THE ALKALOIDS

Chemistry and Pharmacology

Volume 29

THE ALKALOIDS

Chemistry and Pharmacology

A list of contents of volumes in this treatise is available from the publisher on request.

THE ALKALOIDS

Chemistry and Pharmacology

Edited by
Arnold Brossi
National Institutes of Health
Bethesda, Maryland

VOLUME 29

1986

ACADEMIC PRESS, INC.
Harcourt Brace Jovanovich, Publishers

Orlando • San Diego • New York • Austin
Boston • London • Sydney • Tokyo • Toronto

ACADEMIC PRESS, INC.
Orlando, Florida 32887

United Kingdom Edition published by
ACADEMIC PRESS INC. (LONDON) LTD.
24–28 Oval Road, London NW1 7DX

LIBRARY OF CONGRESS CATALOG CARD NUMBER: 50-5522

ISBN 0–12–469529–9

PRINTED IN THE UNITED STATES OF AMERICA

86 87 88 89   9 8 7 6 5 4 3 2 1

CONTENTS

PREFACE vii

Chapter 1.   Chemotaxonomy of Papaveraceae and Fumariaceae

V. PREININGER

   I. Introduction 1
  II. Chemotaxonomic Evaluation of the Papaveraceae 3
 III. Chemotaxonomic Evaluation of the Fumariaceae 51
 IV. Alphabetic Listing of the Papaveraceae and Fumariaceae Alkaloids and Their
      Formulas 66
      References 92

Chapter 2.   Quinazoline Alkaloids

SIEGFRIED JOHNE

   I. Introduction 99
  II. Simple Substituted Quinazolin-4-ones 99
 III. The Pyrroloquinazolines 111
 IV. The Pyrido[2,1-b]quinazolines 121
  V. The Indoloquinazolines 123
 VI. Biosynthesis 125
VII. Biological Activity of Quinazoline Alkaloids and Their Analogs 129
      References 135

Chapter 3.   The Naphthyl Isoquinoline Alkaloids

GERHARD BRINGMANN

   I. Introduction 141
  II. Alkaloids from Ancistrocladaceae 142
 III. Alkaloids from Dionchophyllaceae 145
 IV. Biosynthesis of Naphthyl Isoquinoline Alkaloids 154
  V. The Biomimetic Synthesis of Acetogenin Isoquinoline Alkaloids 163
 VI. Concluding Remarks 181
      References 182

Chapter 4. Alkaloidal Substances from *Aspergillus* Species

Yuzuru Yamamoto and Kunizo Arai

I. Introduction 185
II. Pyrazine Metabolites 185
III. Diketopiperazine Metabolites Derived from Tryptophan and Alanine 192
IV. Diketopiperazines Derived from Tryptophan and Proline 201
V. Other Diketopiperazines Derived from Tryptophan 209
VI. Diketopiperazines Derived from Phenylalanine 211
VII. Metabolites Derived from Tryptophan and Anthranilic Acid 218
VIII. Indole–Mevalonate Metabolites 225
IX. Bisindolylbenzoquinone Metabolites 229
X. Miscellaneous Metabolites 236
References 254

Chapter 5. The Daphniphyllum Alkaloids

Shosuke Yamamura

I. Introduction 265
II. Structures and Properties 266
III. Pharmacological Properties 284
References 286

Chapter 6. The Cularine Alkaloids

Luis Castedo and Rafael Suau

I. Introduction 287
II. Occurrence and Classification 288
III. Physical and Chemical Properties 288
IV. Total Synthesis 308
V. Stereochemistry 316
VI. Biosynthesis 317
VII. Pharmacological Properties 322
References 322

Index 325

PREFACE

It is sad to report that Dr. V. Preininger, who wrote the chapter on "Chemo-taxonomy of Papaveraceae and Fumariaceae Plants," and for many years a member of the famous Šantavý team at the Medical Faculty of the Palacky University in Olomouc, Czechoslovakia, passed away shortly after the writing of his article. His review, the first of its kind in this treatise, will for this reason remain a memorable event. "Quinazoline Alkaloids," discussed in Vols. 3 and 7 of this treatise, have now been updated, and more than 60 members are being presented under the new heading "Simple Substituted Quinazoline Alkaloids," including pharmacological details for several representatives. "Naphthyl Iso-quinoline Alkaloids," not made in nature by conventional isoquinoline biosyn-thetic pathways including Mannich condensation, but of polyketide origin, are presented here for the first time. This group of alkaloids was first recognized to exist by Professor T. R. Govindachari, and the review is honoring his contribu-tion. "Alkaloidal Substances from *Aspergillus* Species," derived from amino acids rather than amines, is an excursion into the field of fungal metabolites, often produced besides antibiotics during fermentation. "Daphniphyllum Al-kaloids," now comprising 14 members, were first discussed in Vol. 15 of this treatise. It is amazing that during the 10 years which have elapsed none of these alkaloids has been synthesized, making this group a challenge to chemists. The cularine alkaloids, reviewed earlier in Vols. 4 and 10 of this treatise, now comprise more than 30 individual alkaloids, which are presented here with details on their structures and synthesis.

Arnold Brossi

CHAPTER 1

CHEMOTAXONOMY OF PAPAVERACEAE AND FUMARIACEAE

V. PREININGER*

*Institute of Chemistry, Medical Faculty, Palacký University,
Olomouc, Czechoslovakia*

I. Introduction

The families Papaveraceae and Fumariaceae are well known as taxonomic groups, and both are very rich in specific alkaloids. Some of them are important in medicine and some others can be considered as promising in this sense (*1,2*). Therefore it is not surprising that in the past and especially at present these alkaloids belong to those that have been most intensively studied (*3–9*). A vast quantity of material has been accumulated, which represents a rich potential source of medicinal compounds of plant origin and a significant taxon in plant systematics. In this connection, attempts are made by both botanists and chemists to find a correlation between chemical properties and botanical features.

Alkaloids with one chiral carbon are shown as racemic mixtures, but their enantiomeric characterization is given in the text with the prefix (−) or (+). In the formulas of alkaloids with more than one chiral center the relative and absolute configurations are omitted. However, these structural details can be found in the literature [Refs. *3–8, 16,* and previous volumes of this treatise: Simple Isoquinolines, Vol. 21 (1983), p. 255; Aporphines, Vol. 24 (1985), p. 153; Phthalideisoquinolines, Vol. 24 (1985), p. 253; Benzophenanthridines, Vol. 26 (1985), p. 185].

The taxonomic delineation of the families Papaveraceae and Fumariaceae is not uniform. Fedde (*10*), Tutin *et al.* (*11*), or Melchior (*12*) classified the fumariaceous plants in the Papaveraceae, as the subfamily Fumarioideae, whereas Hutchinson (*13*) considered the Fumariaceae to be an independent family. The occurrence of isoquinoline alkaloids in those plants indicates a close relationship of the Papaveraceae and Fumariaceae to the Polycarpicae and to the orders Ranales and Magnoliales (*14*). Formerly, the order Rhoeadales included the Papaveraceae and the Fumariaceae, as well as the Brassicaceae, Capparaceae,

---

* Deceased December 22, 1985.

THE ALKALOIDS, VOL. 29

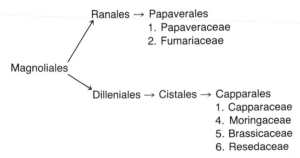

Fig. 1A. Position of the families of the order Rhoeadales in the system of Hutchinson (*13, 14*).

Moringaceae, Resedaceae, and Tovariaceae. More recently, the order Rhoeadales has been replaced by the Papaverales and the Capparales. The Papaverales includes the Papaveraceae and the Fumariaceae, while the remaining
five families are grouped into the Capparales (Fig. 1, a and b). Species of the
Capparales contain glucosinolates rather than alkaloids, whereas species of the
Papaverales bear alkaloids and not glucosinolates. These clear chemical distinc-
tions closely follow the systematics of Hutchinson (*13*) and Takhtajan (*15*) based
on botanical characteristics.

The family Papaveraceae Juss. is subdivided into the subfamilies Hypecoideae
Prantl and Kündig and Papaveroideae A. Br. It embraces 23 genera and ~430
species. The Fumariaceae DC. include 7 genera with ~350 species. In the
literature, the alkaloids of about 168 species of the Papaveraceae and 90 species
of the Fumariaceae have been described. A characteristic feature of these two
families is (with a few exceptions, e.g., Section IV) the presence of alkaloids
derived from tetrahydroisoquinolines. These alkaloids arise from phenylalanine
by condensation of dopamine with 3,4-dihydroxyphenylacetaldehyde and subsequent Mannich condensation to yield norlaudanosoline as the primary condensation product (Fig. 2). Other intermediates formed during the biosynthesis of
more complicated constitutional types of isoquinoline alkaloids are (*R*)-(−)- or

Fig. 1B. Position of the families of the order Papaverales in the system of Takhtajan (*14, 15*).

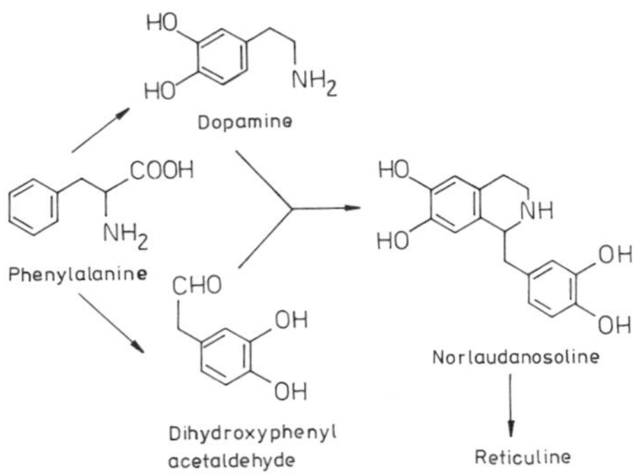

FIG. 2. First steps in the biosynthesis of 1-benzyltetrahydroisoquinoline alkaloids.

(S)-$(+)$-reticuline (**20**) and orientaline (**25**). The alkaloids of the Papaveraceae and Fumariaceae can be subdivided into several constitutional types (Fig. 3), viz., simple isoquinolines, benzylisoquinolines, pavines, isopavines, cularines, proaporphines, aporphines, promorphinanes, morphinanes, protoberberines, retroprotoberberines, secoberbines, benzophenanthridines, protopines, phthali-deisoquinolines, secophthalideisoquinolines, indenobenzazepines, spirobenzyl-isoquinolines, and rhoeadines.

For chemotaxonomic conclusions there are of importance the combination of certain alkaloid types and their biosynthetic interrelationships. One recent rearrangement of the Papaveraceae genera leads to the conclusion that meconic acid is a distinctive taxonomic feature (*17*).

II. Chemotaxonomic Evaluation of the Papaveraceae

The most widespread alkaloids of the Papaveraceae Juss. are protoberberines, benzophenanthridines, and protopines, which, with a few exceptions, are present in all genera. Subsequently, they can be considered as a chemotaxonomically significant characteristic of this family. There should be added the quaternary highly polar alkaloids, which do not produce any form soluble in nonpolar solvents (*16*). The first such finding was the quaternary aporphine alkaloid magnoflorine (**138**) in opium (*18*). Hegnauer (*9,14*) has already drawn attention to the general distribution of magnoflorine in the Polycarpicae. According to recent papers (*16*), quaternary alkaloids of this type are almost ubiquitous in the Papaveraceae. This author isolated about 22 various quaternary alkaloids in some cases as the dominant alkaloid of these plants (Table I). In some instances, the quaternary alkaloid is the *N*-methyl derivative of the main tertiary base. The quarternary tetrahydroprotoberberines were found to play a role of intermediates in the biosynthetic process (*19–23*).

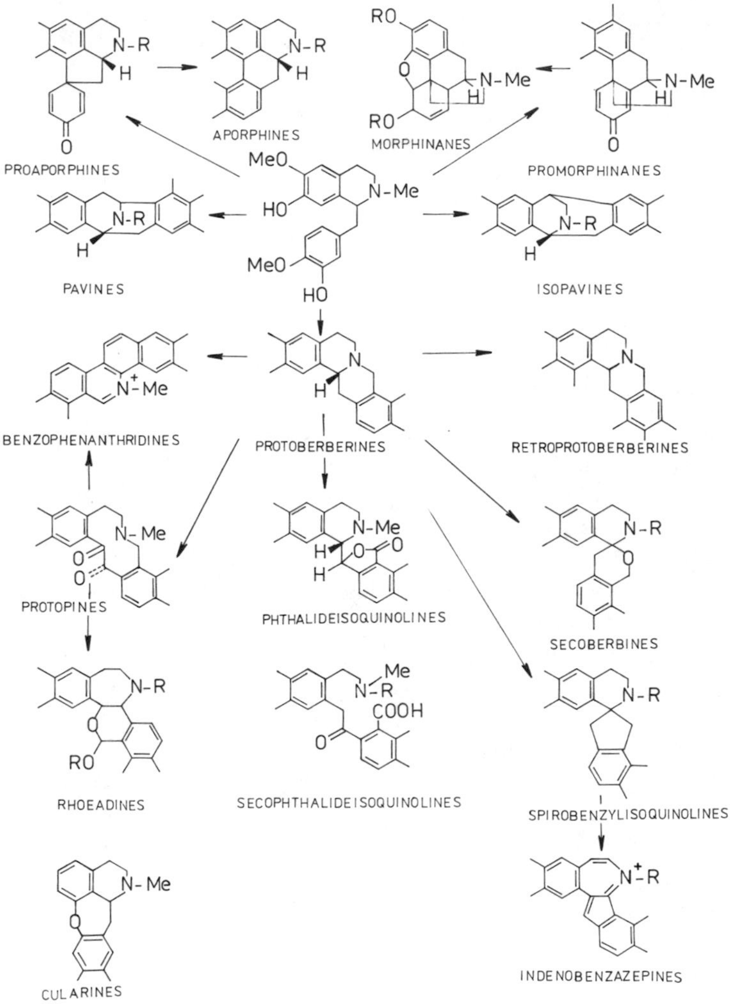

FIG. 3. Biosynthetic relationships between the isoquinoline alkaloids derived from reticuline (**20**).

Most of the chemotaxonomic findings have been made with regard to the genera *Papaver* (Tourn.) L., *Glaucium* (Tourn.) Adans., *Roemeria* Medic., *Meconopsis* Viguier, *Argemone* (Tourn.) L., *Eschscholtzia* Cham., *Dicranostigma* Hook. et Thoms., *Chelidonium* (Tourn.) L., *Bocconia* (Plum.) L., and *Hypecoum* (Tourn.) L.

TABLE I

Occurrence of the Quaternary Alkaloids in Papaveraceae[a]

Benzylisoquinolines

Eschscholamidine (**34**): *Eschscholtzia oregana*
Escholamine (**33**): *Eschscholtzia oregana*
Escholinine (**22**): *Eschscholtzia californica*

Pavines

Argemonine methohydroxide (**48**): *Argemone gracilenta, A. platyceras*
Californidine (**51**): *Eschscholtzia californica, E. douglasii, E. glauca, E. oregana*
Platycerine methohydroxide (**59**): *Argemone gracilenta, A. platyceras*

Isopavines

Amurensinine methohydroxide (**62**): *Meconopsis betonicifolia, M. horridula, M. nepalensis, M. rudis*
Remrefine (**66**): *Roemeria refracta*

Aporphines

Aporheine methohydroxide (**116**): *Papaver albiflorum* ssp. *albiflorum, P. albiflorum* ssp. *austromoravicum, P. caucasicum, P. dubium, P. fugax, P. litwinovii, P. rhoeas*
Bulbocapnine methohydroxide (**130**): *Corydalis cava*
Corytuberine methohydroxide (**138**): *Dicranostigma franchetianum, D. leptopodum*
Isothebaine methohydroxide (**123**): *Papaver pseudoorientale*
Magnoflorine (**138**): *Argemone platyceras, Chelidonium majus, Corydalis cava, Dicranostigma franchetianum, D. lactucoides, D. leptopodum, Eschscholtzia californica, E. douglasii, E. glauca, Glaucium fimbrilligerum, G. flavum, G. squamigerum, Meconopsis cambrica, M. nepalensis, M. paniculata, M. robusta, M. rudis, Papaver albiflorum* ssp. *austromoravicum, P. atlanticum, P. confine, P. dubium, P. dubium* ssp. *lecoquii, P. oreophilum, P. rupifragum, P. somniferum, Pteridophyllum racemosum, Stylophorum diphyllum*
Menisperine (**133**): *Dicranostigma franchetianum, D. lactucoides, D. leptopodum, Papaver oreophilum*
Remrefidine (**116**): *Roemeria refracta*

Morphinanes

Thebaine methohydroxide (**193**): *Papaver albiflorum* ssp. *albiflorum, P. bracteatum, P. decaisnei*

Protoberberines

(−)-α-Canadine methohydroxide (**244**): *Argemone ochroleuca, A. platyceras, Bocconia frutescens, Eschscholtzia californica, E. douglasii, E. glauca, Glaucium corniculatum, G. grandiflorum* var. *torquatum, G. squamigerum, P. albiflorum* ssp. *austromoravicum*
(−)-β-Canadine methohydroxide (**244**): *Corydalis cheilanthifolia, C. thalictrifolia, Glaucium corniculatum, G. squamigerum*
(−)-Cyclanoline (**245**): *Argemone platyceras, Hunnemannia fumariaefolia*
Escholidine (**246**): *Eschscholtzia californica, E. douglasii, E. glauca, Hunnemannia fumariaefolia*
(−)-β-Isocorypalmine methohydroxide (**247**): *Glaucium squamigerum*
N-Methylsinactine (**248**): *Fumaria officinalis*
(−)-α-Scoulerine methohydroxide (**245**): *Argemone albiflora, A. mexicana, A. platyceras*
(−)-β-Scoulerine methohydroxide (**245**): *Argemone albiflora, A. mexicana, A. ochroleuca, A. platyceras, Chelidonium majus, Eschscholtzia californica, E. oregana*

(*continued*)

TABLE I   (*Continued*)

(+)-α-Stylopine methohydroxide (**249**): *Argemone mexicana, A. ochroleuca, A. platyceras, Chelidonium majus, Corydalis cava, Eschscholtzia californica, E. oregana, Glaucium squamigerum, Hypecoum erectum, Papaver albiflorum* ssp. *austromoravicum, P. comutatum, P. dubium* var. *lecoquii, Stylophorum diphyllum*

(−)-β-Stylopine methohydroxide (**249**): *Argemone mexicana, A. ochroleuca, A. platyceras, Chelidonium majus, Fumaria vaillantii, Glaucium corniculatum, G. squamigerum, Papaver comutatum, P. rhoeas, P. syriacum, Stylophorum diphyllum*

(−)-α-Tetrahydropalmatine methohydroxide (**250**): *Argemone ochroleuca*

β-Tetrahydropalmatine methohydroxide (**250**): *Glaucium squamigerum*

(+)-α-Tetrahydrocorysamine methohydroxide (**251**): *Corydalis cava*

Retroprotoberberines
Mecambridine methohydroxide (**258**): *Papaver pseudoorientale*

Protopines
Protopine methohydroxide (**324a**): *Fumaria indica*

Phthalideisoquinolines
Bicuculline methohydroxide (or adlumidine/capnoidine methohydroxide): *Corydalis lutea* (?), *C. ochroleuca, C. sempervirens* (?), *P. rhoeas* (?)

Hydrastine methohydroxide: *Corydalis lutea* (?), *Fumaria parviflora* (?), *F. schleicheri* (?), *F. vaillantii* (?)

(−)-*N*-Methyladlumine (**335**): *Fumaria parviflora, F. vaillantii*

Narcotine methohydroxide: *Papaver somniferum* (?), *P. setigerum* (?), *Fumaria indica* (?)

---

[a] After Ref. *16*.

A. The Genus *Papaver*

The chemotaxonomy of plants of the genus *Papaver* L. was reviewed in Ref. *24*. This genus embraces ~120 species; ~75 of them (inclusive of their varieties) yield about 170 alkaloids (Tables II and III).

The genus *Papaver* was originally divided by Bernhardi (*25*) into five sections—*Lasiotrachyphylla, Oxytona, Miltantha, Rhoeades,* and *Mecones*. Elkan (*26*) introduced the section *Horrida,* and Prantl and Künding (*27*) included the section *Pilosa*. To the section *Rhoeades,* those species were classified which Fedde (*10*) had placed in the sections *Orthorhoeades* and *Argemonorhoeades;* furthermore, he excluded the section *Carinatae*. The division of the genus *Papaver* into nine sections is still used by many authors. Günther (*28*) divided the section *Pilosa,* and to the genus *Papaver* he added the genus *Roemeria* as a special, eleventh section. Later, the section *Mecones* was divided into the section *Papaver* and the section *Glauca* (*29,30*) (Table IV).

TABLE II

Types of Alkaloids Found in the Sections of the Genus *Papaver*

	Benzylisoquinolines	Isopavines	Proaporphines	Aporphines	Promorphinanes	Morphinanes	Protoberberines	Retroprotoberberines	Benzophenanthridines	Protopines	Phthalideisoquinolines	Secophthalideisoquinolines	Rhoeadines
Orthorhoeades	+	–	+	+	–	+	+	–	+	+	–	+	+
Argemonorhoeades	–	–	–	+	–	–	+	–	+	+	–	–	+
Carinatae	–	–	–	+	–	–	–	–	+	+	–	–	+
Papaver (syn. *Mecones*)	+	–	–	+	+	+	+	–	+	+	+	+	+
Glauca	–	–	+	–	–	–	+	–	+	+	–	+	+
Miltantha	+	–	+	+	+	+	+	+	+	+	+	–	+
Pilosa	+	–	+	+	+	+	+	–	+	+	–	–	+
Macrantha	+	–	+	+	+	+	+	+	+	+	–	–	+
Scapiflora	–	+	+	–	+	–	+	+	+	+	–	–	+

TABLE III
Alkaloids Found in Plants of the Genus *Papaver*

Simple isoquinolines and benzylisoquinolines
  Corypalline, hydrocotarnine, *N*-methylcorydaldine, *N*-methyl-6,7-
    dimethoxytetrahydroisoquinolone-1, *O*-methylcorypalline, oxyhydrastinine; arenine, (−)-
    armepavine, codamine, dehydronormacrostomine, glycomarine, latericine, laudanidine,
    laudanine, laudanosine, macrantaline, macrantoridine, macrostomine, macrostomine *N*-oxide,
    orientaline, palaudine, papaveraldine, papaveramine, papaverine, reticuline, sevanine,
    tetrahydropapaverine

Pavines and isopavines
  Amurensine, amurensinine, *O*-methylthalisopavine, reframidine

Proaporphines
  Amuroline, amuronine, dihydroorientalinone, *O,N*-dimethyloridine, glaziovine,
    hexahydromecambrine B, isooridine, mecambrine, *N*-methylcrotonosine, *N*-methyloridine,
    O-methyloridine, oridine, orientalinone (= bracteine), pronuciferine

Aporphines
  Alkaloid PO-3, aporheine, aporheine methohydroxide, bracteoline, corydine, corytuberine,
    dehydroglaucine, 6a,7-dehydroisothebaine, dehydroroemerine, *O*-demethylnuciferine,
    floripavidine, glaucine, isoboldine, isocorydine, isothebaidine, isothebaine, isothebaine
    methohydroxide, lirinidine, liriodenine, magnoflorine, mecambroline, menisperine, *N*-
    methylasimilobine, *O*-methyl-6a,7-dehydroisothebaine, *O*-methylorientine, nantenine,
    norcorydine, nuciferine, nuciferoline, orientine, remrefidine, roemerine, roemerine
    methohydroxide

Promorphinanes
  Amurine, amurinine, dihydronudaurine, epiamurinine, flavinantine, *O*-methylflavinantine, *O*-
    methylsalutaridine, nudaurine, salutaridine, sinoacutine

Morphinanes
  Codeine, codeine *N*-oxide, codeinone, 10-hydroxycodeine, 14β-hydroxycodeine, 14β-
    hydroxycodeinone, 16-hydroxythebaine, 6-methylcodeine, morphine, morphine *N*-oxide,
    neopine, normorphine, oripavidine, oripavine, pseudomorphine, thebaine, thebaine
    methohydroxide, thebaine *N*-oxide

Protoberberines
  Canadine, canadine methohydroxide, cheilanthifoline, coreximine, isocorypalmine
    (= tetrahydrocolumbamine), scoulerine, sinactine, stepholidine, stylopine, β-stylopine
    methohydroxide; berberine, coptisine, corysamine, palmatine

Retroprotoberberines
  Alkaloid PO-4, alkaloid PO-5 (= alborine), aryapavine, mecambridine, mecambridine
    methohydroxide, orientalidine

Benzophenanthridines
  6-Acetonyldihydrosanguinarine, chelerythrine, chelirubine, dihydrosanguinarine,
    norsanguinarine, oxysanguinarine, sanguinarine

Protopines
  Allocryptopine, cryptopine, dihydroprotopine, 1-methoxyallocryptopine, 1-methoxy-13-
    oxoallocryptopine, muramine, 13-oxocryptopine, 13-oxomuramine, 13-oxoprotopine,
    protopine

TABLE III   (*Continued*)

Phthalideisoquinolines
  Gnoscopine, narcotine, narcotoline
Secophthalideisoquinolines
  Adlumidiceine, narceine, narceine imide, nornarceine
Rhoeadines
  Alpinigenine, alpinine, epialpinine, epiglaudine, glaucamine, glaudine, glaupavine
  (= epiglaudine?), isorhoeadine, isorhoeagenine, isorhoeagenine glycoside, *N*-demethyl-14-*O*-
  demethylepiporphyroxine, oreodine, oreogenine, papaverrubines A, B, C, D, E, F, G,
  rhoeadine, rhoeagenine
Alkaloids of unknown structure
  Bractamine, floribundine, glaucidine, glaupavine, lanthopine, nudicaulinole, pahybrine,
  papaveramine, roemeridine

1. Section *Orthorhoeades* (Syn. *Rhoeadium*)

A survey of the plants of this section and their alkaloids is given in Table V (*31–37*). The major alkaloids are rhoeadine (**389**) and protopine (**309**), the minor alkaloids are isorhoeadine (**389**) and the papaverrubines (**396–400, 406**). However, plants have also been found whose major alkaloids are aporphines. The presence of these alkaloids can be considered as a significant characteristic of the section *Orthorhoeades*.

The plant *P. dubium* differs in that rhoeadine alkaloids or proaporphine and aporphine alkaloids or both types together are present therein. *Papaver litwinovii* Fedde resembles this type, containing aporphine alkaloids. The major alkaloids found therein are aporheine (**115**) and its methohydroxide (**116**). In *P. dubium* L. var. *lecoquii* (Lamotte) Fedde, some authors (*38,39*) found berberine (**209b**) as the major alkaloid, whereas in the material studied by us (*40*) the major alkaloids were rhoeadine (**389**) and isorhoeadine (**389**), though in some instances berberine was also detected. The plant *P. commutatum* also was shown to contain either alkaloids of the rhoeadine or those of the aporphine type and papaverine or all of these alkaloids simultaneously. Noticeable is the high content of isorhoeagenine glycoside (**390**) (*41*).

Two herbal specimens of *P. rhoeas* have been studied (*34*) that do not differ morphologically from the typical *P. rhoeas*. One of the specimens is typical, with yellow latex, and it corresponds to Kuntze's description (*10,42*) of the taxon *P. rhoeas* var. *chelidonioides* O. Ktze., while the other has a white latex, turning pink in air. The composition of the tertiary fraction, where rhoeadine (**389**) and aporheine (**115**) are the dominant alkaloids in approximately equal amounts, characterizes the chemical type, which forms a transition between the typical *P. rhoeas* and *P. dubium*. In contrast to other populations of *P. rhoeas,* the

TABLE IV

Retrospective Historical Survey of the Sections of the Genus *Papaver*

Bernhardi (25)	Elkan (26)	Prantl (27)	Fedde (10)	Günther (28)	Present state
Lasiotrachyphylla Bernh.	*Scapiflora* Reich.	*Lasiotrachyphylla* Bernh.	*Scapiflora* Reich.	*Lasiotrachyphylla* Bernh.	*Meconella* Spach.
Oxytona Bernh.	*Macrantha* Elk.	*Macrantha* Elk.	*Macrantha* Elk.	*Oxytona* Bernh.	*Macrantha* Elk.
		Pilosa Prantl	*Pilosa* Prantl	*Pilosa* Prantl	*Pilosa* Prantl
Miltantha Bernh.	*Pyramistigmata* Elk.	*Miltantha* Elk.	*Miltantha* Bernh.	*Miltantha* Bernh.	*Miltantha* Bernh.
Rhoeades Bernh.	*Rhoeades* Bernh.	*Rhoeades* Bernh.	*Argemonorhoeades* Fedde	*Argemonidium* Spach.	*Argemonidium* Spach.
			Carinatae Fedde	*Carinatae* Fedde	*Carinatae* Fedde
			Orthorhoeades Fedde	*Rhoeades* Bernh.	*Rhoeadium* Spach.
Mecones Bernh.	*Mecones* Bernh.	*Mecones* Bernh.	*Mecones* Bernh.	*Papaver*	*Papaver*
					Glauca J. Nov. et V. Prein.
	Horrida Elk.	*Horrida* Elk.	*Horrida* Elk.	*Horrida* Elk.	*Horrida* Elk.
				Pseudo-pilosa M. Pop.	
				Roemeria (Medic.) Günther	*Roemeria* (Medic.) Günther

TABLE V

Plants of *Papaver* Section *Orthorhoeades* and Their Alkaloids

Papaver albiflorum Pacz. ssp. *albiflorum*
Aporheine methohydroxide (**116**), berberine (**209b**), corytuberine (**137**), protopine (**309**), thebaine (**192**), thebaine methohydroxide (**193**) (*31*)

P. albiflorum ssp. *austromoravicum* Kubát
Allocryptopine (**310**), aporheine (**115**), aporheine methohydroxide (**116**), berberine (**209b**), canadine (**209a**), (−)-α-canadine methohydroxide (**244**), coptisine (**217b**), corydine (**136**), corysamine (**234b**), corytuberine (**137**), isocorydine (**132**), magnoflorine (**138**), mecambrine (**110**), protopine (**309**), papaverrubine A, C, D, E (**396, 398, 400**), rhoeadine (**389**), scoulerine (**212**), α-stylopine methohydroxide (**249**), thebaine (**192**) (*31*)

P. arenarium Marsch. et Bieb.
Glycomarine (**38**), isorhoeadine (**391**), macrostomine (**40**), papaverrubine A (**396**), rhoeadine (**389**), rhoeagenine (**404**) (*3,32,33*)

P. californicum A. Gray
Coptisine (**217b**), cryptopine (**311**), latericine (**15**), muramine (**314**), papaverrubine A, B, D, E (**396–398**), protopine (**309**), rhoeadine (**389**), rhoeagenine (**404**) (*3*)

P. comutatum Fisch. et Mey
Coptisine (**217b**), corytuberine (**137**), cryptopine (**311**), isocorydine (**132**), isorhoeadine (**389**), isorhoeagenine (**391**), isorhoeagenine glycoside (**390**), papaverine (**35**), papaverrubine A, B, C, D, E, F (**396–398, 400**), roemerine (**115**), rhoeadine (**389**), rhoeagenine (**404**), sanguinarine (**296**), (−)-stylopine methohydroxide (**249**) (*3*)

P. confine Jord. (syn. *P. dubium*)
Magnoflorine (**138**) (*34*)

P. dubium L.
Aporheine methohydroxide (**116**), berberine (**209b**), coptisine (**217b**), mecambrine (**110**), oxysanguinarine (**281**), papaverrubine A, B, E (**396, 397**), protopine (**309**), rhoeadine (**389**), rhoeagenine (**404**), roemerine (**115**) (*3,34*)

P. dubium ssp. *albiflorum* (Boiss.) Dost.
Allocryptopine (**310**), berberine (**209b**), coptisine (**217b**) (*3*)

P. dubium L. var. *glabrum*
Oxyhydrastinine (**6**) (*35*)

P. dubium ssp. *lecoquii* (Lamotte)
Allocryptopine (**310**), berberine (**209b**), canadine (**209a**), coptisine (**217b**), corydine (**136**), corytuberine (**137**), cryptopine (**311**), (−)-hexahydromecambrine B (**98**), isocorydine (**132**), isorhoeadine (**389**), magnoflorine (**138**), mecambrine (**110**), oxysanguinarine (**281**), papaverrubine A, C, D, E (**396, 398, 400**), protopine (**309**), rhoeadine (**389**), scoulerine (**212**), stylopine (**217b**), α-stylopine methohydroxide (**249**) (*3,31*)

P. intermedium Bedker O. Ktze. (syn. *P. rhoeas*)
Oxysanguinarine (**281**), rhoeadine (**389**), thebaine (**192**) (*3*)

P. lacerum Popov (syn. *P. laevigatum* auct. *non* Bieb.)
Mecambrine (**110**), *N*-methylasimilobine (**119**), pronuciferine (**112**), roemerine (**115**) (*36*)

P. litwinovii Fedde
Aporheine (**115**), aporheine methohydroxide (**116**), oxysanguinarine (**281**) (*3,34*)

(*continued*)

TABLE V   (*Continued*)

P. rhoeas L.

Adlumidiceine (**340**), allocryptopine (**310**), asimilobine methohydroxide (**119**), aporheine
methohydroxide (**116**), berberine (**209b**), coptisine (**217b**), corydine (**136**), cryptopine (**311**),
glaucamine (**393**), glaudine (**392**), isorhoeadine (**389**), isocorydine (**132**), isorhoeagenine
(**391**), isorhoeagenine glycoside (**390**), oxysanguinarine (**281**), papaverrubine A, B, C, D, E
(**396–398, 400**), rhoeadine (**389**), rhoeagenine (**404**), sanguinarine (**296**), (−)-sinactine
(**215a**), stylopine (**217a**), β-stylopine methohydroxide (249), thebaine (**192**), *N*-
methyltetrahydronorharmane (**409**) (*3,34,37*)

P. rhoeas var. *decaisnei*

Coptisine (**217b**), isorhoeadine (**389**), isorhoeagenine glycoside (**390**), oxysanguinarine (**281**),
protopine (**309**), rhoeadine (**389**), rhoeagenine (**404**), stylopine (**217a**) (*3*)

P. rhoeas var. *flore albo*

Isorhoeadine (**389**), oxysanguinarine (**281**), papaverrubine A (**396**), rhoeadine (**389**) (*3*)

P. rhoeas var. *flore pleno*

Coptisine (**217b**), isorhoeadine (**389**), oxysanguinarine (**281**), papaverrubine A (**396**), protopine
(**309**), rhoeadine (**389**), rhoeagenine (**404**), stylopine (**217a**) (*3*)

P. strigosum Schur. (syn. *P. rhoeas*)

Coptisine (**217b**), papaverrubine A, B, D, E (**396–398**), protopine (**309**), rhoeadine (**389**),
rhoeagenine (**404**), thebaine (**192**) (*3*)

---

chelidonioides variety contains a substantially higher content of coptisine
(**217b**), which is undoubtedly responsible for the yellow color of the latex. The
presence of aporheine (**115**), aporheine methohydroxide (**116**), or mecambrine
(**110**) has not been detected in the populations of the typical *P. rhoeas* as yet.

Studies of the section *Orthorhoeades* have shown the occurrence of at least
two chemotypes in this section. One chemotype contains rhoeadine alkaloids and
the other, proaporphine and aporphine alkaloids. This is consistent with the
theory postulated by Mothes about chemical races (*43*). The question also arises
how firmly some hereditary features are fixed to the plant, consequently, how
easily the change in the geographical or climatic conditions may affect the
spectrum of the alkaloids.

Interesting is the finding of the alkaloid glycoside latericine (**15**) in *P. californicum* (*44*) and of the secophthalideisoquinoline alkaloid adlumidiceine (**341**) in
P. rhoeas (*45*) because latericine is characteristic of plants of the section *Pilosa*
and adlumidiceine of the genera *Fumaria* and *Corydalis*. Noticeable is the presence of thebaine (**192**) in *P. strigosum* Schur. and in *P. intermedium* O. Ktze.

2. Section *Argemonorhoeades* (Syn. *Argemonidium*)

Studies of plants of the section *Argemonorhoeades* Fedde, in which Fedde
(*10*) classifies six species that are similar to the plants of the section *Orthorhoe-*

TABLE VI

Plants of *Papaver* Section *Argemonorhoeades* and Their Alkaloids

---

Papaver apulum Ten.
  Coptisine (**217b**) (*3*)

P. argemone L.
  Coptisine (**217b**), isorhoeadine (**389**), oxysanguinarine (**281**), papaverrubine D, E (**398, 396**),
    protopine (**309**), rhoeadine (**389**), rhoeagenine (**404**) (*3*)

P. hispidum Lam. (syn. *P. hybridum* L.)
  Berberine (**209b**), coptisine (**217b**), cryptopine (**311**), 13-oxoprotopine (**319**), oxysanguinarine
    (**281**), protopine (**309**), rhoeadine (**389**), rhoeagenine (**404**) (*3*)

P. hybridum L.
  Berberine (**209b**), coptisine (**217b**), pahybrine (?), papaverrubine A, D, E (**396, 398**),
    sanguinarine (**296**) (*3*)

P. pavoninum Fisch. et Mey
  Allocryptopine (**310**), coptisine (**217b**), papaverrubine D, E, (**398, 396**), protopine (**309**),
    rhoeadine (**389**), roemeridine (?), sanguinarine (**296**) (*3*)

---

ades, show that these plants contain alkaloids of the protoberberine, benzophe-nanthridine, protopine, rhoeadine, and papaverrubine types. These alkaloids resemble those found in the section *Orthorhoeades*, but the alkaloid content is markedly lower (Table VI).

3. Section *Carinatae*

Of the six species classified by Fedde (*10*) in this section, only *P. macrostomum* Boiss. et Huet. has been studied (Table VII). The other species are morphologically closely related (*46*). Plants cultivated in central Europe (ČSSR, GDR), were found to contain rhoeadine, papaverrubine, and protopine alkaloids in the same way as the species of the sections *Orthorhoeades* and *Argemonor-hoeades* (*40,47*). Plants collected in the vicinity of Lake Sevan (Armenia) yielded macrostomine (**40**) as the major alkaloid together with dehydronor-macrostomine (**43**) and sevanine (**37**) (benzylisoquinolines of the papaverine type) but no rhoeadines or papaverrubines (*48*). These differences in the alkaloid content may be explained by the observation that *P. macrostomum* from Armenia

TABLE VII

Plants of *Papaver* Section *Carinatae* and Their Alkaloids

Papaver macrostomum Boiss. et Huet.
  Dehydronormacrostomine (**43**), macrostomine (**40**), papaverrubine A, B, D, E (**396–398**),
    protopine (**309**), rhoeadine (**389**), sevanine (**37**) (*3*)

TABLE VIII
Types of Alkaloids Contained in *Papaver* Sections *Papaver* and *Glauca*

	Tetrahydroisoquinolines and benzyltetrahydroisoquinolines	Benzylisoquinolines	Aporphines	Promorphinanes	Morphinanes	Protoberberines	Protopines	Phthalideisoquinolines	Secophthalideisoquinolines	Rhoeadines	Benzophenanthridines
Papaver somniferum L.	+	+	+	+	+	+	+	+	+	Trace	+
P. setigerum DC.	+	+	−	−	+	+	+	+	+	Trace	+
P. glaucum Boiss. et Hausskn.	−	−	−	−	−	+	−	−	−	+	+
P. gracile Auch.	−	−	−	−	−	−	−	−	−	+	+
P. decaisnei Hochst. et Steud.	−	−	−	−	−	+	+	−	−	+	−

TABLE IX

Rhoeadine–Papaverrubine Alkaloids Contained in *Papaver* Sections *Papaver* and *Glauca*

	Rhoeadines		Papaverrubines
	C-14—OH	C-14—OMe	
Papaver somniferum L.	*N*-Methyl-14-*O*-demethylepiporphyroxine	Glaudine	Papaverrubine C, D
P. *setigerum* DC.	—	—	Papaverrubine A, B, C, D
P. *glaucum* Boiss. et Hausskn.	Glaucamine	Rhoeadine, glaudine	Papaverrubine A, B, C, D
P. *gracile* Auch.	—	Rhoeadine	?
P. *decaisnei* Hochst. et Steud.	Rhoeagenine	Rhoeadine, isorhoeadine	Papaverrubine A, C, D, E

15

grows in its natural habitat, whereas the plants obtained from central Europe are cultivated artificially. The different findings reported (40,47,48) can also be explained by taking into account that in 1962 and 1964 the isolation technique was not sufficiently developed and the quantity of the initial material under investigation was very small. The existence of two chemical races cannot be excluded, one containing rhoeadine and protopine and the other macrostomine. Macrostomine (40) has also been found in *P. arenarium (Orthorhoeades)*.

4. Section *Papaver* (Syn. *Mecones*)

According to the new classification (29,30), the section *Papaver* includes the species *P. somniferum* L. and *P. setigerum* DC. Systematic investigations have shown that the original section *Papaver* (syn. *Mecones* Bernh.) is rather heterogeneous in its chemical composition. The presence of morphinane alkaloids, thebaine (192), codeine (198), and morphine (204), together with the secophthalideisoquinoline alkaloids narceine (343), nornarceine (342), and narceine imide (350) and the phthalideisoquinoline alkaloids narcotine (332) and narcotoline (333), is characteristic of *P. somniferum* L. and *P. setigerum'* DC. *In contrast, rhoeadines and papaverrubines predominate in P. glaucum Boiss. et Hausskn., P. gracile Auch., and P. decaisnei Hochst. et Steud.* (Tables VIII–XI). Traces of rhoeadines are also found in *P. somniferum* and *P. setigerum*,

TABLE X

Plants of *Papaver* Section *Papaver* (syn. *Mecones*) and Their Alkaloids

---

Papaver somniferum L.

    6-Acetonyldihydrosanguinarine (285), allocryptopine (310), berberine (209b), canadine (209a), codamine (14), codeine (198), codeine *N*-oxide (199), codeinone (206), coptisine (217b), coreximine (221a), corytuberine (137), cryptopine (311), dihydroprotopine (323), dihydrosanguinarine (280), glaudine (392), gnoscopine (332), hydrocotarnine (4), 10-hydroxycodeine (200), 16-hydroxythebaine (197), isoboldine (147), isocorypalmine (214a), lanthopine (?), laudanidine (16), laudanine (16), laudanosine (17), magnoflorine (138), 6-methylcodeine (202), *N*-methyl-14-*O*-demethylepiporphyroxine (401), morphine (204), morphine *N*-oxide (205), narceine (343), narceine imide (350), narcotine (332), narcotoline (333), neopine (208), normorphine (203), nornarceine (342), norsanguinarine (292), orientaline (25), oripavine (195), 13-oxocryptopine (317), oxysanguinarine (281), palaudine (36), papaveraldine (39), papaveramine (?), papaverine (35), papaverrubine C, D (400,398), protopine (309), pseudomorphine (204), reticuline (20), salutaridine (188), sanguinarine (296), scoulerine (212), stepholidine (216a), thebaine (192), thebaine *N*-oxide (194), tetrahydropapaverine (21) (3,49,50)

P. setigerum DC.

    Codeine (198), coptisine (217b), cryptopine (311), laudanine (16), laudanosine (17), morphine (204), narceine (343), narcotine (332), narcotoline (333), papaverine (35), papaverrubine A, B, D, E (396–398), protopine (309), sanguinarine (296), thebaine (192) (3,49)

TABLE XI

Plants of *Papaver* Section *Glauca* and Their Alkaloids

Papaver glaucum Boiss. et Hauskn.

Coptisine (**217b**), dehydroroemerine (**167**), glaucamine (**393**), glaudine (**392**), glaupavine (**402**), liriodenine (**157**), oxysanguinarine (**281**), papaverrubine A, B, C, D (**396–398, 400**), rhoeadine (**389**), rhoeagenine (**404**), roemerine (**115**), roemerine *N*-oxide, sanguinarine (**296**) (*3,49,51*)

P. gracile Auch.

Oxysanguinarine (**281**), rhoeadine (**389**) (*3,49*)

P. decaisnei Hochst

Codeine (**198**), coptisine (**217b**), corytuberine (**137**), isorhoeadine (**389**), morphine (**204**), narcotine (**332**), papaverine (**35**), papaverrubine A, C, D, E (**396, 398, 400**), protopine (**309**), rhoeadine (**389**), rhoeagenine (**404**), thebaine (**192**), thebaine methohydroxide (**193**) (*29,30,52*)

---

whereas morphinanes have not been found in *P. glaucum* and *P. gracile*. In one publication (*52*), morphine from a species other than *P. somniferum* or *P. setigerum* has been reported as a minor alkaloid [in addition to rhoeadine (**389**) and coptisine (**217b**)] from the poppy heads of *P. decaisnei*. Its major alkaloid is papaverine (**35**) (Table XI).

Thebaine (**192**) is found in numerous plants of the sections *Orthorhoeades, Pilosa, Miltantha,* and *Macrantha.* The phenolic ortho–para oxidation leading to the formation of thebaine is shown to be a common phenomenon, whereas the specific ability to demethylate the methoxy groups to produce morphine (**204**) or codeine (**198**) appears to be a rare characteristic (Fig. 4).

On the basis of these findings, it has been proposed to divide the original

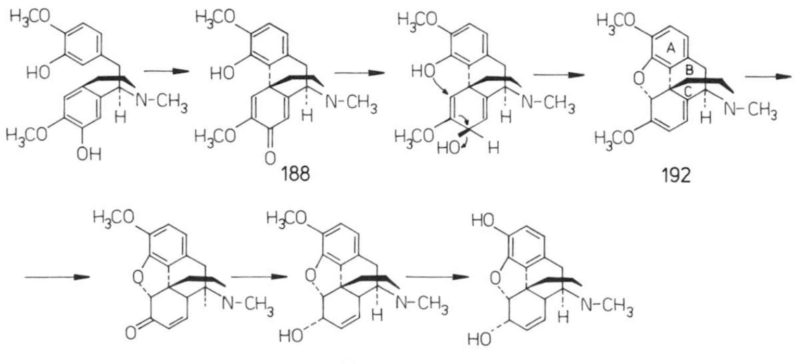

FIG. 4. Biosynthesis of the morphinan alkaloids thebaine (**192**), codeine (**198**), and morphine (**204**).

section *Papaver* (syn. *Mecones*) into the sections *Papaver*, in which *P. somniferum* and *P. setigerum* are classified, and *Glauca* J. Novák et V. Preininger, a new section which includes *P. glaucum, P. gracile,* and *P. decaisnei* (*29*). This conclusion was substantiated by studies of the morphological features and their correlation with the basic chromosome numbers (*29*). Species of the section *Papaver* have the chromosome number $x = 11$, whereas the species of the section *Glauca* have $x = 7$. In *P. somniferum* and *P. setigerum* there mostly occur diploid individuals ($2n = 22$) and very rarely tetraploid individuals ($2n = 44$). Our findings do not confirm the assumption that the occurrence of tetraploids is more frequent in *P. setigerum*. Even in this species, diploid individuals occur much more frequently.

5. Section *Glauca*

The new division of the section *Papaver* (syn. *Mecones*) (*29,30*) assigns the species *P. glaucum* Boiss. et Hausskn., *P. gracile* Auch., and *P. decaisnei* Hochst. et Steud. to the new section *Glauca* J. Novák et V. Preininger (Table XI). The species of the section *Glauca* are, to a certain extent, morphologically similar to some species of the section *Orthorhoeades,* but they differ in their habitat. Some authors (*53*) assume that these plants are very closely related since they actually contain very similar alkaloids. The basic chromosome number of the section *Glauca* is $x = 7$. Within this section, the diploid number of chromosomes is found to be $2n = 14$ (*54*).

As already mentioned, this section is characterized by the presence of rhoeadine–papaverrubine alkaloids. Chemical investigations of the herbarium samples of Iraqi *P. glaucum* indicate that chemical races exist in this section (*51*). In two samples, no alkaloids were detected, one contained rhoeagenine (**404**) as the major alkaloid, one unidentified rhoeadines, one unidentified aporphines, while the remaining sample yielded aporphine alkaloids including liriodenine (**157**), roemerine (**115**), and its *N*-oxide (**116**).

6. Section *Miltantha*

Some of the species of the section *Miltantha* Bernh. differ morphologically. The external habitat and the form of growth of *P. polychaetum* and *P. libanoticum* are unlike those of the other species. Nomenclatural difficulties also exist, particularly for chemists.

Detailed studies have been made of the alkaloids contained in *P. armeniacum* (L.) DC., *P. caucasicum* Marsch.-Bieb., *P. fugax* Poir., *P. persicum* Lindl., *P. floribundum* Desf., *P. polychaetum* Schott et Kotschy, and *P. triniaefolium* Boiss., cultivated from seeds obtained from different botanical gardens in the ČSSR and GDR. On the basis of the results obtained from these studies, the chemotaxonomical conclusion was drawn (55) that the significant alkaloids of the section are armepavine (12), mecambrine (110) and protopine (309) (Table XII) (56–60). The alkaloid armepavine was not found in any other section of the genus *Papaver*, which makes it a chemotaxonomic characteristic of the section *Miltantha*.

TABLE XII

Plants of *Papaver* Section *Miltantha* and Their Alkaloids

Papaver acrochaetum Born
Protopine (309), rhoeagenine (404) (51)

P. armeniacum L. (DC.)
Armepavine (12), coptisine (217b), cryptopine (311), floripavidine (120), glaudine (392), lirinidine (118), mecambrine (110), palmatine (218b), protopine (309), rhoeadine (389), rhoeagenine (404), sanguinarine (296), thebaine (192) (3,49,51,56)

P. caucasicum Marsch.-Bieb
Aporheine methohydroxide (116), armepavine (12), coptisine (217b), N-methylcrotonosine (111), glaziovine (109), mecambrine (110), mecambroline (124), nuciferine (117), nuciferoline (125), palmatine (218b), papaverrubine A, B, C, D, E (396–398, 400), (−)-pronuciferine (112), protopine (309), roemerine (115), salutaridine (188), sanguinarine (296) (3,34,49)

P. curviscapum Nabk.
Allocryptopine (310), alpinigenine (403), berberine (209b), flavinantine (185), 1-methoxyallocryptopine (316), 1-methoxy-13-oxoallocryptopine (320), protopine (309) (51,51a)

P. floribundum Desf.
Armepavine (12), floribundine (?), floripavidine (120), salutaridine (188) (3,49)

P. cylindricum Cullen
Armepavine (12), cheilanthifoline (210a), floripavidine (120), N-methylasimilobine (119), narcotine (332), oripavine (195), papaverine (35), rhoeadine (389), salutaridine (188), scoulerine (212), thebaine (192) (57)

P. fugax Poir.
Alpinigenine (403), amurensinine (61), aporheine (115), armepavine (12), chelerythrine (295), coptisine (217b), glaudine (392), glaucamine (393), mecambrine (110), N-methylasimilobine (119), narcotine (332), nuciferine (117), oreogenine (405), oreodine (392), palmatine (218b), papaverrubine B, C, D, E (397, 398, 400, 396), protopine (309), pronuciferine (112), remrefidine (116), roemerine (115), rhoeadine (389), salutaridine (188), sanguinarine (296), thebaine (192) (3,49,56)

P. persicum Lindl.
Alpinigenine (403), armepavine (12), coptisine (217b), floripavidine (120), glaudine (392), mecambrine (110), mecambridine (256a), nuciferine (117), O-demethylnuciferine (118), N-methylasimilobine (119), narcotine (332), palmatine (218b), papaverrubine B, D, (397, 398), (−)-pronuciferine (112), protopine (309), rhoeadine (389), roemerine (115), sanguinarine (296) (3,49,51)

(*continued*)

TABLE XII   *(Continued)*

P. polychaetum Schott. et Kotschy

Armepavine (**12**), mecambrine (**110**), nuciferine (**117**), palmatine (**218b**), papaverrubine B, D, E (**397, 398, 396**), pronuciferine (**112**), protopine (**309**), roemerine (**115**), sanguinarine (**296**) *(3,49)*

P. tauricolum Boiss.

Amurensinine (**61**), armepavine (**12**), cryptopine (**311**), epiglaudine (**402**), glaucamine (**393**), glaudine (**392**), lirinidine (**118**), mecambrine (**110**), narcotine (**332**), nantenine (**144**), nuciferine (**117**), oreodine (**392**), oreogenine (**405**), palmatine (**218b**), papaverrubine E (**396**), pronuciferine (**112**), protopine (**309**), rhoeadine (**389**), rhoeagenine (**404**), roemerine (**115**), scoulerine (**212**), sinoacutine (**188**), thebaine (**192**) *(3,49,56,58)*

P. triniaefolium Boiss.

Amurine (**183**), aporheine (**115**), armepavine (**12**), cheilanthifoline (**210a**), coptisine (**217b**), floripavidine (**120**), mecambrine (**110**), nuciferine (**117**), oreodine (**392**), oxysanguinarine (**281**), palmatine (**218b**), papaverine (**35**), papaverrubine B, D (**397, 398**), pronuciferine (**112**), protopine (**309**), rhoeadine (**389**), roemerine (**115**), salutaridine (**188**), sanguinarine (**296**), scoulerine (**212**), sinactine (**215a**), thebaine (**192**) *(3,49,59)*

P. urbanianum Fedde

Aporheine (**115**), armepavine (**12**), dehydroroemerine (**167**), floripavidine (**120**), mecambrine (**110**), *O*-methylsalutaridine (**187**), muramine (**314**), *N*-methyl-6,7-dimethoxytetrahydroisoquinolone-1 (**5**), palmatine (**218b**), papaverrubine B (**397**), protopine (**309**), salutaridine (**188**) *(3,49,60)*

However, the investigations of the plants of the section *Miltantha*, obtained from their natural habitats (Turkey, Iraq), showed that some of the species produce thebaine (**192**) (or morphinanes), whereas some others, rhoeadine alkaloids *(56,61)* (Table XII). *Papaver cylindricum* was found to contain the morphinane alkaloid oripavine (**195**) *(57)*, which is characteristic of *P. orientale* (section *Macrantha*), and papaverine (**35**) and narcotine (**332**), which are characteristic of *P. somniferum* and *P. setigerum* (section *Papaver*). Narcotine was also found in *P. fugax, P. persicum,* and *P. tauricolum,* and papaverine in *P. triniaefolium.* In the plants *P. fugax,* cultivated in England from seeds collected in Turkey, thebaine (**192**) was predominant, whereas in the plants collected in eastern Turkey, the major alkaloids were glaudine (**392**), glaucamine (**393**), and rhoeadine (**389**) *(56)*. The latter alkaloid was also found to be the major alkaloid of *P. armeniacum* collected in Turkey. In *P. tauricolum* [in three different batches collected in Anatolia (Turkey)], the major alkaloids were those of the rhoeadine type *(58)*. The isolation of different alkaloids from separate samples of the same species may be explained by genetic, climatic, or edaphic factors. Of interest is the analogy between the alkaloids present in this plant [rhoeadine (**389**), rhoeagenine (**404**), glaucamine (**393**), glaudine (**392**), and sinactine (**215a**)] and those found in *P. rhoeas* (section *Orthorhoeades*). This was the first time that rhoeadine alkaloids were found as major alkaloids in the section *Miltantha*, although the presence of rhoeadine as a trace alkaloid was demonstrated by us in *P. caucasicum* earlier (*62*).

The investigation of the herbarium samples from Iraq was carried out (*63*). *Papaver acrochaetum* yielded rhoeagenine (*404*) as the major alkaloid. The major alkaloids of *P. curviscapum* were protopine (**309**) and allocryptopine (**310**). The major alkaloid of Iraqi *P. persicum* and of *P. armeniacum* was identified as the glycosidic alkaloid floripavidine (**119**), which had been first reported from *P. floribundum*. For the infraspecific variation of the alkaloids of the section *Miltantha*, see also Ref. *64*.

In the section *Miltantha*, the basic chromosome number $x = 6$ and 7. All the studied cases are diploid species (*54*). In the species *P. acrochaetum, P. urbanianum*, and *P. cylindricum*, the number found is $2n = 12$; in the species *P. tauricolum, P. persicum, P. fugax, P. armeniacum, P. triniaefolium*, and *P. polychaetum*, it is $2n = 14$. In the other species, chromosome numbers were not determined.

On the basis of the results of the studies of alkaloids of the section *Miltantha*, it is concluded that there exist at least three different chemical races in which the major alkaloidal types are either benzylisoquinolines and proaporphines, or morphinanes or rhoeadines.

7. Section *Pilosa*

The section *Pilosa* Prantl. was systematically studied by several authors and almost all of the species classified in it by Fedde were investigated for their alkaloid content (Table XIII) (*65–68*).

In all of the studied species, rhoeadine–papaverrubine and protopine alkaloids were constantly present. Protopine alkaloids, having an oxygen at C-13, were found in *P. atlanticum* and *P. oreophilum*. This type of alkaloids was also isolated from *P. somniferum* (section *Papaver*), *P. hispidum* (section *Argemonorhoeades*), and from some species of the section *Scapiflora*. Three studied plants (*P. lateritium, P. pilosum*, and *P. monanthum*) contained the alkaloid glycoside latericine (**15**), which was also demonstrated in *P. californicum* (section *Orthorhoeades*). The plant species *P. oreophilum* contained the alkaloids oreodine (**392**) and oreogenine (**405**), which are related to the rhoeadine alkaloids glaudine (**392**) and glaucamine (**393**) found in the sections *Papaver* and *Glauca*. This indicates that they differ from other plants of the section *Pilosa*. Aporphine alkaloids were found in seven species (*P. atlanticum, P. feddei, P. heldreichii, P. oreophilum, P. pannosum, P. pilosum*, and *P. rupifragum*) and the promorphinane alkaloids (especially amurine) in six species (*P. feddei, P. heldreichii, P. pannosum, P. pilosum, P. spicatum*, and *P. strictum*) (Table XIV).

TABLE XIII

Plants of *Papaver* Section *Pilosa* and Their Alkaloids

Papaver atlanticum Ball.
  Coptisine (**217b**), cryptopine (**311**), magnoflorine (**138**), muramine (**314**), 13-oxocryptopine (**317**), 13-oxoprotopine (**319**), oxysanguinarine (**281**), papaverrubine A, B, C, D, E (**396–398, 400**), protopine (**309**), rhoeadine (**389**), rhoeagenine (**404**), sanguinarine (**296**), (−)-stylopine (**217a**) (*3,34,49*)

P. feddei Schwarz.
  Amurine (**183**), glaucine (**145**), papaverrubine B, D, E (**397, 398, 396**), roemerine (**115**) (*3,49*)

P. heldreichii Boiss.
  Amurine (**183**), aporheine (**115**), coptisine (**217b**), glaucine (**145**), liriodenine (**157**), papaverrubine A, B, D, E (**396–398**), sanguinarine (**296**) (*3,49*)

P. lateritium C. Koch.
  Latericine (**15**), papaverrubine A, B, D, E (**396–398**), protopine (**309**), rhoeadine (**389**), rhoeagenine (**404**) (*3,49*)

P. monanthum Trautv.
  Coptisine (**217b**), latericine (**15**), oxysanguinarine (**281**), protopine (**309**), rhoeadine (**389**), sanguinarine (**296**) (*3,49*)

P. oreophilum Rupr.
  Alborine (alkaloid PO-5) (**256b**), allocryptopine (**310**), berberine (**209b**), *O,N*-dimethyloridine (**97**), chelirubine (**300**), corydine (**136**), corysamine (**234b**), isocorydine (**132**), isorhoeadine (**389**), magnoflorine (**138**), mecambridine (**256a**), (−)-isooridine (**99**), menisperine (**133**), 1-methoxy-13-oxoallocryptopine (**320**), *N*-methyloridine (**100**), *O*-methyloridine (**101**), nuciferine (**117**), oreodine (**392**), oreogenine (**405**), oridine (**102**), oxysanguinarine (**281**), papaverrubine A, B, C, D, E, F (**396–398, 400**), protopine (**309**), rhoeadine (**389**), rhoeagenine (**404**), sanguinarine (**296**), thebaine (**192**) (*3,16,49,65,66*)

P. pannosum Schw.
  Amurine (**183**), dihydronudaurine (**189**), glaucine (**145**), mecambrine (**110**), nudaurine (**191**), roemerine (**115**) (*3,49*)

P. pilosum Sibth. et Smith
  Amurine (**183**), (+)-amurinine (**190**), coptisine (**217b**), dehydroglaucine (**172**), dehydroroemerine (**167**), (+)-dihydronudaurine (**189**), (−)-epiamurinine (**190**), (−)-glaucine (**145**), latericine (**15**), (−)-mecambrine (**110**), muramine (**314**), papaverrubine A, B, D, E (**396–398**), protopine (**309**), rhoeadine (**389**), rhoeagenine (**404**), (+)-roemerine (**115**), sanguinarine (**296**) (*3,49,67*)

P. rupifragum Boiss. et Reut.
  Allocryptopine (**310**), coptisine (**217b**), corysamine (**234b**), corytuberine (**137**), cryptopine (**311**), isorhoeadine (**389**), magnoflorine (**138**), papaverrubine A, B, C, D, E (**396–398, 400**), protopine (**309**), rhoeadine (**389**), rhoeagenine (**404**), (−)-stylopine (**217a**) (*3,49,68*)

P. spicatum Boiss. et Ball.
  Amurine (**183**) (*3,49*)

P. strictum Boiss. et Ball.
  Amurine (**183**) (*3,49*)

TABLE XIV
Chemotaxonomically Significant Alkaloids in Species of *Papaver* Section *Pilosa*

Species	Promorphinanes	Aporphines	13-Oxoprotopines	Latericine
Papaver pilosum	+	+	+	+
P. oreophilum	+	+	+	−
P. feddei	+	+	−	−
P. heldreichii	+	+	−	−
P. pannosum	+	+	−	−
P. spicatum	+	−	−	−
P. strictum	+	−	−	−
P. atlanticum	−	+	+	−
P. rupifragum	−	+	−	−
P. lateritium	−	−	−	+
P. monanthum	−	−	−	+

The basic chromosome numbers of the section *Pilosa* are $x = 6$ and 7 (*54*). All the studied plants were diploid species. In *P. pilosum*, *P. strictum*, *P. atlanticum*, *P. oreophilum*, and *P. monanthum*, $2n = 14$; in *P. rupifragum*, $2n = 12$.

The presence/absence of the promorphinane alkaloid amurine (**183**) has shown the difference between the species *P. feddei*, *P. heldreichii*, *P. pannosum*, *P. pilosum*, *P. spicatum*, and *P. strictum* and the species *P. atlanticum*, *P. lateritium*, *P. monanthum*, *P. oreophilum*, and *P. rupifragum* (*24,69*). These conclusions and the results from studies of the morphological features indicate the possibility of a subdivision within the section *Pilosa* (Table XIV).

8. Section *Macrantha* (Syn. *Oxytona* Bernh.)

Plants of the section *Macrantha* Elkan have been intensively studied, particularly since 1973 (*70–73*), because some varieties of *P. bracteatum* produce considerable yields of thebaine (**192**), which can readily be converted to codeine (**198**) (a milder analgesic) but not so readily to morphine (**204**) (see Fig. 4). The illegal use of opiates and especially the easy conversion of morphine (**204**) to heroin shows that opium poppy (*P. somniferum*)is a dangerous plant to cultivate on a large scale. The severity of the problem is demonstrated by the U.N. Division of Narcotic Drugs' approval and financing of the project "Scientific Research of *P. bracteatum*" (1974). No case of thebaine abuse has been reported to date. However, since thebaine (**192**) is the major alkaloid of *P. bracteatum* and since it is the starting material for 6,14-*endo*-ethenotetrahydrothebaines (Bentley compounds) (*74*), which are powerful analgesics but also cause strong dependency, it was felt that *P. bracteatum* should be controlled as well. Attempts have been made to obtain thebaine from tissue cultures or to increase its natural production (75–79).

The genus *Macrantha* was reinvestigated by Goldblatt (80) in 1974 who recognized three species: *P. bracteatum, P. orientale,* and *P. pseudoorientale.* Each of these species can be distinguished morphologically by its petal marking, the shape of the capsules, and by the chromosome number (54). Studies of the chromosome numbers have shown that, in most of the species of the genus *Papaver* and also in plants of the section *Macrantha,* the basic chromosome number is $x = 7$ (71,81,82). It seems that all three species are evidently determined on the basis of somatic numbers of chromosomes, and karyologic conditions provide confirmatory evidence for the elaborated classification of this section. In *P. bracteatum,* the diploid number of chromosomes is $2n = 14$. In comparison with the other two species, they have fewer metaphase structures, the cells and interkinetic nuclei are smaller, and the chromosomes are usually at a shorter distance. *Papaver orientale* is a tetraploid species with $2n = 28$. In the species *P. pseudoorientale,* the hexaploid number of chromosomes is $2n = 42$. Thus the assumption has been confirmed that it is an allohexaploid species that has arisen by hybridization between the diploid *P. bracteatum* and the tetraploid *P. orientale.* Also found were species of *P. pseudoorientale* having other chromosome numbers. Their development might have been due to various causes, e.g., crossbreeding.

Papaver bracteatum Halle III (83) and *P. bracteatum* of Iranian origin (83a) contain thebaine (**192**) as the major alkaloid and only small quantities of alpinigenine (**403**), orientalidine (**257a**), and isothebaine (**122**) (Table XV) (85–90). It has been assumed that the variety Halle III exists in two different races, one producing only thebaine and the other thebaine and alpinigenine (73). In *P. bracteatum* of Turkish origin, the major alkaloids are thebaine and salutaridine (**188**) (61). The three existing chemotypes of *P. bracteatum* are thebaine, thebaine–alpinigenine, and thebaine–salutaridine. The presence of other chemotypes cannot be excluded (55). Interesting are the ontogenetic studies of *P. bracteatum* Halle III (83) and *P. bracteatum* from Iran (84) (Fig. 5). The bound form of thebaine occurs in the pericarp of *P. bracteatum* (91).

Analyses of *P. orientale* from natural habitats showed that the dried latex contained 20% of oripavine (**195**) and 9% of thebaine (**192**) (92). Traces of isothebaine (**122**) were found in plants collected from Khalkhala in northwest Iran. Six different chemotypes of *P. orientale* were reported (64,93), namely, oripavine, oripavine–thebaine, oripavine–isothebaine, oripavine–alpinigenine, oripavine–thebaine–alpinigenine, and mecambridine–orientalidine–salutaridine. In all of them, with the exception of one, oripavine was the major alkaloid (82). Feeding experiments in *P. orientale* with radioactively labeled reticuline and thebaine have demonstrated that oripavine is derived from reticuline and thebaine (93a). Reticuline undergoes racemization in this plant as has been earlier shown for *P. somniferum* and *P. bracteatum.*

TABLE XV

Plants of *Papaver* Section *Macrantha* and Their Alkaloids

Papaver bracteatum Lindl.

Alpinigenine (**403**), alpinine (**394**), bractamine (?), bracteine (**113**), bracteoline (**142**), codeine (**198**), coptisine (**217b**), corypaline (**1**), floripavidine (**119**), 14β-hydroxycodeine (**201**), 14β-hydroxycodeinone (**207**), isothebaine (**122**), mecambridine (**256a**), *O*-methylcorypaline (**2**), nuciferine (**117**), *N*-methylcorydaldine (**5**), neopine (**208**), norcorydine (**135**), orientalidine (**257a**), oripavine (**195**), oxysanguinarine (**281**), papaverrubine B, D, E (**397, 398, 396**), protopine (**309**), salutaridine (**188**), thebaine (**192**), thebaine *N*-oxide (**194**), thebaine methohydroxide (**193**), α-thebaol, *O*-methylflavinanthine (**184**) (*3,85–87*)

P. orientale L.

Alkaloids PO-3 (**162**), PO-4 (**257b**), PO-5 (= alborine) (**256b**), bracteine (**113**), bracteoline (**142**), coptisine (**217b**), 6a,7-dehydroisothebaine (**169**), glaucidine (?), isothebaidine (**121**), isothebaine (**122**), mecambridine (**256a**), *O*-methyl-6a,7-dehydroisothebaine (**168**), *O*-methylorientine (**126**), nuciferine (**117**), (−)-orientalinone (**113**), oripavidine (**196**), orientalidine (**257a**), orientaline (**25**), orientine (**128**), oripavine (**195**), oxysanguinarine (**281**), dihydroorientalinone (**108**), papaverrubine B, C, D, E (**397, 398, 400, 396**), salutaridine (**188**), sanguinarine (**296**), thebaine (**192**) (*3,88–90*)

P. pseudoorientale (Fedde)

Alkaloids PO-4 (**257b**), PO-5 (**256b**), aryapavine (**255a**), bracteoline (**142**), macrantaline (**261**), macranthoridine (**262**), isothebaine (**122**), isothebaine methohydroxide (**123**), orientalidine (**257a**), salutaridine (**188**) (*3,34*)

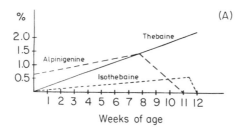

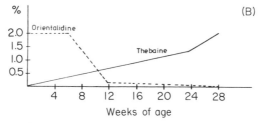

FIG. 5. Alkaloid content in *Papaver bracteatum* L. during ontogenesis [according to Nyman and Bruhn (*71*); (A) Böhm (*83*), (B) Cheng (*84*)].

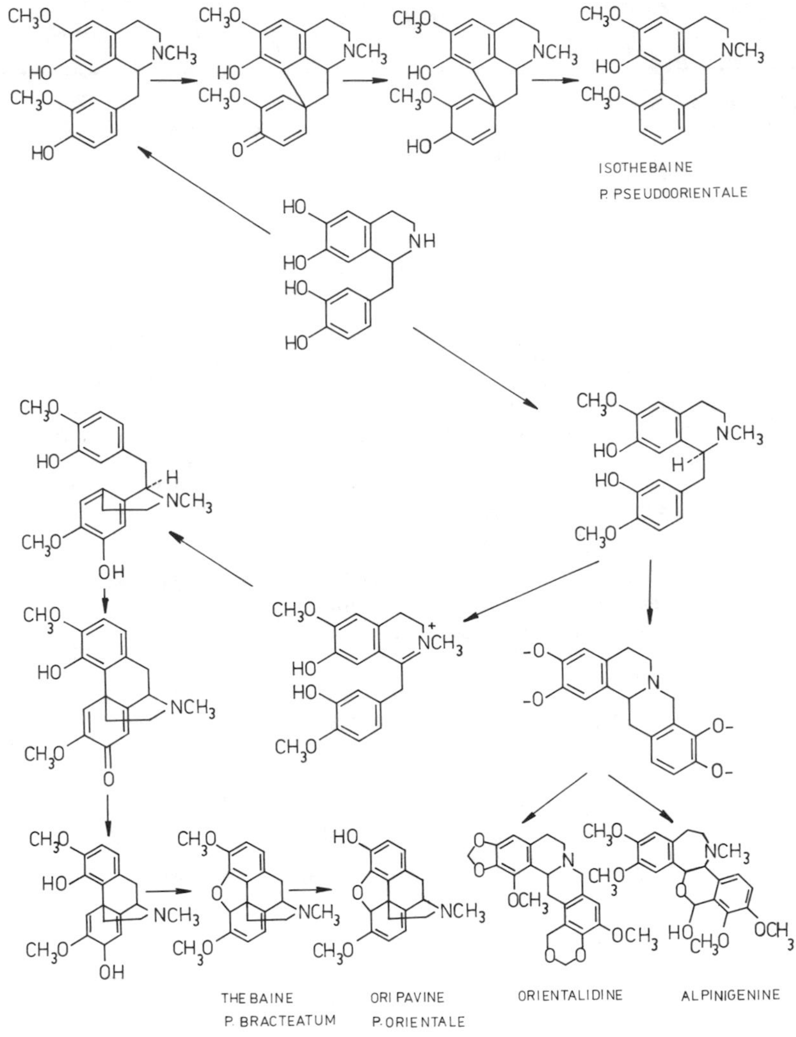

FIG. 6. Biosynthesis of the major alkaloids of section *Macrantha*.

The dry latex of *P. pseudoorientale* from Iran yielded isothebaine (**122**) (11.7%) as the major alkaloid as well as orientalidine (**257a**) (0.5%) and some other alkaloids (*92*). Samples of *P. pseudoorientale* from Turkey contained the major alkaloids salutaridine (**188**) and macrantaline (**261**) [the minor alkaloid macrantoridine (**262**)], or salutaridine and thebaine, or only salutaridine (*94*). Consequently, *P. orientale* exists in at least four different chemical races, one containing isothebaine, the second, salutaridine and thebaine, the third, salutaridine, and the fourth, salutaridine and macrantaline as the major alkaloids. Figure 6 shows the biosynthesis of the major alkaloids of the section *Macrantha*.

Thebaine (**192**) and oripavine (**195**) were isolated as the dominant alkaloids from hybrids of *P. bracteatum* and *P. orientale*, whereas isothebaine (**122**), with lesser quantities of oripavine, was obtained from hybrids of *P. orientale* and *P. pseudoorientale*. Morphine (**204**), thebaine (**192**), and alpinigenine (**403**) were isolated from dried capsules of hybrids of *P. bracteatum* and *P. somniferum* (*83,95*).

Isothebaine (**122**) and orientalidine (**257a**) do not appear in any other sections of the genus *Papaver* or in the other genera of the Papaveraceae and Fumariaceae. Therefore, their presence can be considered as a chemotaxonomic characteristic of the section *Macrantha*. Since the plants of the section *Macrantha* form isothebaine from (*S*)-(+)-orientaline (**25**), thebaine and oripavine from (*R*)-(−)-reticuline (**20**), and orientalidine and alpinigenine from (*S*)-(+)-reticuline, it follows that they contain stereospecific enzymes.

9. Section *Scapiflora* (Syn. *Meconella* Spach)

This section has been subdivided into nine series on the basis of a morphologicogeographical analysis (*96*). The presence of isopavines, promorphinanes, protopines, and rhoeadines–papaverrubines can be considered as chemotaxonomically significant for this section. The isopavines amurensine (**60**) and amurensinine (**61**) were detected in almost all of the studied species of this section (Table XVI) (*92–98*). Promorphinanes amurine (**183**) and nudaurine (**191**) were found in all subspecies of *P. alpinum*. Amurine (**183**) could be demonstrated in the majority of the var. *P. nudicaule*. All subspecies of *P. alpinum* contain the rhoeadine-type alkaloids alpinine (**394**) and alpinigenine (**403**) but not rhoeadine (**389**). In contrast, rhoeadine (**389**) is present in some of the varieties of *P. nudicaule*. Aporphine alkaloids have not yet been found in species of the section *Scapiflora* except in *P. lisae*, which is considered to be a relic of the sedimentary rock and endemic to the Caucasus. Originally, it was described as a variety of the species *P. nudicaule*, which was later classified in the section *Pilosa*. (*96*).

B. THE GENUS *Glaucium*

The species of the genus *Glaucium* Tourn. are characterized by the presence of aporphine and protopine alkaloids in addition to protoberberines and benzophenanthridines (Tables XVII and XVIII). Platonova *et al.* (*110*) isolated corydine (**136**) from *G. squamigerum* as the major alkaloid in addition to a smaller amount of protopine (**309**) and allocryptopine (**310**). In *G. oxylobum*, the major alkaloid was corydine together with protopine and allocryptopine (*107*). In contrast, Slavík *et al.* (*108*) isolated allocryptopine (**310**) and protopine (**303**) as the

TABLE XVI

Plants of *Papaver* Section *Scapiflora* (syn. *Meconella*) and Their Alkaloids

---

Papaver alboroseum Hulten
  Alborine (**256b**), mecambridine (**256a**), papaverrubine C, D (**398, 400**) (*3*)

P. alpinum spp. *alpinum*
  Alborine (**256b**), alpinigenine (**403**), alpinine (**394**), amurensine (**60**), amurensinine (**61**),
    amurine (**183**), cryptopine (**311**), mecambridine (**256a**), nudaurine (**191**), papaverrubine C,
    D (**398, 400**), protopine (**309**), sanguinarine (**296**) (*3*)

P. alpinum ssp. *burseri* (Crantz) Fedde
  Amurensine (**60**), amurensinine (**61**), amurine (**183**), coptisine (**217b**), cryptopine (**311**) (*3*)

P. alpinum ssp. *kerneri* (Hayek) Fedde
  Alborine (**256b**), alpinigenine (**403**), alpinine (**394**), amurensinine (**61**), amurensine (**60**),
    amurine (**183**), cryptopine (**311**), mecambridine (**256a**), muramine (**314**), nudaurine (**191**),
    papaverrubine B, D, E, G (**396–399**), protopine (**309**), sanguinarine (**296**) (*3*)

P. alpinum ssp. *tatricum* Nyár
  Epialpinine (**395**), oxysanguinarine (**281**), sanguinarine (**296**) (*3*)

P. alpinum ssp. *rhaeticum*
  Alborine (**256b**), alpinigenine (**403**), alpinine (**394**), amurensine (**60**), amurensinine (**61**),
    amurine (**183**), cryptopine (**311**), mecambridine (**256a**), muramine (**314**), nudaurine (**191**),
    papaverrubine D, G (**398, 399**), protopine (**309**), sanguinarine (**296**) (*3*)

P. alpinum ssp. *sendtneri*
  Alborine (**256b**), alpinigenine (**403**), alpinine (**394**), amurensine (**60**), amurensinine (**61**),
    amurine (**183**), cryptopine (**311**), mecambridine (**256a**), muramine (**314**), nudaurine (**191**),
    papaverrubine D, G (**398, 399**), protopine (**309**), sanguinarine (**296**) (*3*)

P. anomalum Fedde
  Alborine (**256b**), allocryptopine (**310**), amurensine (**60**), amurensinine (**61**), amurine (**183**),
    nudaurine (**191**), coptisine (**217b**), cryptopine (**311**), glaucamine (**393**), mecambridine
    (**256a**), oxysanguinarine (**281**), papaverrubine C, D, G (**398–400**), pavanoline (?), protopine
    (**309**), reframidine (**64**), rhoeadine (**389**), sanguinarine (**296**) (*3*)

P. lisae Reichenb.
  (+)-Isocorydine (**132**), macrantaline (**261**), (−)-mecambridine (**256a**), *N*-methyloridine (**100**),
    oridine (**102**), protopine (**309**) (*97,98*)

P. nudicaule L.
  Papaverrubine B, D (**397, 398**) (*3*)

P. nudicaule L. var. *amurense* N. Busch.
  Amurensine (**60**), amurine (**183**), amuroline (**104**), amuronine (**106**), muramine (**314**) (*3*)

P. nudicaule L. var. *aurantiacum* Loisel
  Amurine (**183**), nudaurine (**191**) (*3*)

P. nudicaule L. var. *croceum* Ledeb
  Amurine (**183**), muramine (**314**), 13-oxomuramine (**318**), protopine (**309**), amuronine (**106**) (*3*)

P. nudicaule L. var. *leiocarpum* (Turez.) Fedde
  Amurensine (**60**), amurensinine (**61**), cryptopine (**311**), glaucamine (**193**), mecambridine
    (**256a**), oxysanguinarine (**281**), papaverrubines, protopine (**309**), rhoeadine (**389**) (*3*)

TABLE XVI   *(Continued)*

P. nudicaule L. ssp. *radicatum* Rottb.
   Allocryptopine (**310**), amurensinine (**61**), amurine (**183**), berberine (**209b**), cryptopine (**311**),
      O-methylthalisopavine (**63**), papaverrubine B, D, E (**397, 398, 396**), protopine (**309**),
      sanguinarine (**296**) (*96a*)

P. nudicaule L. ssp. *rubroaurantiacum* (DC.) Fedde
   Amurensine (**60**), amurensinine (**61**), amurine (**183**), amuronine (**106**), muramine (**314**),
      nudicaulinole (?), oxysanguinarine (**281**), papaverrubines, protopine (**309**), rhoeadine (**389**),
      rhoeagenine (**404**), sanguinarine (**296**) (*3*)

P. nudicaule L. ssp. *xanthopetalum* (Trautv.) Fedde
   Amurensine (**60**), amurensinine (**61**), amurine (**183**), coptisine (**217b**), cryptopine (**311**),
      mecambridine (**256a**), muramine (**314**), oxysanguinarine (**281**), papaverrubines, protopine
      (**309**), rhoeadine (**389**), sanguinarine (**296**) (*3*)

P. pseudocanescens M. Pop.
   Alborine (**256b**), amurensine (**60**), amurensinine (**61**), cryptopine (**311**), mecambridine (**256a**),
      mecambridine methohydroxide (**258**), papaverrubine C, D, E (**398, 400, 396**), protopine
      (**309**), rhoeadine (**389**) (*3*)

P. pyrenaicum (L.) Kerner
   Allocryptopine (**310**), amurensine (**60**), amurensinine (**61**), coptisine (**217b**), muramine (**314**),
      oxysanguinarine (**281**), papaverrubines, sanguinarine (**296**) (*3*)

P. suaveolens Lap.
   Amurensinine (**61**), coptisine (**217b**), oxysanguinarine (**281**), papaverrubines, protopine (**309**),
      sanguinarine (**296**) (*3*)

TABLE XVII
Plants of the Genus *Glaucium* and Their Alkaloids

Glaucium contortuplicatum Boiss.
   Dicentrine (**140**), sinoacutine (**188**) (*3*)

G. corniculatum Curt.
   Allocryptopine (**310**), berberine (**209b**), (−)-α-canadine methohydroxide (**244**), chelerythrine
      (**295**), (−)-chelidonine (**266**), chelirubine (**300**), coptisine (**217b**), (+)-corydine (**136**),
      dehydrocorydine (**170**), glaufidine (**155**), heliotrine (**410**), (+)-isocorydine (**132**),
      norbracteoline (**141**), predicentrine (**151**), protopine (**309**), reticuline (**20**), sanguinarine
      (**296**), (−)-β-stylopine methohydroxide (**249**), thalicmidine (**153**) (*3,99*)

G. corniculatum ssp. *refractum*
   α-Allocryptopine (**310**), (±)-chelidonine (**266**), dehydrodicentrine (**171**), dicentrine (**140**),
      dicentrinone (**159**), glaucine (**145**), *N*-methyllaurotetanine (**149**), protopine (**309**) (*100*)

G. elegans Fisch. et Mey.
   Allocryptopine (**310**), chelerythrine (**295**), (±)-chelidonine (= diphylline) (**266**), chelirubine
      (**300**), coptisine (**217b**), corydine (**136**), dihydrochelerythrine (**278**), glaucine (**145**), glauvine
      (**178**), isoboldine (**147**), isocorydine (**132**), *O*-methylatheroline (**160**), protopine (**309**),
      sanguinarine (**296**) (*3*)

(*continued*)

TABLE XVII   (*Continued*)

G. fimbrilligerum Boiss.
Allocryptopine (**310**), berberine (**209b**), chelerythrine (**295**), chelidonine (**266**), chelirubine (**300**), coptisine (**217b**), (+)-corydine (**136**), corytuberine (**137**), glaufidine (**155**), glaufine (**136**), glaunine (**158**), glaunidine (**163**), isoboldine (**147**), (+)-isocorydine (**132**), (−)-isocorypalmine (**214a**), magnoflorine (**138**), *N*-methyllindcarpine (**139**), *O*-methylarmepavine (**19**), *N*-methylcoclaurine (**18**), protopine (**309**), sanguinarine (**296**) (*3,101,102*)

G. flavum Crantz.
Allocryptopine (**310**), (−)-aurotensine (= isoboldine) (**147**), bocconoline (**279**), bulbocapnine (**129**), chelerythrine (**295**), (−)-chelidonine (**266**), chelirubine (**300**), coptisine (**217b**), (+)-corydine (**136**), corytuberine (**137**), dehydroglaucine (**172**), dehydronorglaucine (**176**), dicentrine (**140**), (+)-glaucine (**145**), glauflavine (= corytuberine) (**137**), glauvine (**178**), 1-hydroxy-2,9,10-trimethoxyaporphine (**146**), 6-iminosanguinarine (**283**), (+)- and (±)-isoboldine (**147**), (+)-isocorydine (**132**), luguine (**270**), magnoflorine (**138**), *O*-methylatheroline (**160**), (−)-norchelidonine (**267**), protopine (**309**), sanguinarine (**296**), scoulerine (**212**), thaliporphine (= *O*-methylisoboldine (**153**) (*3,103,104*)

G. flavum ssp. *fulvum* (Smith) Fedde
(+)- and (±)-Isoboldine (**147**) (*3*)

G. flavum Crantz. pop. Ghom.
Bulbocapnine (**129**), dicentrine (**140**), salutaridine (**188**) (*3*)

G. flavum Crantz. var. *leiocarpum*
Oxoglaucine (= *O*-methylatheroline) (**160**) (*3*)

G. flavum Crantz. var. *vestitum*
Arosine (**163**), arosinine (**164**), cataline (**156**), corunnine (**175**), glaucine (**145**), pontevedrine (**182**), thalicmidine (= thaliporphine) (**153**) (*3,105*)

G. grandiflorum Boiss. et Huet.
Berberine (**209b**), (+)-glaucine (**145**), glauvine (**178**), isoboldine (**147**), isocorydine (**132**), *O*-methylatheroline (**160**), protopine (**309**), sanguinarine (**296**), thalicmidine (**153**) (*3,51*)

G. grandiflorum Boiss. et Huet. var. *torquatum* Cullen
Allocryptopine (**310**), canadine methohydroxide (**244**), corydine (**136**), cryptopine (**311**), (+)-glaucine (**145**), isocorydine (**132**), protopine (**309**) (*106*)

G. leiocarpum Boiss.
Dehydroglaucine (**172**), glaucine (**145**), (+)- and (±)-isoboldine (**147**), protopine (**309**) (*3*)

G. oxylobum Boiss. et Buhse
Allocryptopine (**310**), berberine (**209b**), chelerythrine (**295**), chelilutine (**201**), chelirubine (**300**), coptisine (**217b**), corydine (**136**), dehydrocorydine (**170**), domesticine (**143**), glaufidine (**155**), (+)- and (±)-isoboldine (**147**), isocorytuberine (**134**), *N*-methylcoclaurine (**18**), norisocorydine (**131**), protopine (**309**), sanguinarine (**296**) (*3,107*)

G. pulchrum Stafp. pop. Elika
Bulbocapnine (**129**), corydine (**136**), isocorydine (**132**), *N*-methyllindcarpine (**139**), protopine (**309**) (*3*)

G. squamigerum Kar. et Kir.
Allocryptopine (**310**), berberine (**209b**), canadine (**209a**), (−)-α-canadine methohydroxide (**244**), (−)-β-canadine methohydroxide (**244**), chelerythrine (**295**), (−)- and (±)-chelidonine

TABLE XVII   (*Continued*)

(266), chelirubine (300), coptisine (217b), corydine (136), corysamine (234b), corytuberine (137), (−)-β-isocorypalmine methohydroxide (247), magnoflorine (138), protopine (309), sanguinarine (296), scoulerine (212), stylopine (217a), α-stylopine methohydroxide (249), (−)-β-stylopine methohydroxide (249), β-tetrahydropalmatine methohydroxide (250) (*3,108*)

G. *vitelinum* Boiss. et Buhse pop. Seerjan
  Allocryptopine (310), bulbocapnine (129), corydine (136), chelidonine (266), dehydrodicentrine (171), dicentrine (140), dicentrinone (159), dihydrochelerythrine (278), dihydrosanguinarine (280), glaucine (145), isocorydine (137), 4-hydroxybulbocapnine (154), *N*-methyllaurotetanine (149), *N*-methyllindcarpine (139), muramine (314), neolitsine (150), protopine (309), salutaridine (188), tetrahydropalmatine (218a) (*3,109*)

TABLE XVIII
Alkaloids Found in Plants of the Genus *Glaucium*

Benzylisoquinolines
  O-Methylarmepavine, *N*-methylcoclaurine, reticuline

Aporphines
  Arosine, arosinine, bulbocapnine, cataline, corunnine, (+)-corydine, corytuberine (= glauflavine), dehydrocorydine, dehydroglaucine, dehydrodicentrine, dehydronorglaucine, dicentrine, dicentrinone, domesticine, (+)-glaucine, glaufine, glaufidine, glaunine, glanidine, glauvine, 4-hydroxybulbocapnine, 1-hydroxy-2,9,10-trimethoxyaporphine, (+)- and (±)-isoboldine, (+)-isocorydine, isocorytuberine, magnoflorine, *N*-methyllaurotetanine, *N*-methyllindcarpine, *O*-methylatheroline (= oxoglaucine), neolitsine, norbracteoline, norisocorydine, pontevedrine, predicentrine, thalicmidine (= thaliporphine), thaliporphine (= *O*-methylisoboldine)

Promorphinanes
  Salutaridine, sinoacutine

Protoberberines
  Canadine, (−)-α-canadine methohydroxide, (−)-β-canadine methohydroxide, (−)-isocorypalmine, (−)-β-isocorypalmine methohydroxide, scoulerine, stylopine, α-stylopine methohydroxide, (−)-β-stylopine methohydroxide, tetrahydropalmatine, β-tetrahydropalmatine methohydroxide, berberine, coptisine, corysamine

Benzophenanthridines
  Bocconoline, chelerythrine, chelilutine, (−)- and (±)-chelidonine, chelirubine, dihydrochelerythrine, dihydrosanguinarine, luguine, (−)-norchelidonine, 6-iminosanguinarine, sanguinarine

Protopines
  Allocryptopine, cryptopine, muramine, protopine

dominant alkaloids from *G. squamigerum*. They found that, in comparison with other species of the genus *Glaucium, G. squamigerum* displays a relatively high content of quaternary alkaloids (*16*). After allocryptopine, (−)-β-canadine methohydroxide (**244**) represents the second dominant alkaloid of the plant. The herbal specimens of *Glaucium* species of Iraq contain the same major alkaloids—allocryptopine and protopine (*51*); the other alkaloids identified therein are corydine (**136**), isocorydine (**132**), and berberine (**209b**). Thus chemical races also exist in the genus *Glaucium*.

From some species of the genus *Glaucium*, (−)-chelidonine (**266**) was isolated, while its enantiomer (+)-chelidonine is the dominant alkaloid in *Chelidonium majus* L. and *Stylophorum diphyllum* (Michx.) Nutt.

C. The Genus *Roemeria*

The genus *Roemeria* Medic. comprises approximately 10 species, of which only *R. hybrida* (L.) DC. and *R. refracta* (Stev.) DC. have been studied for their alkaloid content (Tables XIX and XX). From *R. hybrida,* Platonova *et al.* (*110*) isolated protopine (**309**) as the major alkaloid together with roemeridine. In contrast to Soviet authors, Slavík *et al.* (*111*) were not able to detect the presence of protopine (**309**) in their plant material. They isolated roemeridine as the main component. In two herbal samples of *R. hybrida* from Iraq, the major alkaloid was roehybrine (**103**) (*51*).

From *R. refracta,* Konovalova *et al.* (*112*) isolated (−)-roemerine (**115**) as the major alkaloid and a smaller quantity of L-(−)-ephedrine and D-(+)-pseudoephedrine. Yunusov *et al.* (*113*) isolated, in addition to roemerine, a low yield of anonaine (**114**), liriodenine (**157**), and remrefine (**66**). On the contrary, anonaine, liriodenine, and ephedrine could not be proved by Slavík *et al.* (*114*), who found reframine (**65**) and roemerine as the major alkaloids.

A comparison of the two species shows that biochemically they differ considerably. Both of them contain proaporphines, which in the herbal specimens of *R. hybrida* native to Iraq are the major alkaloids. They differ by the presence of

TABLE XIX

Plants of the Genus *Roemeria* and Their Alkaloids

Roemeria hybrida (L.) DC. (syn. *R. violacea* Medic.)
 Coptisine (**217b**), (−)-isocorypalmine (**214a**), roehybridine (?), roehybrine (**103**), roemeridine (?) (*3,51*)

R. refracta (Stev.) DC. (syn. *R. rhoeadiflora* Boiss.)
 Anonaine (**114**), liriodenine (**157**), (+)-mecambrine (**110**), (−)-mecambroline (**124**), protopine (**309**), reframidine (**64**), reframine (**65**), reframoline (**67**), remrefidine (**116**), remrefine (= reframine methohydroxide) (**66**), roehybrine (**103**), roemeramine (**105**), (−)-roemerine (**115**), roemeroline (**127**), roemeronine (**107**) (*3,51*)

TABLE XX

Alkaloids Found in Plants of the Genus *Roemeria*

Pavines and isopavines
  Reframidine, reframine, reframoline, remrefine (= reframine methohydroxide)

Proaporphines
  (+)-Mecambrine, roehybrine, roemeramine, roemeronine

Aporphines
  Anonaine, liriodenine, (−)-mecambroline, remrefidine, (−)-roemerine, roemeroline

Protoberberines
  (−)-Isocorypalmine; coptisine

Protopines
  Protopine

Alkaloids of unknown structure
  Roehybridine, roemeridine

---

isopavines in *R. refracta* and their absence in *R. hybrida.* In Papaveraceae, the presence of isopavines is very rare. With the exception of single findings in the individual plants of the genus *Papaver* and *Meconopsis,* isopavines are tax-onomically significant for the *Papaver* section *Scapiflora.*

D. THE GENUS *Meconopsis*

Meconopsis Vig. is the fourth largest genus (after *Eschscholtzia, Papaver,* and *Platystemon*) of the family Papaveraceae, including some 40–45 species that are indigenous mainly to the Himalayas and China. It has been concluded that the common biochemical feature of all investigated Asian species of the genus *Meconopsis* is the presence of protopine (**309**) or amurensinine (**61**), which in many instances are also the major alkaloids of the plant (*115*) (Tables XXI and XXII). They are often accompanied by cryptopine (**311**) or allocryptopine (**310**). The occurrence of 6-methoxy-2-methyl-1,2,3,4-tetrahydro-β-carboline in sever-al species is also remarkable. The only West European polycarpic species, *M. cambrica,* occupies a rather isolated position by its geographic occurrence and the content of its alkaloids. It yields the major tertiary alkaloids mecambrine (**110**) and mecambridine (**256b**) and the minor alkaloids roemerine (**115**) and corytuberine (**137**), which have not been detected in other *Meconopsis* species as yet. A reinvestigation of *M. cambrica* grown in England (*116*) revealed that its major alkaloid was the quaternary aporphine magnoflorine (**138**). The general occurrence of papaverrubine, or of small amounts of rhoeadine (**389**) and iso-rhoeadine (**389**), indicates a close biochemical relationship with the *Papaver* genus. The genus *Meconopsis* is the only one that together with the genus *Papaver* bears rhoeadine alkaloids. The presence of amurensinine (**61**) as the

TABLE XXI
Plants of the Genus *Meconopsis* and Their Alkaloids

Meconopsis aculeata Royle
    Protopine (**309**), sanguinarine (**296**) (*116*)

M. betonicifolia Franch.
    Allocryptopine (**310**), amurensinine methohydroxide (**62**), berberine (**209b**), coptisine (**217b**),
        corysamine (**234b**), cryptopine (**311**), isorhoeadine (**389**), papaverrubine A, C, D, E (**396,
        398, 400**), protopine (**309**), rhoeadine (**389**), sanguinarine (**296**) (*3,115,116*)

M. cambrica (L.) Vig.
    Amurine (**183**), berberine (**209b**), coptisine (**217b**), corytuberine (**137**), flavinantine (**185**),
        magnoflorine (**138**), mecambridine (**256a**), mecambrine (**110**), *N*-methylcrotonosine (**111**),
        papaverrubine D (**398**), (−)-pronuciferine (**112**), roemerine (**115**), sanguinarine (**296**) (*3,116*)

M. dhwojii G. Taylor
    Sanguinarine (**296**) (*116*)

M. horridula Hook. f. et Thoms.
    Allocryptopine (**310**), amurensinine (**61**), amurensinine methohydroxide (**62**), coptisine (**217b**),
        isorhoeadine (**389**), papaverrubine C, D, E (**398, 400, 396**), protopine (**309**), rhoeadine
        (**389**), sanguinarine (**296**) (*3,115,116*)

M. latifolia Prain.
    Protopine (**309**), sanguinarine (**296**) (*116*)

M. nepalensis DC
    Allocryptopine (**310**), amurensinine (**61**) amurensinine methohydroxide (**62**), coptisine (**217b**),
        corysamine (**234b**), cryptopine (**311**), magnoflorine (**138**), isorhoeadine (**389**), *N*-
        methyltetrahydronorharmane (**409**), papaverrubine A, C (**396, 400**), protopine (**309**),
        rhoeadine (**389**) (*3,115*)

M. paniculata (D. Don) Prain.
    Allocryptopine (**310**), coptisine (**217b**), corysamine (**234b**), cryptopine (**311**), magnoflorine
        (**138**), papaverrubine A, C, D, E (**396, 398, 400**), protopine (**309**), rhoeadine (**389**) (*3,115*)

M. robusta Hook. f. et Thoms.
    Allocryptopine (**310**), coptisine (**217b**), corysamine (**234b**), cryptopine (**311**), magnoflorine
        (**138**), papaverrubine C, D, E (**398, 400, 396**), protopine (**309**), rhoeadine (**389**) (*3,115*)

M. rudis Prain.
    Allocryptopine (**310**), amurensine (**60**), amurensinine (**61**), amurensinine methohydroxide (**62**),
        coptisine (**217b**), isorhoeadine (**389**), magnoflorine (**138**), papaverrubine A, C, D, E (**396,
        398, 400**), protopine (**309**), rhoeadine (**389**), sanguinarine (**296**) (*3,115,116*)

M. sinuata Prain.
    Allocryptopine (**310**), amurensinine (**61**), coptisine (**217b**), cryptopine (**311**), papaverrubine D
        (**398**), protopine (**309**) (*3,115*)

TABLE XXII
Alkaloids Found in Plants of the Genus *Meconopsis*

Pavines and isopavines
  Amurensine, amurensinine, amurensinine methohydroxide

Proaporphines
  Mecambrine, (−)-*N*-methylcrotonosine, (−)-pronuciferine

Aporphines
  Corytuberine, magnoflorine, roemerine

Promorphinanes
  (−)-Amurine, (−)-flavinantine

Protoberberines
  Berberine, coptisine, corysamine

Retroprotoberberines
  Mecambridine

Benzophenanthridines
  Sanguinarine

Protopines
  Allocryptopine, cryptopine, protopine

Rhoeadines
  Isorhoeadine, papaverrubine A, C, D, E, rhoeadine

---

major alkaloid in the species *M. rudis* and *M. horridula* indicates the relationship with the species of the section *Scapiflora* (genus *Papaver*). Thus chemical races also appear to exist in *Meconopsis*.

The isolated promorphinane alkaloids were (−)-amurine (**183**) and (−)-flavinantine (**185**), but no (+)-isomers were isolated from other natural sources. It is not known as yet whether this feature will be of chemotaxonomic significance for this genus.

E. The Genus *Argemone*

Much attention has been paid to the genus *Argemone* Tourn. from the viewpoints of classification (*10,117*) and content of alkaloids and taxonomic alliances (summarized in Ref. *118*) (Tables XXIII and XXIV). Fedde (*10*) adopted a conservative approach and recognized only nine species of *Argemone*, although he included many varieties. A comprehensive reevaluation of the classification was accomplished by Ownbey (*117*). Altogether, 28 species in North America and the West Indies were described. The discrepancy between the numbers of species described by Fedde and by Ownbey perhaps overstates the difference between the two classifications. Ownbey added a number of species that were

TABLE XXIII
Plants of the Genus *Argemone* and Their Alkaloids

Argemone aenea Ownb.
  Allocryptopine (**310**), berberine (**209b**), protopine (**309**) (*3*)

A. alba Lestib.
  Allocryptopine (**310**), berberine (**209b**), chelerythrine (**295**), coptisine (**217b**), protopine (**309**), sanguinarine (**296**) (*3*)

A. albiflora Hornem. (syn. *A. mexicana* L.)
  Allocryptopine (**310**), berberine (**209b**), chelerythrine (**295**), norchelerythrine (**291**), norsanguinarine (**292**), protopine (**309**), sanguinarine (**296**), (−)-scoulerine (**212**), (−)-β-scoulerine methohydroxide (**245**) (*3*)

A. albiflora Hornem. ssp. *texana* Ownb.
  Allocryptopine (**310**), berberine (**209b**), coptisine (**217b**), protopine (**309**), sanguinarine (**296**) (*3*)

A. aurantiaca Ownb.
  Coptisine (**217b**), protopine (**309**) (*3*)

A. brevicornuta Ownb.
  Berberine (**209b**), norargemonine (**55**) (*3*)

A. chisosensis Ownb.
  Allocryptopine (**310**), berberine (**209b**), protopine (**309**) (*3*)

A. corymbosa Greene ssp. *arenicola*
  Allocryptopine (**310**), berberine (**209b**), protopine (**309**), sanguinarine (**296**) (*3*)

A. echinata Ownb.
  Berberine (**209b**), cryptopine (**311**) (*3*)

A. fruticosa Thurber
  Allocryptopine (**310**), cryptopine (**311**), hunnemannine (**312**) (*3*)

A. glauca var. *glauca*
  Allocryptopine (**310**), berberine (**209b**), chelerythrine (**295**), protopine (**309**), sanguinarine (**296**) (*3*)

A. gracilenta Greene
  (−)-Argemonine (**47**), (−)-argemonine methohydroxide (**48**), (−)-argemonine *N*-oxide (**49**), (−)-isonorargemonine (**54**), (+)-laudanidine (**16**), (−)-munitagine (**57**), muramine (**314**), (−)-platycerine methohydroxide (**59**), protopine (**309**), (+)-reticuline (**20**) (*3*)

A. grandiflora Sweet ssp. *grandiflora*
  α-Allocryptopine (**310**), berberine (**209b**), (−)-cheilanthifoline (**210a**), chelerythrine (**295**), (+)-codamine (**14**), corypalmine (**211a**), (+)-laudanosine (**17**), protopine (**309**), sanguinarine (**296**) (*3*)

A. hispida Gray
  (−)-Argemonine (**47**), (−)-bisnorargemonine (**50**), (−)-norargemonine (**55**), (+)-reticuline (**20**) (*3*)

A. mexicana L.
  Allocryptopine (**310**), berberine (**209b**), (−)-cheilanthifoline (**210a**), chelerythrine (**295**), coptisine (**217b**), cryptopine (**311**), dihydrosanguinarine (**280**), norchelerythrine (**291**), norsanguinarine (**292**), oxyhydrastinine (**6**), protopine (**309**), sanguinarine (**296**), (−)-β-scoulerine methohydroxide (**245**), (−)-α- and (−)-β-stylopine methohydroxide (**249**), 6-acetonyldihydrosanguinarine (**285**) (*3,119*)

TABLE XXIII   (*Continued*)

A. munita Dur. et Hilg. ssp. *argentea* Ownb.
Allocryptopine (**310**), (−)-argemonine (**47**), (−)-isonorargemonine (**54**), protopine (**309**) (*3*)

A. munita Dur. et Hilg. ssp. *rotundata* (Rydb.) Ownb.
(−)-Bisnorargemonine (**50**), cryptopine (**311**), 2,9-dimethoxy-3-hydroxypavine (**56**), (−)-munitagine (**57**), muramine (**314**), (+)-reticuline (**20**) (*3*)

A. ochroleuca Sweet [syn. *A. mexicana* var. *ochroleuca* (Sweet) L.]
Allocryptopine (**310**), berberine (**209b**), (−)-α-canadine methohydroxide (**244**), (−)-cheilanthifoline (**210a**), chelerythrine (**295**), coptisine (**217b**), sanguinarine (**296**), (−)-stylopine methohydroxide (**249**), (−)-α-tetrahydropalmatine methohydroxide (**250**) (*3*)

A. platyceras Link et Otto
Allocryptopine (**310**), argemonine (**47**), (−)-argemonine methohydroxide (**48**), berberine (**209b**), (−)-α-canadine methohydroxide (**244**), chelerythrine (**295**), coptisine (**217b**), corysamine (**234b**), cyclanoline [= (−)-α-scoulerine methohydroxide] (**245**), magnoflorine (**138**), norargemonine (**55**), platycerine (**58**), (−)-platycerine methohydroxide (**59**), protopine (**309**), sanguinarine (**296**), (−)-stylopine methohydroxide (**249**) (*3*)

A. platyceras ssp. *pinnatisecta*
Bisnorargemonine (**50**), munitagine (**57**) (*3*)

A. platyceras ssp. *pleiacantha*
Allocryptopine (**310**), berberine (**209b**), bisnorargemonine (**50**), cryptopine (**311**), munitagine (**57**), protopine (**309**) (*3*)

A. pleiacantha Greene ssp. *ambigua* Ownb.
Allocryptopine (**310**), berberine (**209b**), bisnorargemonine (**50**), cryptopine (**311**), munitagine (**57**), protopine (**309**) (*3*)

A. polyanthemos (Fedde) Ownb.
Allocryptopine (**310**), berberine (**209b**), chelerythrine (**295**), coptisine (**217b**), norchelerythrine (**291**), protopine (**309**), sanguinarine (**296**), (−)-scoulerine (**212**) (*3*)

A. sanguinea
Allocryptopine (**310**), argemonine (**47**), berberine (**209b**), muramine (**314**) (*3*)

A. squarrosa Greene
Berberine (**209b**), muramine (**314**) (*3*)

A. subfusiformis
Allocryptopine (**310**), berberine (**209b**), chelerythrine (**295**), protopine (**309**), sanguinarine (**296**) (*3*)

A. subfusiformis ssp. *subfusiformis* Ownb.
Allocryptopine (**310**), berberine (**209b**), chelerythrine (**295**), protopine (**309**), sanguinarine (**296**) (*3*)

A. subfusiformis ssp. *subinermis* Ownb.
Allocryptopine (**310**), berberine (**209b**), chelerythrine (**295**), protopine (**309**), sanguinarine (**296**) (*3*)

A. subintegrifolia Ownb.
Allocryptopine (**310**), berberine (**209b**), protopine (**309**) (*3*)

A. turnerae Powel
(+)-Armepavine (**12**), (−)-tetrahydropalmatine (**218a**) (*3*)

TABLE XXIV
Alkaloids Found in Plants of the Genus *Argemone*

Simple isoquinolines and benzylisoquinolines
 Oxyhydrastinine; (+)-armepavine, (+)-codamine, (+)-laudanidine, (+)-laudanosine, (+)-
 reticuline

Pavines and isopavines
 (−)-Argemonine, (−)-argemonine methohydroxide, (−)-argemonine *N*-oxide, (−)-
 bisnorargemonine, 2,9-dimethoxy-3-hydroxypavine, (−)-isonorargemonine, (−)-munitagine,
 (−)-norargemonine, platycerine, (−)-platycerine methohydroxide

Aporphines
 Magnoflorine

Protoberberines
 (−)-α-Canadine methohydroxide, (−)-cheilanthifoline, corypalmine, cyclanoline [= (−)-α-
 scoulerine methohydroxide], (−)-scoulerine, (−)-β-scoulerine methohydroxide, (−)-α- and
 (−)-β-stylopine methohydroxide, (−)-tetrahydropalmatine, (−)-α-tetrahydropalmatine
 methohydroxide; berberine, coptisine, corysamine

Benzophenanthridines
 6-Acetonyldihydrosanguinarine, chelerythrine, dihydrosanguinarine, norchelerythrine,
 norsanguinarine, sanguinarine

Protopines
 Allocryptopine, cryptopine, hunnemannine, muramine, protopine

completely unknown to Fedde, but in many instances he raised to species level taxa that were recognized as varieties by Fedde. This double classification is also found in chemical publications. An example is the designation *A. albiflora* Hornem. (Ownbey) which is now in use for *A. alba* Lestib. (Fedde); The same plant has evidently been studied under these two names (Table XXIII). In the literature, the data concerning 30 species (or their varieties) of the genus *Argemone* are given. With six exceptions, these are all those that Ownbey classifies in this genus.

In the Papaveraceae, it is the genus *Argemone* next to the genus *Eschscholtzia* where pavines have been found. In view of the chemical characteristics, it has been suggested (*120*) to subdivide the genus *Argemone* into two groups. The first includes six species containing pavine alkaloids, and the second is characterized by the presence of alkaloids of the protopine type (Table XXIII). Stermitz (*118*) goes still further in his conclusion. On the basis of his extensive studies, including morphological signs and alkaloid chemistry, he suggests the subdivision of the genus *Argemone* into five groups.

Interesting is the finding of (+)-armepavine (**12**) in *A. turnerae*. This is the only finding of a (+)-enantiomer of armepavine in the Papaveraceae. The presence of the (−)-enantiomer is characteristic of the section *Miltantha* (genus *Papaver*).

F. The Genus *Eschscholtzia*

Fedde (*10*) divided the genus *Eschscholtzia* Cham. and its 123 species into two sections. This division is also consistent with the chemical characteristics of those species that have been investigated for their alkaloid content (Tables XXV and XXVI). *Eschscholtzia californica* Cham., *E. glauca* Greene, and *E. douglasii* (Hook. et Arn.) Walp., which belong to the section *Eurycraspedontae,* are characterized by the presence of pavine alkaloids and a high content of quaternary alkaloids. While the composition of the nonquaternary alkaloid fraction of these species does not differ appreciably, the composition of the quaternary fraction from the roots shows remarkable differences (*16,123*) (Table XXVII). It

TABLE XXV
Plants of the Genus *Eschscholtzia* and Their Alkaloids

Eschscholtzia californica Cham.
  Allocryptopine (**310**), (−)-bisnorargemonine (**50**), californidine (**51**), (−)-α-canadine
    methohydroxide (**244**), chelerythrine (**295**), chelirubine (**300**), chelilutine (**301**), coptisine
    (**217b**), dihydrochelerythrine (**278**), dihydrochelirubine (**288**), dihydromacarpine (**289**),
    dihydrosanguinarine (**280**), escholidine (= tetrahydrothalifendine methohydroxide) (**246**),
    escholine (= magnoflorine) (**138**), escholinine [= (+)-romneine methohydroxide] (**32**),
    eschscholtzidine (**52**), eschscholtzine (= californine) (**53**), glaucine (**145**), isocorydine (**132**),
    lauroscholtzine (= laurotetanine methohydroxide) (**149**), laurotetanine (**148**), norargemonine
    (**55**), protopine (**309**), sanguinarine (**296**), (−)-α-stylopine methohydroxide (**249**)
    (*3,121,122*)

E. douglasii (Hook. et Arn.) Walp.
  Allocryptopine (**310**), berberine (**209b**), (−)-bisnorargemonine (**50**), californidine (**51**), (−)-α-
    canadine methohydroxide (**244**), chelilutine (**301**), chelirubine (**300**), coptisine (**217b**),
    escholidine (**246**), escholine (**138**), eschscholtzidine (**52**), eschscholtzine (**53**), macarpine
    (**302**), (−)-norargemonine (**55**), protopine (**309**) (*3*)

E. glauca Greene
  Allocryptopine (**310**), berberine (**209b**), (−)-bisnorargemonine (**50**), californidine (**51**), (−)-α-
    canadine methohydroxide (**244**), chelerythrine (**295**), chelirubine (**300**), coptisine (**217b**),
    escholidine (**246**), escholine (**138**), eschscholtzidine (**52**), eschscholtzine (**53**), macarpine
    (**302**), norargemonine (**55**), protopine (**309**), sanguinarine (**296**) (*3*)

E. lobbii Greene
  Allocryptopine (**310**), berberine (**209b**), chelerythrine (**295**), chelirubine (**300**), coptisine
    (**217b**), corydine (**136**), corysamine (**234b**), corytuberine (**137**), macarpine (**302**), protopine
    (**309**), sanguinarine (**296**), (−)-scoulerine (**212**) (*3*)

E. oregana Greene
  Allocryptopine (**310**), berberine (**209b**), californidine (**51**), chelerythrine (**295**), chelilutine
    (**301**), chelirubine (**300**), coptisine (**217b**), corydine (**136**), escholamidine (**34**), escholamine
    (**33**), escholinine (**22**), protopine (**309**), sanguinarine (**296**), scoulerine (**212**), (−)-α-
    stylopine methohydroxide (**249**) (*3*)

TABLE XXVI
Alkaloids Found in Plants of the Genus *Eschscholtzia*

Simple isoquinolines
Escholamidine, escholamine, escholinine [= (+)-romneine methohydroxide]

Pavines and isopavines
(−)-Bisnorargemonine, californidine, eschscholtzidine, eschscholtzine (= californine), (−)-norargemonine

Aporphines
Corydine, corytuberine, glaucine, isocorydine, escholine (= magnoflorine), laurotetanine, lauroscholtzine (= laurotetanine methohydroxide)

Protoberberines
(−)-α-Canadine methohydroxide, escholidine (= tetrahydrothalifendine methohydroxide), (−)-scoulerine, (−)-α-stylopine methohydroxide; berberine, coptisine, corysamine

Benzophenanthridines
Chelerythrine, chelirubine, chelilutine, dihydrochelerythrine, dihydrochelirubine, dihydromacarpine, macarpine, sanguinarine

Protopines
Allocryptopine, protopine

seems that the occurrence of pavines in the Papaveraceae is limited to the *Eschscholtzia* and *Argemone* genera growing in North America.

Species belonging to the section *Stenocraspedontae* are characterized by the presence of benzylisoquinoline alkaloids, for example, *E. oregana* Greene contains escholamine (**33**) as the major alkaloid. *Eschscholtzia lobbii* Greene appears to be exceptional since neither pavine nor benzylisoquinoline alkaloids have been detected therein. The presence of aporphine alkaloids is characteristic for all the species.

G. THE GENUS *Dicranostigma*

The genus *Dicranostigma* Hook. f. et Thoms. comprises three species (*10*) (Tables XXVIII and XXIX). They are *D. franchetianum* (Prain.) Fedde, *D.*

TABLE XXVII
Significant Quaternary Alkaloids in the Roots of Three *Eschscholtzia* Species[a]

	E. californica	E. douglasii	E. glauca
Escholidine	+	+	+
(−)-α-Canadine methohydroxide	+	+	+
Magnoflorine	+	+	−
Escholinine	+	−	−
(−)-α-Stylopine methohydroxide	+	−	−

[a] After Ref. *16*.

TABLE XXVIII

Plants of the Genus *Dicranostigma* and Their Alkaloids

Dicranostigma franchetianum (Prain.) Fedde

Allocryptopine (**310**), berberine (**209b**), chelerythrine (**295**), chelirubine (**300**), coptisine (**217b**), corydine (**136**), corysamine (**234b**), corytuberine methohydroxide (**138**), isocorydine (**132**), magnoflorine (**138**), menisperine [= (+)-isocorydine methohydroxide] (**133**), protopine (**309**), sanguinarine (**296**) (*3,124*)

D. lactucoides Hook. f. et Thoms.

Allocryptopine (**310**), berberine (**209b**), chelerythrine (**295**), chelirubine (**300**), coptisine (**217b**), corytuberine (**137**), isocorydine (**132**), magnoflorine (**138**), menisperine (**133**), oxysanguinarine (**281**), protopine (**309**), sanguinarine (**296**) (*3*)

D. leptopodum (Maxim.) Fedde

Allocryptopine (**310**), berberine (**209b**), chelerythrine (**295**), chelirubine (**300**), coptisine (**217b**), corysamine (**234b**), corytuberine (**137**), corytuberine methohydroxide (**138**), isocorydine (**132**), isocorypalmine (**214a**), magnoflorine (**138**), menisperine (**133**), protopine (**309**), sanguinarine (**296**) (*3,124*)

lactucoides Hook. f. et Thoms. and *D. leptopodum* (Maxim.) Fedde, occurring in the region stretching from the eastern Himalayas to central China. Cullen studied (*125*) taxonomy of *Dicranostigma* and found that the names of the species cultivated in botanical gardens are frequently incorrect. *Dicranostigma leptopodum* is often called *D. franchetianum* and it is usually given the name *Glaucium vitellinum*. Much attention has been paid to the quarternary alkaloids of the aforementioned three species of *Dicranostigma,* and chemotaxonomical conclusions have been drawn (*124*). For *Dicranostigma* the predominance of aporphines is a characteristic feature, which indicates a close relationship with the genus *Glaucium*. The results of morphological (*125*), cytological, and chemotaxonomic (*126*) studies are consistent with this viewpoint, and the conclusion has been drawn that the genera *Dicranostigma* and *Glaucium* are much more closely related than is indicated by their classification into different tribes of Chelidonieae and Papavereae. Therefore it was proposed to classify both genera

TABLE XXIX

Alkaloids Found in Plants of the Genus *Dicranostigma*

Aporphines

Corydine, corytuberine, corytuberine methohydroxide, isocorydine, magnoflorine, menisperine [= (+)-isocorydine methohydroxide]

Protoberberines

Isocorypalmine; berberine, coptisine, corysamine

Benzophenanthridines

Chelerythrine, chelirubine, oxysanguinarine, sanguinarine

Protopines

Allocryptopine, protopine

TABLE XXX
Plants of the Genus *Chelidonium* and Their Alkaloids

Chelidonium majus L.

Allocryptopine (**310**), chelamidine (= 10-hydroxyhomochelidonine) (**269**), chelamine (= 10-hydroxychelidonine) (**268**), chelidamine [= (−)-stylopine] (**217a**), chelerythrine (**295**), chelidonine (**266**), chelilutine (**301**), chelirubine (**300**), coptisine (**217b**), corysamine (**234b**), *N*-demethyl-9,10-dihydroxysanguinarine (**303**), dihydrochelerythrine (**278**), dihydrochelilutine (**286**), dihydrochelirubine (**288**), dihydrosanguinarine (**280**), magnoflorine (**138**), oxysanguinarine (**281**), protopine (**309**), sanguinarine (**296**), sparteine (**407**), stylopine (**217a**), (−)-α-stylopine methohydroxide (**249**), (−)-β-stylopine methohydroxide (**249**) (*3*)

C. japonicum Thunb. (syn. *Stylophorum japonicum*)

Bocconoline (**279**), 6-hydroxymethyldihydrosanguinarine (**284**), chelerythrine (**295**), norchelerythrine (**291**), norsanguinarine (**292**), sanguinarine (**296**) (*3,127*)

---

together in the tribe Chelidonieae (*125*) or with respect to biochemical features in the tribe Papavereae (*126*). Some species of *Glaucium*, containing (−)- or (±)-chelidonine (**266**) and (−)-norchelidonine (**267**) may thus be considered as a connecting link to the "Chelidonium Gruppe" (*10*) in the tribe Chelidonieae.

H. The Genus *Chelidonium*

Of the genus *Chelidonium* Tourn., two species have been studied (Tables XXX and XXXI), of which *Chelidinium majus* L.—a medicinal plant employed from time immemorial—is one of the most often investigated plants of the Papaveraceae. It is known (*128*) that the root and the aerial part of this plant differ substantially both in content and composition of their alkaloids. It is also

TABLE XXXI
Alkaloids Found in Plants of the Genus *Chelidonium*

Aporphines
  Magnoflorine

Protoberberines
  Chelidamine [= (−)-stylopine], (−)-stylopine, (−)-α-stylopine methohydroxide, (−)-β-stylopine methohydroxide; coptisine, corysamine

Benzophenanthridines
  Bocconoline, 6-hydroxymethyldihydrosanguinarine, chelamidine (= 10-hydroxyhomochelidonine), chelamine (= 10-hydroxychelidonine), chelerythrine, chelidonine, chelilutine, chelirubine, *N*-demethyl-9,10-dihydrooxysanguinarine, dihydrochelerythrine, dihydrochelilutine, dihydrochelirubine, norchelerythrine, norsanguinarine, oxysanguinarine, sanguinarine

Other alkaloids
  Sparteine

---

evident that the occurrence and the relative amounts of individual alkaloids are considerably dependent on geographic and climatic conditions and on the vegetation stage of the plant. While the major alkaloid in the material collected at the period of seed ripening was chelidonine (**266**) (*129*), coptisine (**217b**) was the predominant alkaloid in the aboveground parts of the plant collected at the beginning of the flowering period. These two groups of alkaloids (benzophenanthridines and protoberberines) were the only ones (except for magnoflorine) found in the genus *Chelidonium*.

I. THE GENUS *Bocconia*

The genus *Bocconia* Plum. of the tribe Chelidonieae comprises some 10 species that are native to tropical and subtropical areas of Central and South America. *Bocconia*, which is closely related to the east Asian genus *Macleaya*, does not appear to contain alkaloids that are taxonomically significant (Tables XXXII and XXXIII). In these two genera the major alkaloids are allocryptopine

TABLE XXXII
Plants of the Genus *Bocconia* and Their Alkaloids

Bocconia arborea
   1,3-Bis(6-hydroxychelerythrinyl)acetone (= chelerythridimerine) (**307**), dihydrosanguinarine (**280**), 6-*O*-methyldihydrochelerythrine (= angoline) (**290**), oxysanguinarine (**281**) (*3*)

B. cordata Wild.
   Allocryptopine (**310**), bocconine (= chelirubine) (**300**), bocconoline (**279**), chelerythrine (**295**), cryptopine (**311**), dehydrocheilanthifolinc (**210b**), 9,10-demethylen-9,10-dihydrosanguinarine (**294**), 9,10-demethylensanguinarine (**297**), dihydrochelerythrine (**278**), dihydrosanguinarine (**280**), oxysanguinarine (**281**), protopine (**309**), protopine *N*-oxide (**324**), sanguinarine (**296**) (*3,130–132*)

B. frutescens L.
   Allocryptopine (**310**), berberine (**209b**), chelerythrine (**295**), columbamine (**214b**), coptisine (**217b**), corysamine (**234b**), (−)-isocorypalmine (**214a**), papaverrubine E (**396**), protopine (**309**), rhoeadine (**389**), sanguinarine (**296**) (*3*)

B. latisepala Wats.
   Allocryptopine (**310**), chelerythrine (**295**), oxysanguinarine (**281**), protopine (**309**), sanguinarine (**296**) (*3*)

B. laurine
   Protopine (**309**) (*3*)

B. microcarpa Maxim.
   Allocryptopine (**310**), chelerythrine (**295**), 9,10-demethylene-9,10-dihydrosanguinarine (**294**), 9,10-demethylenesanguinarine (**297**), protopine (**309**), sanguinarine (**296**) (*3,130*)

B. pearcei Hutsch.
   Allocryptopine (**310**), berberine (**209b**), chelerythrine (**295**), chelirubine (**300**), coptisine (**217b**), protopine (**309**), sanguinarine (**296**) (*3*)

TABLE XXXIII
Alkaloids Found in Plants of the Genus *Bocconia*

Protoberberines
  (−)-Isocorypalmine; berberine, columbamine, coptisine, corysamine, dehydrocheilanthifoline

Benzophenanthridines
  1,3-Bis(6-hydroxychelerythrinyl)acetone (= chelerythrindimerine), bocconine (= chelirubine),
    bocconoline, chelerythrine, 9,10-demethylensanguinarine, 9,10-demethylen-9,10-
    dihydrosanguinarine, dihydrochelerythrine, dihydrosanguinarine, 6-*O*-
    methyldihydrochelerythrine (= angoline), oxysanguinarine, sanguinarine

Protopines
  Allocryptopine, cryptopine, protopine, protopine *N*-oxide

Rhoeadines
  Papaverrubine E, rhoeadine

(**310**) and protopine (**309**). These plants may represent rich sources of these alkaloids. Interesting is the finding of rhoeadine (**389**) and papaverrubine E (**396**) in *B. frutescens* L. (*133*). Thus the genus *Bocconia* is the third genus (in addition to the genera *Papaver* and *Meconopsis*) in which the presence of rhoeadines has been demonstrated.

J. THE GENUS *Hypecoum*

Plants of the genus *Hypecoum* Tourn. (subfamily Hypecoideae) differ chemotaxonomically from the plants of the other genera of the subfamily Papaveroideae. Species of the genus *Hypecoum* contain secoberbines in addition to protopines [protopine (**309**) predominates], protoberberines, and benzophenanthridines (Tables XXXIV and XXXV). The presence of secoberbines is indicative of a close relationship between the genus *Hypecoum* and some species of Fumariaeceae.

K. OTHER GENERA OF THE PAPAVERACEAE

1. The Genus *Stylophorum*

Of the genus *Stylophorum* Nutt., only *Stylophorum diphyllum* (Michx.) Nutt. of the tribe Chelidonieae, indigenous to North America, has been studied. The alkaloid fraction from the root and the aerial part differ both in total content and in relative proportions of individual alkaloids (*137*). From the root, (−)-stylopine (**217a**) and (+)-chelidonine (**266**) have been isolated as the main components (Tables XXXVI and XXXVII). The major component from the aerial part is stylopine, which represents a mixture of (−)-and (±)-forms. In contrast to the root, the aerial part contains small amounts of (+)-chelidonine (**266**) and pro-

TABLE XXXIV

Plants of the Genus *Hypecoum* and Their Alkaloids

Hypecoum erectum L.

Allocryptopine (**310**), (−)-canadine methohydroxide (**244**), coptisine (**217b**), corydamine (**306**), hypecorine (**264**), hypecorinine (**263**), hyperectine (**388**), protopine (**309**) (*3,134,135*)

H. imberbe Sibth. et Smith

Allocryptopine (**310**), chelerythrine (**295**), protopine (**309**), sanguinarine (**296**) (*3*)

H. lactiflorum

Allocryptopine (**310**), hypecorine (**264**), hypecorinine (**263**), protopine (**309**) (*135*)

H. leptocarpum Hook. f. et Thoms.

Chelerythrine (**295**), chelirubine (**300**), coptisine (**217b**), protopine (**309**), sanguinarine (**296**) (*3*)

H. parviflorum Kar. et Kir.

Peshawarine (**265**), protopine (**309**) (*3*)

H. pendulum L.

Protopine (**309**) (*3*)

H. procumbens L.

8-Acetonyldihydrosanguinarine (**285**), allocryptopine (**310**), chelerythrine (**295**), chelirubine (**300**), coptisine (**217b**), (−)-corydalisol (**260**), (+)-glaucine (**145**), (±)-hypecorinine (**263**), (+)-isocorydine (**132**), 8-methoxydihydrosanguinarine (**282**), norsanguinarine (**292**), oxysanguinarine (**281**), protopine (**309**), sanguinarine (**296**), (−)-scoulerine (**212**) (*3,136*)

H. trilobum Trautv.

Chelerythrine (**295**), protopine (**309**), sanguinarine (**296**) (*3*)

TABLE XXXV

Alkaloids Found in Plants of the Genus *Hypecoum*

Aporphines

(+)-Glaucine, (+)-isocorydine

Protoberberines

(−)-Canadine methohydroxide, (−)-scoulerine; coptisine

Secoberbines

(−)-Corydalisol, (±)-hypecorinine, (±)-hypecorine, (−)-peshawarine

Benzophenanthridines

8-Acetonyldihydrosanguinarine, chelerythrine, chelirubine, corydamine, 8-methoxydihydrosanguinarine, norsanguinarine, oxysanguinarine, sanguinarine

Protopines

Allocryptopine, protopine

Spirobenzylisoquinolines

Hyperectine

TABLE XXXVI
Plants of the Genus *Stylophorum* and Their Alkaloids

Stylophorum diphyllum (Michx.) Nutt.
  Allocryptopine (**310**), berberine (**209b**), chelerythrine (**295**), (+)- and (±)-chelidonine (**266**), chelirubine (**300**), coptisine (**217b**), corysamine (**234b**), corytuberine (**137**), cryptopine (**311**), isoboldine (**147**), macarpine (**302**), magnoflorine (**138**), protopine (**309**), sanguinarine (**296**), scoulerine (**212**), (−)- and (±)-stylopine (**217a**), (−)-α- and (−)-β-stylopine methohydroxide (**249**) (*3,137*)

---

topine (**309**), while (±)-chelidonine and quaternary benzophenanthridines are completely absent. The occurrence of (−)-α- and (−)-β-stylopine methohydroxide (**249**) and magnoflorine (**138**) is evidence of the close relationship between *S. diphyllum* and *Chelidonium majus,* in both of which the same quaternary alkaloids have been detected. As in *Chelidonium majus,* the content and ratio of individual alkaloids in *S. diphyllum* is largely dependent on the vegetation period.

2. The Genus *Hunnemannia*

The only representative of the genus *Hunnemannia* A. Juss., which is closely related to the genera *Eschscholtzia, Dendromecon,* and *Petromecon,* is Hunnemannia fumariaefolia Sweet, native to Central America. In the aerial part of this plant, the major alkaloid is hunnemanine (**312**) (Tables XXXVIII and XXXIX) and in the roots, protopine (**309**), hunnemanine, and allocryptopine (**310**). Cyclanoline (**245**) represents the main component of the quaternary alkaloid fraction of the aerial part and the roots.

TABLE XXXVII
Alkaloids Found in Plants of the Genus *Stylophorum*

Aporphines
  Corytuberine, isoboldine, magnoflorine

Protoberberines
  Scoulerine, (−)- and (±)-stylopine, (−)-α- and (−)-β-stylopine methohydroxide; berberine, coptisine, corysamine

Benzophenanthridines
  Chelerythrine, (+)- and (±)-chelidonine, chelirubine, macarpine, sanguinarine

Protopines
  Allocryptopine, cryptopine, protopine

TABLE XXXVIII

Plants of the Genus *Hunnemannia* and Their Alkaloids

Hunnemannia fumariaefolia Sweet

Allocryptopine (**310**), berberine (**209b**), chelerythrine (**295**), chelilutine (**301**), chelirubine (**300**), coptisine (**217b**), corysamine (**234b**), cyclanoline [= (−)-α-scoulerine methohydroxide] (**245**), escholidine (**246**), hunnemannine (**312**), oxyhydrastinine (**6**), protopine (**309**), sanguinarine (**296**), scoulerine (**212**) (*3,138,139*)

---

3. The Genus *Stylomecon*

Stylomecon heterophylla (Benth.) G. Tayl. (syn. *Meconopsis heterophylla* Benth.), indigenous to North America, contains small quantities of alkaloids (Tables XL and XLI). Its major alkaloid is cryptopine (**311**) together with allocryptopine (**310**), protopine (**309**) and (−)-α-hydrastine (**331**) (*140*).

4. The Genus *Hylomecon*

In comparison with the other species belonging to the tribe Chelidonieae, *Hylomecon vernalis* Maxim., indigenous to the Far East, contains only small quantities of alkaloids (*140*). In the roots, ~65% are quaternary benzophenanthridines; the aerial part contains protoberberines, protopines, and benzophenanthridines (Tables XLII and XLIII).

5. The Genus *Sanguinaria* Dill.

Sanguinaria canadensis L. is indigenous to North America. In the roots, the major alkaloids are the benzophenanthridines, which form ~75% of the total

TABLE XXXIX

Alkaloids Found in Plants of the Genus *Hunnemannia*

Simple isoquinolines
  Oxyhydrastinine

Protoberberines
  Cyclanoline [= (−)-α-scoulerine methohydroxide], escholidine, scoulerine; berberine, coptisine, corysamine

Benzophenanthridines
  Chelerythrine, chelilutine, chelirubine, sanguinarine

Protopines
  Allocryptopine, hunnemannine, protopine

TABLE XL

Plants of the Genus *Stylomecon* and Their Alkaloids

Stylomecon heterophylla (Benth.) G. Tayl. (syn. *Meconopsis heterophylla)*
  Allocryptopine (**310**), berberine (**209b**), chelerythrine (**295**), coptisine (**217b**), cryptopine
    (**311**), protopine (**309**), sanguinarine (**296**), stylophylline [= (−)-α-hydrastine) (**331**] (*3*)

TABLE XLI

Alkaloids Found in Plants of the Genus *Stylomecon*

Protoberberines
  Berberine, coptisine

Benzophenanthridines
  Chelerythrine, sanguinarine

Protopines
  Allocryptopine, cryptopine, protopine

Phthalideisoquinolines
  Stylophylline [= (−)-α-hydrastine]

TABLE XLII

Plants of the Genus *Hylomecon* and Their Alkaloids

Hylomecon vernalis Maxim.
  Allocryptopine (**310**), berberine (**209b**), canadine (**209a**), chelerythrine (**295**), chelidonine
    (**266**), chelilutine (**301**), chelirubine (**300**), coptisine (**217b**), isocorydine (**132**), protopine
    (**309**), sanguinarine (**296**), stylopine (**217a**) (*3*)

TABLE XLIII

Alkaloids Found in Plants of the Genus *Hylomecon*

Aporphines
  Iscorydine

Protoberberines
  Canadine, stylopine; berberine, coptisine

Benzophenanthridines
  Chelerythrine, chelidonine, chelilutine, chelirubine, sanguinarine

Protopines
  Allocryptopine, protopine

TABLE XLIV

Plants of the Genus *Sanguinaria* and Their Alkaloids

Sanguinaria canadensis L.

Allocryptopine (**310**), berberine (**209b**), canadine (**209a**), chelerythrine (**295**), chelerythrindimerine (**307**), chelilutine (**301**), chelirubine (**300**), coptisine (**217b**), dihydrosanguilutine (**287**), oxysanguinarine (**281**), protopine (**309**), sanguidimerine (**308**), sanguilutine (**299**), sanguinarine (**296**), sanguirubine (**298**), stylopine (**217a**) (*3*)

TABLE XLV

Alkaloids Found in Plants of the Genus *Sanguinaria*

Protoberberines

Canadine, stylopine; berberine, coptisine

Benzophenanthridines

Chelerythrine, chelerythridimerine, chelilutine, chelirubine, dihydrosanguilutine, oxysanguinarine, sanguidimerine, sanguilutine, sanguinarine, sanguirubine

Protopines

Allocryptopine, protopine

TABLE XLVI

Plants of the Genus *Romneya* and Their Alkaloids

Romneya coulteri var. *trichocalyx* Jepson

Coulteropine (**316**), dihydrosanguinarine (**280**), norromneine (**23**), romneine (**24**), protopine (**309**), (+)-reticuline (**20**), sanguinarine (**296**) (*3*)

---

alkaloid content in addition to protopines (*140*) (Tables XLIV and XLV). Also in the aerial part, the benzophenanthridines are the major alkaloids in addition to protoberberines, which are not found in the roots. *Sanguinaria canadensis* represents a good source of benzophenanthridine alkaloids, especially of sanguinarine (**296**) and chelerythrine (**295**).

6. The Genus *Romneya* Harv.

From *Romneya coulteri* var. *trichocalyx* Jepson, benzylisoquinolines, protopines, and benzophenanthridines have been isolated (Tables XLVI and XLVII).

7. The Genus *Platystemon* Benth.

Platystemon californicus Benth. contains quaternary protoberberines, benzophenanthridines, and protopine (**309**) (Tables XLVIII and XLIX).

50   V. PREININGER

TABLE XLVII
Alkaloids Found in Plants of the Genus *Romneya*

Benzylisoquinolines
Norromneine, (+)-reticuline, romneine

Benzophenanthridines
Dihydrosanguinarine, sanguinarine

Protopines
Coulteropine, protopine

TABLE XLVIII
Plants of the Genus *Platystemon* and Their Alkaloids

Platystemon californicus Benth.
Berberine (**209b**), chelerythrine (**295**), chelirubine (**300**), coptisine (**217b**), protopine (**309**),
sanguinarine (**296**) (*3*)

TABLE XLIX
Alkaloids Found in Plants of the Genus *Platystemon*

Protoberberines
Berberine, coptisine

Benzophenanthridines
Chelerythrine, chelirubine, sanguinarine

Protopines
Protopine

TABLE L
Plants of the Genus *Meconella* and Their Alkaloids

Meconella oregana Nutt. var. *californica*
Protopine (**309**)

TABLE LI
Alkaloids Found in Plants of the Genus *Meconella*

Protopines
Protopine

8. The Genus *Meconella* Nutt.

Meconella oregana Nutt. var. *californica* yields protopine (**309**) (Tables L and LI).

No mention is to be found in the literature of the other genera of the Papaveraceae, that is, *Artomecon* Torr. et Frém., *Cynbya* Parry, *Dendromecon* Benth., *Eomecon* Hance, and *Pteridophyllum* Sieb. et Zucc.

III. Chemotaxonomic Evaluation of the Fumariaceae

The Papaveraceae and Fumariaceae DC. are closely related families. Their relationship as well as their independence are well demonstrated by the external structure of the flowers, the various anatomical characteristics of their different organs, and alkaloid content. As in the Papaveraceae, the protoberberines, benzophenanthridines, and protopines are chemotaxonomically significant for the Fumariaceae. Characteristic for the Fumariaceae is the presence of spirobenzylisoquinolines, indenobenzazepines, secophthalideisoquinolines, phthalideisoquinolines, and cularines. The Fumariaceae also contain quaternary highly polar alkaloids, insoluble in nonpolar solvents (*16*).

A. THE GENUS *Corydalis*

In the literature, the number of species classified in the genus *Corydalis* Vent. differs considerably. The *Index Kewensis* (*141*) mentions 194 items under the name *Corydalis,* 78 of them synonyms. Popov (*142*), who subdivides the genus *Corydalis* into 12 sections, assigns to it only 61 items, Ryberg (*143*) 26 in five sections. Sixty-three of *Corydalis* species have been studied for their alkaloid content. The alkaloids found in the genus *Corydalis* are given in Tables LII (*144–193*) and LIII.

Almost all of the studied plants have been found to contain protoberberines, protopines, and phthalideisoquinolines. The occurrence of aporphines and benzophenanthridines is frequent. The secoberbines have been isolated from *C. incisa* and *C. ochotensis.* The cularines, which are characteristic of some plants of the genus *Corydalis,* are also found in the genera *Dicentra* and *Sarcocapnos.*

B. THE GENUS *Fumaria*

The chemotaxonomy of plants of the genus *Fumaria* Tourn. has been reviewed (*194,195*). The number of taxons classified in the genus *Fumaria* differs

TABLE LII
Plants of the Genus *Corydalis* and Their Alkaloids

Corydalis ambigua Cham. et Schlecht.
  α-Allocryptopine (**310**), cavidine (**233a**), coptisine (**217b**), (+)-corybulbine (**230a**), (+)-
  corydaline (**232a**), (+)-corydalmine (**230a**), (−)-corypalmine (**211a**), dehydrocorydaline
  (**232b**), dehydrothalictrifoline (**233b**), (+)-glaucine (**145**), (+)-1-methylcorypalline (**2**),
  noroxyhydrastinine (**7**), protopine (**309**), (−)-scoulerine (**212**), (−)-tetrahydrocolumbamine
  [= (−)-isocorypalmine] (**214a**), (−)- and (±)-tetrahydrocoptisine [= (−)- and (±)-
  stylopine] (**217a**), (+)-tetrahydrojatrorrhizine [= (+)-corypalmine] (**211a**), (±)-
  tetrahydropalmatine (**218a**), thalictrifoline (**233a**) (*3*)

C. ambigua Cham. et Schlecht. var. *amurensis* Maxim.
  (+)-Corybulbine (**230a**), (+)-corydaline (**232a**), dehydrocorybulbine (**230b**), dehydrocorydaline
  (**232b**), palmatine (**218b**), protopine (**309**), sinactine (?) (**215a**), tetrahydroberberine (=
  canadine) (**209a**), (±)-tetrahydropalmatine (**218a**) (*144*)

C. angustifolia (Marsch.-Bieb.) DC.
  Protopine (**309**), (±)-tetrahydropalmatine (**218a**) (*145*)

C. aurea Willd.
  Allocryptopine (**310**), aurotensine (= isoboldine) (**147**), bicucine (?), bicuculline (**325**),
  capauridine (**219a**), capaurine (**219a**), cordrastine (**327**), corpaverine (= capaurine +
  sendaverine), corydaline (**232a**), corypalline (**1**), dehydrocorydaline (**232b**), protopine (**309**),
  (−)- and (±)-tetrahydropalmatine (**218a**) (*3*)

C. bulbosa (L.) DC.
  (−)-Adlumidine (**325**), α-allocryptopine (**310**), (+)-bicuculline (**325**), bicucullinine (**345**), (+)-
  bulbocapnine (**129**), bulbodione (**177**), columbamine (**214b**), coptisine (**217b**), (+)-
  corybulbine (**230a**), (+)-corydaline (**232a**), (+)-corydine (**136**), corydione (**174**),
  dehydrocorydaline (**232b**), dehydroglaucine (**172**), dehydronantenine (**173**),
  dihydrosanguinarine (**280**), (−)-domesticine (**143**), (+)-glaucine (**145**), glaucinone (**165**),
  (+)-isoboldine (**147**), nandazurine (**166**), (+)-nantenine (**144**), noroxyhydrastinine (**7**),
  oxonantenine (**161**), protopine (**309**), predicentrine (**151**), (−)-scoulerine (**212**), (±)-sinactine
  (**215a**), (−)-tetrahydrocolumbamine (**214a**), (−)-tetrahydrocoptisine [= (−)-stylopine]
  (**217a**), (+)-tetrahydrojatrorrhizine [= (+)-corypalmine] (**211a**), tetrahydropalmatine (**218a**),
  thaliporphine (**153**) (*3,146,147*)

C. campulicarpa Hayata
  α-Allocryptopine (**310**), berberine (**209b**), ophiocarpine (**240**), protopine (**309**) (*3*)

C. caseana A. Gray
  Allocryptopine (**310**), bicuculline (**325**), caseadine (**226**), caseamine (**227**), caseanadine (**228**),
  corypalmine (**211a**), (−)- and (±)-isocorypalmine (**214a**), protopine (**309**), (−)-scoulerine
  (**212**), (−)-tetrahydropalmatine (**218a**) (*3*)

C. caucasica
  Allocryptopine (**310**), chelerythrine (**295**), protopine (**309**), sanguinarine (**296**) (*3*)

C. cava Schweigg. et Korte (syn. *C. tuberosa* DC.)
  Adlumidiceine (**340**), allocryptopine (**310**), apocavidine (**229a**), berberine (**209b**), bulbocapnine
  (**129**), (+)-bulbocapnine methohydroxide (**130**), (+)-canadine (**209a**), capnoidine (**325**),
  columbamine (**214b**), coptisine (**217b**), corybulbine (**230a**), corycavamine (**321**),
  corycavidine (**322**), corycavine (**322**), corydaline (**232a**), corydine (**136**), corypalmine

TABLE LII (*Continued*)

(**211a**), corysamine (**234b**), corytuberine (**137**), dehydroapocavidine (**229b**), dehydrocorybulbine (**230b**), dehydrocorydaline (**232b**), dehydrothalictricavine (**235b**), domesticine (**143**), domestine (**144**), glaucine (**145**), hydrastinine (**8**), isoboldine (**147**), isocorybulbine (**236a**), isocorydine (**132**), (+)- and (±)-isocorypalmine (**214a**), jatrorrhizine (**211b**), magnoflorine (**138**), 1,2-methylenedioxy-6a,7-dehydroaporphine-10,11-quinone (**179**), 13-methyltetrahydroprotoberberine (= thalictricavine) (**235a**), narcotine (**332**), 8-oxocoptisine (**254**), palmatine (**218b**), predicentrine (**151**), protopine (**309**), reticuline (**20**), (−)-scoulerine (**212**), sinoacutine (**188**), (+)-stylopine (**217a**), (+)-α-stylopine methohydroxide (**249**), tetrahydrocoptisine (= stylopine) (**217a**), (+)-tetrahydrocorysamine (**234a**), α-tetrahydrocorysamine methohydroxide (**251**), (+)- and (±)-tetrahydropalmatine (**218a**) (*3,148,149*)

C. cheilanthifolia Hemsl.
Allocryptopine (**310**), berberine (**209b**), (−)-canadine (**209a**), (−)-β-danadine methohydroxide (**244**), cheilanthifoline (**210a**), coptisine (**217b**), (−)-corypalmine (**211a**), (−)-13β-hydroxystylopine (**239**), ophiocarpine (**240**), protopine (**309**), sanguinarine (**296**), (−)- and (±)-stylopine (**217a**) (*150,151*)

C. claviculata DC.
Berberine (**209b**), coptisine (**217b**), (+)-claviculine (**78**), crassifolazonine (**411**), crassifoline (**27**), culacorine (**68**), (+)-cularicine (**76**), (+)-cularidine (**72**), (+)-cularine (**74**), limousamine (**81**), norcularicine (**77**), norcularidine (**73**), norsecocularine (**91**), noyaine (**93**), oxocularine (**84**), protopine (**309**), ribasine (**305**), secocularidine (**89**), (−)- and (±)-stylopine (**217a**), viguine (**46**) (*152–159*)

C. conspersa
Acetylcorynoline (**272**), consperine [= 6-acetonylacetylcorynoline] (**276**), corynoline (**271**), corynoloxine (**304**) (*160*)

C. cornuta Royle
Protopine (**309**), (+)-stylopine (**217a**) (*161*)

C. crystalina Engelm.
Bicuculline (**325**), capnoidine (**325**), protopine (**309**) (*162*)

C. decumbens Pers.
Adlumidine (**325**), bulbocapnine (**129**), dehydrocorydaline (**232b**), protopine (**309**), (+)- and (−)-tetrahydropalmatine (**218a**) (*3*)

C. fimbrillifera Korsh.
(+)-Hydrastine (**331**), protopine (**309**) (*3*)

C. gigantea Trautv. et Mey.
(−)-Adlumidine (**325**), (−)-adlumine (**326**), (+)-bicuculline (**325**), (−)-cheilanthifoline (**210a**), dihydrosanguinarine (**280**), ophiocarpine (**240**), protopine (**309**), sanguinarine (**296**), (−)-scoulerine (**212**) (*3,163*)

C. gortschakovii Schrenk.
Adlumidine (**325**), (−)-adlumine (**326**), (+)-bicuculline (**325**), bracteoline (**142**), cheilanthifoline (**210a**), corgoine (**44**), corydine (**136**), cryptopine (**311**), domesticine (**143**), gortschakoine (**28**), isoboldine (**147**), isocorydine (**132**), N-methylcoclaurine (**18**), protopine (**309**), reticuline (**20**), scoulerine (**212**), sendaverine (**45**), stylopine (**217a**) (*3,164*)

(*continued*)

TABLE LII (*Continued*)

C. govaniana Wall.
Bicuculline (**325**), corlumine (**326**), corygovanine (**222a**), *S*-govadine (**224a**), *S*-govanine (**225a**), isocorydine (**132**), protopine (**309**) (*3*)

C. incisa (Thunb.) Pers.
Acetylcorynoline (**272**), acetylisocorynoline (**272**), adlumidine (**325**), bicuculline (**325**), coreximine (**221**), corycavine (**322**), corydalisol (**260**), corydalispiron (**263**), corydalmine (**230a**), corydamine (**306**), corynolamine (**277a**), (+)- and (±)-corynoline (**271**), (+)-corynoline 11-*O*-sulfate (**273**), corynoloxine (**304**), corypalmine (**211a**), corysamine (**234b**), coptisine (**217b**), (+)-11-epicorynoline (**277**), 12-hydroxycorynoline (**275**), (−)-cheilanthifoline (**210a**), chelidonine (**266**), isocorynoline (**277**), 6-oxocorynoline (**274**), pallidine (**186**), protopine (**309**), (+)-reticuline (**20**), sanguinarine (**296**), (−)-scoulerine (**212**), sinoacutine (**188**), (−)-tetrahydrocorysamine (**251**) (*3,165,166*)

C. intermedia Méret (syn. *C. fabacea* Pers.)
α-Allocryptopine (**310**), isoboldine (**147**), protopine (**309**) (*3*)

C. koidzumiana Ohwi
Allocryptopine (**310**), capaurine (**219a**), cheilanthifoline (**210a**), corydalidzine (**231a**), corydaline (**232a**), (+)-corybulbine [= (+)-corydalmine] (**230a**), corynoxidine (**241**), dihydrosanguinarine (**280**), epicorynoxidine (**241**), (−)-isocorypalmine (**214a**), oxysanguinarine (**281**), protopine (**309**), (+)-reticuline (**20**), sanguinarine (**296**), (−)-scoulerine (**212**), stylopine (**217a**), tetrahydropalmatine (**218a**) (*3*)

C. ledebouriana Kar. et Kir.
Allocryptopine (**310**), bulbocapnine (**129**), (±)-cavidine (**233a**), corledine (**329**), corydaline (**232a**), cryptopine (**311**), dihydrochelerythrine (**278**), dihydrosanguinarine (**280**), ledeborine (**378**), ledeboridine (**383**), ledecorine (**31**), lederine (**385**), oxysanguinarine (**281**), protopine (**309**), (±)-raddeanine (**381**), (+)-tetrahydrocorysamine (**234a**), tetrahydropalmatine (**218a**) (*3,167–170*)

C. lineariloba var. *papilligera*
Corydaline (**232a**), glaucine (**145**), protopine (**309**), (−)-scoulerine (**212**), (−)-tetrahydrocolumbamine [= (−)-isocorypalmine] (**214a**), tetrahydrocoptisine [= (−)-stylopine] (**217a**), tetrahydropalmatine (**218a**) (*3*)

C. linearioides Maxim.
Acetylcorynoline (**272**), (+)-adlumine (**326**), adlumidine (**325**), (+)-bicuculline (**325**), corlumidine (**330**), corynoline (**271**), cheilanthifoline (**210a**), hydrastine (**331**), protopine (**309**) (*171*)

C. lutea (L.) DC.
Adlumidiceine (**340**), adlumidiceine enol lactone (**337**, *trans*), berberine (**209b**), (+)-bicuculline (**325**), coptisine (**217b**), (−)-corypalmine (**211a**), corysamine (**234b**), isocorydine (**132**), (−)-isocorypalmine (**214a**), jatrorrhizine (**211b**), *N*-methylhydrastine (**336**), *N*-methylhydrasteine (**339**), ochrobirine (**379**), 8-oxocoptisine (**254**), oxysanguinarine (**281**), protopine (**309**), (−)-stylopine (**217a**), (−)-tetrahydropalmatine (**218a**) (*3,172*)

C. marschalliana Pers.
Adlumidine (**325**), allocryptopine (**310**), bicuculline (**325**), bulbocapnine (**129**), cheilanthifoline (**210a**), corydaline (**232a**), (+)-corydine (**136**), corydione (**174**), dehydronantenine (**173**), (−)-domesticine (**143**), isoboldine (**147**), isocorypalmine (**214a**), isocorybulbine (**236a**),

TABLE LII (*Continued*)

marschaline (**32**), (+)-nantenine (**144**), oxonantenine (**161**), protopine (**309**), sanguinarine
  (**296**), (±)-sinactine (**215a**), (±)-stylopine (**217a**) (*3,163,173,174*)

C. meifolia Wall.
  Apocavidine (**229a**), (+)-cavidine (**233a**), cheilanthifoline (**210a**), corlumine (**326**),
    dehydrocavidine (**233b**), dihydrosanguinarine (**280**), protopine (**309**), (+)-sinactine (**215a**),
    stylopine (**217a**), yenhusomidine (**375**), yenhusomine (**380**) (*175*)

C. melanochlora Maxim.
  Acetylcorynoline (**272**), adlumidine (**325**), adlumine (**326**), (+)-bicuculline (**325**),
    cheilanthifoline (**210a**), corlumidine (**330**), corynoline (**271**), (+)-hydrastine (**331**), protopine
    (**309**) (*171*)

C. micrantha (Engelm.) Gray
  Capauridine (**219a**), capaurine (**219a**), protopine (**309**), scoulerine (**212**), (−)-
    tetrahydropalmatine (**218a**) (*162*)

C. montana (Engelm.) Britton (syn. *C. aurea* Willd.)
  Capauridine (**219a**), capaurimine (**220a**), capaurine (**219a**), corydaline (**232a**),
    dehydrocorydaline (**232b**), protopine (**309**), scoulerine (**212**), (±)-tetrahydropalmatine (**218a**)
    (*176*)

C. mucronifera
  Acetylcorynoline (**272**), adlumidine (**325**), (+)-adlumine (**326**), (+)-bicuculline (**325**),
    cheilanthifoline (**210a**), corlumidine (**330**), corynoline (**217**), (+)-hydrastine (**331**), protopine
    (**309**) (*171*)

C. nakaii Ishidoya
  Berberine (**209b**), coptisine (**217b**), corydine (?) (**136**), corysamine (?) (**234b**), isocorydine
    (**132**), (+)-norisocorydine (**131**), noroxyhydrastinine (**7**), (−)-reticuline (**20**), scoulerine
    (**212**), (−)-tetrahydrocolumbamine [= (−)-isocorypalmine] (**214a**) (*3*)

C. nobilis Pers.
  Bicuculline (**325**), corlumine (**326**), corydaline (**232a**), corytuberine (**137**), cryptopine (**311**),
    (+)-isocorypalmine (**214a**), protopine (**309**), stylopine (**217a**), (+)- and (±)-
    tetrahydropalmatine (**218a**) (*177*)

C. nokoensis Hayata
  Capaurine (**219a**), coptisine (**217b**), corybulbine (**230a**), dehydrocorybulbine (**230b**),
    dihydrosanguinarine (**280**), nokoensine (**243**), palmatine (**218b**), protopine (**309**),
    sanguinarine (**296**), (±)-tetrahydropalmatine (**218a**) (*178*)

C. ochotensis Turcz.
  Adlumidine (**325**), aobamidine (**337**, *cis*), aurotensine (= isoboldine) (**147**), corytenchine
    (**223**), corytenchirine (**238**), cryptocavine (= cryptopine) (**311**), didehydrocheilanthifoline
    (**210b**), lienkonine (**237**), ochotensimine (**359**), protopine (**309**), raddeanamine (**364**),
    raddeanine (**381**), raddeanone (**376**), yenhusomidine (**375**), yenhusomine (**380**) (*179*)

C. ochotensis var. *raddeana*
  Adlumidine (**325**), aobamidine (**337**, *cis*), aobamine (**259**), bicuculline (**325**), cheilanthifoline
    (**210a**), dihydrosanguinarine (**280**), fumarine (?), ochotensimine (**359**), ochotensine (**358**),
    pallidine (**186**), protopine (**309**), raddeanamine (**364**), raddeanidine (**382**), raddeanine (**381**),
    raddeanone (**376**), scoulerine (**212**), sinoacutine (**188**) (*3*)

(*continued*)

TABLE LII (*Continued*)

C. ochroleuca Koch
  Adlumidine (**325**), bicuculline (**325**), (−)-corypalmine (**211a**), fumarine (?), fumaramine (**347**), (+)-glaucine (**145**), isocorydine (**132**), (−)-isocorypalmine (**214a**), ochrobirine (**379**), protopine (**309**), scoulerine (**212**), sinactine (**215a**), sinoacutine (**188**), (−)-tetrahydropalmatine (**218a**) (*3*)

C. ophiocarpa Hook. et Thoms.
  Allocryptopine (**310**), adlumidine (**325**), adlumine (**326**), berberine (**209b**), bicuculline (**325**), (−)-canadine (**209a**), carpoxidine (= ophiocarpine *N*-oxide) (?) (**242**), chelerythrine (**295**), (−)-corypalmine (**211a**), coptisine (**217b**), corysamine (**234b**), cryptopine (**311**), dehydrocorypalline (**9**), dehydroisocorypalmine (**214b**), fumarine (?), fumaramine (**347**), (+)-glaucine (**145**), 13β-hydroxystylopine (**239**), isocorydine (**132**), noroxyhydrastinine (**7**), ophiocarpine (**240**), (−)-ophiocarpine *N*-oxide (**242**), protopine (**309**), sanguinarine (**296**), scoulerine (**212**), sinoacutine (**188**) (*180–182*)

C. paczoskii N. Bush.
  Corpaine (**360**), corydaine (**374**) (*3*)

C. pallida (Thunb.) Pers.
  Capauridine (**219a**), capaurimine (**220a**), capaurine (**219a**), corydaline (**232a**), cryptopine (**311**), isoboldine (**147**), (+)-isosalutaridine (**186**), kikemanine (= corydalmine) (**230a**), pallidine (**186**), protopine (**309**), (±)-stylopine (**217a**), (−)-tetrahydropalmatine (**218a**) (*3*)

C. pallida Pers. var. *tenuis* Yatabe
  Aurotensine (= isoboldine) (**147**), capauridine (**219a**), capaurimine (**220a**), capaurine (**219a**), coptisine (**217b**), corydalactame (**408**), corydaline (**232a**), corysamine (**234b**), corytuberine (**137**), dehydrocapaurine (**219b**), dehydrocapaurimine (**220b**), dehydrocorydaline (**232b**), dehydrocorydalmine (**253**), dihydrosanguinarine (**280**), kikemanine (= corydalmine) (**230a**), oxysanguinarine (**281**), pallidine [= (+)-isosalutaridine] (**186**), palmatine (**218b**), protopine (**309**), sanguinarine (**296**), (−)-scoulerine (**212**), sinoacutine (**188**), (+)-tetrahydrocorysamine (**234a**), (−)-tetrahydropalmatine (**218a**) (*3*)

C. paniculigera Regel et Schmalh.
  Adlumine (**326**), bicuculline (**325**), coclaurine (**13**), corunnine (**175**), dihydrosanguinarine (**280**), oxysanguinarine (**281**), pancoridine (**180**), pancorine (**293**), pancorinine (**181**), protopine (**309**), sanguinarine (**296**), sibiricine (**373**), stylopine (**217a**), thalicmidine (**153**), wilsonirine (**152**) (*183*)

C. persica Cham. et Schlecht.
  Chelerythrine (**295**), protopine (**309**), sanguinarine (**296**) (*3*)

C. platycarpa Makino
  Aurotensine (= isoboldine) (**147**), bicuculline (**325**), capaurimine (**220a**), capaurine (**219a**), cheilanthifoline (**210a**), corybulbine (**230a**), corydaline (**232a**), corysamine (**234b**), dehydrocapaurimine (**220b**), dehydrocorybulbine (**230b**), isocorydine (**132**), (−)-isocorypalmine (**214a**), jatrorrhizine (**211b**), palmatine (**218b**), protopine (**309**), sanguinarine (**296**), (+)- and (±)-scoulerine (212), (±)-stylopine (**217a**), tetrahydrocolumbamine (= isocorypalmine) (**214a**), (−)-tetrahydropalmatine (**218a**) (*3*)

C. pseudoadunca Popov
  Adlumidine (**325**), (+)- and (±)-bicuculline (**325**), coramine (= coreximine) (?) (**221**), corftaline (**328**), (+)-β-hydrastine (**8**), protopine (**309**), scoulerine (**212**) (*3,184*)

TABLE LII   (*Continued*)

C. pumila Koch
    Corydaline (**232a**), protopine (**309**) (*145*)

C. racemosa Pers.
    Protopine (**309**), (±)-tetrahydropalmatine (**218a**) (*3*)

C. remota Fisch. (syn. *C. bulbosa* DC.)
    (±)-Adlumidine (**325**), (±)-adlumine (**326**), allocryptopine (**310**), dihydrosanguinarine (**280**), sanguinarine (**296**) (*163*)

C. rosea Zeyh.
    (−)- and (±)-Adlumidine (**325**), (−)- and (±)-adlumine (**326**), protopine (**309**), sanguinarine (**296**) (*3,163*)

C. scouleri Hook.
    (−)-Adlumine (**326**), allocryptopine (**310**), bicuculline (**325**), capnoidine (**325**), cheilanthifoline (**210a**), corlumidine (**338**), corlumine (**326**), cryptopine (**311**), protopine (**309**), (−)-scoulerine (**212**) (*185*)

C. sempervirens (L.) Pers. (syn. *C. glauca* Pursh)
    Adlumiceine (**341**), adlumidiceine (**340**), adlumidiceine enol lactone (**337**, *trans*), adlumine (**326**), berberine (**209b**), bicuculline (**325**), capnoidine (**325**), coptisine (**217b**), cryptopine (**311**), oxysanguinarine (**281**), protopine (**309**) (*186*)

C. sewerzovi Regel
    Allocryptopine (**310**), bicuculline (**325**), chelerythrine (**295**), corlumine (**326**), cryptopine (**311**), dihydrosanguinarine (**280**), protopine (**309**), sanguinarine (**296**), severcinine (**384**) (*3*)

C. sibirica (L.) Pers.
    Adlumidine (**325**), bicuculline (**325**), cheilanthifoline (**210a**), corlumine (**326**), cryptopine (**311**), ochotensine (**358**), ochrobirine (**379**), protopine (**309**), sibiricine (**373**), scoulerine (**212**) (*3*)

C. slivensis [syn. *C. solida* (L.) Swartz. subsp. *slivensis* (Velen.) Hayek]
    Allocryptopine (**310**), berberine (**209b**), bulbocapnine (**129**), (−)-canadine (**209a**), cavidine (**233a**), corydine (**136**), dehydronantenine (**173**), domesticine (**143**), isoboldine (**147**), isocorydine (**132**), (+)-nantenine (**144**), predicentrine (**151**), protopine (**309**), (±)-sinactine (**215a**), (−)-stylopine (**217a**), (+)-tetrahydrocorysamine (**234a**) (*187*)

C. solida (L.) Swartz. (syn. *C. bulbosa* DC.)
    Allocryptopine (**310**), aurotensine (= isoboldine) (**147**), berberine (**209b**), bulbocapnine (**129**), (+)- and (−)-canadine (**209a**), columbamine (**214b**), coptisine (**217b**), corydaline (**232a**), (+)-corydine (**136**), dehydrocorydaline (**232b**), domesticine (**143**), (+)-isocorydine (**132**), nantenine (**144**), ochotensine (**358**), predicentrine (**151**), protopine (**309**), (±)-sinactine (**215a**), solidaline (**252**), (+)- and (−)-stylopine (**217a**), (−)-tetrahydrocolumbamine [= (−)-isocorypalmine] (**214a**), (±)-tetrahydrocorysamine (**234a**), (+)-, (−)-, and (±)-tetrahydropalmatine (**218a**) (*188,189*)

C. speciosa Maxim. (syn. *C. pallida* Pers.)
    Allocryptopine (**310**), capaurimine (**220a**), capaurine (**219a**), corypalline (**1**), protopine (**309**), tetrahydropalmatine (**218a**) (*3*)

(*continued*)

TABLE LII   (*Continued*)

C. stewartii Fedde
Coptisine (**217b**), corycidine (?), corydicine (?), corydinine (= protopine) (**309**), (+)-tetrahydrocoptisine (**217a**) (*3*)

C. stricta Steph.
Acetylcorynoline (**272**), adlumidine (**325**), (+)-adlumine (**326**), (+)-bicuculline (**325**), cheilanthifoline (**210a**), corlumidine (**330**), corynoline (**271**), (+)-hydrastine (**331**), protopine (**309**), sanguinarine (**296**) (*3,171*)

C. taliensis Fr.
Acetylcorynoline (**272**), bicuculline (**325**), corycavine (**322**), (−)-corynoline (**271**), corydamine (**277a**), protopine (**309**) (*190*)

C. tashiroi Makino
(−)-Corydalmine (**230a**), dehydrocheilanthifoline (**210b**), dehydrocorydalmine (**253**), dehydrodiscretamine (**213b**), (−)-discretamine (**213a**), palmatine (**218b**), protopine (**309**), sendaverine (**45**), (−)-sinactine (**215a**), stepharanine (**216b**), (−)-stepholidine (**216a**), (−)- and (±)-tetrahydropalmatine (**218a**) (*3,191*)

C. ternata Nakai
Allocryptopine (**310**), (−)-canadine (**209a**), coptisine (**217b**), (−)-corydine (**136**), (−)-glaucine (**145**), isocorydine (**132**), protopine (**309**), stylopine (**217a**) (*192*)

C. thalictrifolia Jameson
Adlumidine (**325**), adlumine (**326**), berberine (**209b**), (−)-canadine (**209a**), (−)-β-canadine methohydroxide (**244**), cavidine (**233a**), coptisine (**217b**), (−)-corypalmine (**211a**), dehydrothalictrifoline (**233b**), (−)-isocorypalmine (**214a**), protopine (**309**), (+)-stylopine (**217a**), (+)-thalictrifoline (**233a**) (*3,193*)

C. vaginans Royle
(±)-Adlumine (**326**), bicuculline (**325**), bulbocapnine (**129**), cheilanthifoline (**210a**), corydaine (**374**), dihydrosanguinarine (**280**), isocorypalmine (**214a**), *O*-methylcorpaine (**361**), ochotensine (**358**), (+)-ochrobirine (**379**), protopine (**309**), sanguinarine (**296**), scoulerine (**212**) (*3,163*)

TABLE LIII
Alkaloids Found in Plants of the Genus *Corydalis*

Simple isoquinolines, benzylisoquinolines, and *N*-benzylisoquinolines
Corypalline, dehydrocorypalline, hydrastinine, (+)-1-methylcorypalline, noroxyhydrastinine; coclaurine, corpaverine (= capaurine + sendaverine), (+)-crassifoline, gortschakoine, ledecorine, marschaline, *N*-methylcoclaurine, (+)- and (−)-reticuline; corgoine, sendaverine, viguine

Cularines
(+)-Claviculine, (+)-culacorine, (+)-cularicine, (+)-cularidine, (+)-cularine, limousamine, (+)-norculacorine, (+)-norcularicine, norcularidine, norsecocularine, noyaine, oxocularine, secocularidine

Aporphines
Bracteoline, bulbocapnine, (+)-bulbocapnine methohydroxide, bulbodione, corunnine, corydine, corydione, corytuberine, dehydroglaucine, dehydronantenine, (−)-domesticine,

TABLE LII *(Continued)*

domestine, (+)-glaucine, glaucinone, isoboldine (= aurotensine), isocorydine, 1,2-methylene dioxy-6a,7-dehydroaporphine-10,11-quinone, magnoflorine, nandazurine, (+)-nantenine, (+)-norisocorydine, oxonantenine, pancoridine, pancorinine, predicentrine, thalicmidine, thaliporphine, wilsonirine

Promorphinanes

Pallidine [= (+)-isosalutaridine], sinoacutine

Protoberberines

Apocavidine, (+)- and (−)-canadine, (−)-β-canadine methohydroxide, (±)-capauridine [= (±)-capaurine], capaurimine, capaurine, caseadine, caseamine, caseanadine, (+)- and (±)-cavidine, (−)-cheilanthifoline, coreximine, (+)-corybulbine, corydalidzine, (+)-corydaline, corydalmine (= kikemanine, corybulbine), corygovanine, corynoxidine, (+)- and (−)-corypalmine (= tetrahydrojatrorrhizine), corytenchine, corytenchirine, discretamine, epicorynoxidine, govadine, govanine, (−)-13β-hydroxystylopine, isocorybulbine, (+)-, (−)-, and (±)-isocorypalmine (= tetrahydrocolumbamine), lienkonine, 13-methyltetrahydroprotoberberine (= thalictricavine), nokoensine, ophiocarpine, ophiocarpine N-oxide, (+)-, (−)-, and (±)-scoulerine, (±)-sinactine, solidaline, stepholidine, (+)-, (−)-, and (±)-stylopine, (±)-α-stylopine methohydroxide, (+)-, (−)-, and (±)-tetrahydrocorysamine, α-tetrahydrocorysamine methohydroxide, (+)-, (−)-, and (±)-tetrahydropalmatine, (+)- and (±)-thalictricavine, (+)-thalictrifoline; berberine, columbamine, coptisine, corysamine, dehydroapocavidine, dehydrocapaurine, dehydrocapaurimine, dehydrocavidine, dehydrocheilanthifoline, dehydrocorybulbine, dehydrocorydaline, dehydrocorydalmine, dehydrodiscretamine, dehydroisocorypalmine, dehydrothalictrifoline, dehydrothalictricavine, didehydrocheilanthifoline, jatrorrhizine, 8-oxocoptisine, palmatine, stepharanine

Secoberbines

Aobamine, corydalisol, corydalispiron

Benzophenanthridines

Acetylcorynoline, acetylisocorynoline, chelerythrine, chelidonine, consperine, corydamine, corynolamine, (+)-, (−)-, and (±)-corynoline, corynoloxine, (+)-corynoline 11-O-sulfate, dihydrochelerythrine, dihydrosanguinarine, (+)-11-epicorynoline, 12-hydroxycorynoline, isocorynoline, 6-oxocorynoline, oxysanguinarine, pancorine, ribasine, sanguinarine

Protopines

Allocryptopine, corycavamine, corycavidine, corycavine, cryptopine, protopine

Phthalideisoquinolines

Adlumidine, adlumine, (+)- and (±)-bicuculline, capnoidine, cordrastine, corftaline, corledine, corlumidine, corlumine, (+)-α-hydrastine, narcotine

Secophthalideisoquinolines

Adlumiceine, adlumiceine enol lactone, adlumidiceine, adlumidiceine enol lactone, aobamidine, bicucullinine, fumaramine, N-methylhydrasteine, N-methylhydrastine

Spirobenzylisoquinolines

Corpaine, corydaine, ledeboridine, ledeborine, lederine, ochotensimine, ochotensine, O-methylcorpaine, ochrobirine, raddeanamine, raddeanidine, (±)- raddeanine, raddeanone, severcinine, sibiricine, yenhusomidine, yenhusomine

Dibenzazonines

Crassifolazonine

Alkaloids of unknown structure

Bicucine, caseamine, corydalactame, corycidine, corydicine, fumarine

60 V. PREININGER

TABLE LIV
Plants of the Genus *Fumaria* and Their Alkaloids

Fumaria agraria Lag.
  Protopine (**309**) (*197*)

F. capreolata
  Coptisine (**217b**), protopine (**309**), sanguinarine (**296**) (*198*)

F. densiflora DC. (syn. *F. micrantha* Lag.)
  Adlumidiceine (**340**), coptisine (**217b**), cryptopine (**311**), densiflorine (**387**), fumaramine (**347**),
    palmatine (**218b**), protopine (**309**), (±)-sinactine (**215a**) (*3,199*)

F. indica Pugsley
  (−)-Adlumidine (**325**), (+)- and (±)-bicuculline (**325**), coptisine (**217b**),
    dehydrocheilanthifoline (**210b**), fumarilicine (?), fumariline (**367**), (−)-8-
    methoxydihydrosanguinarine (**282**), narceine imide (**350**), narceine imine (**345**), narlumidine
    (**352**), oxysanguinarine (**281**), protopine (**309**), protopine methohydroxide (**324a**),
    sanguinarine (**296**), (−)- and (±)-stylopine (**217a**) (*3,200,201*)

F. judaica Boiss.
  Bicuculline (**325**), cheilanthifoline (**210a**), coptisine (**217b**), parfumine (**366**), protopine (**309**),
    stylopine (**217a**) (*202*)

F. kralikii Jord. (syn. *F. anatolica*)
  Adlumidiceine (**340**), (−)-adlumine (**326**), berberine (**209b**), (−)-canadine (**209a**), coptisine
    (**217b**), cryptopine (**311**), fumaritine *N*-oxide (**363**), fumarofine (**355**), fumarophycine (**370**),
    O-methylfumarophycine (**371**), (+)-parfumine (**366**), protopine (**309**), (−)-stylopine (**217a**)
    (*203,204*)

F. macrocarpa Parl.
  Bicuculline (**325**), fumariline (**367**), parfumine (**366**) (*205*)

F. micrantha Lag.
  Fumaramine (**347**), protopine (**309**), sanguinarine (**296**) (*3,206*)

F. microcarpa Boiss.
  Fumarofine (**355**) (*207*)

F. muralis Green. (syn. *F. media*)
  Protopine (**309**) (*197*)

F. officinalis L.
  Canadine (**209a**), cryptocavine (**311**), cryptopine (**311**), dihydrofumariline (**368**), (−)-
    fumaricine (**262**), fumarilicine (?), (+)-fumariline (**367**), fumaritine (**363**), fumaritine *N*-
    oxide (**386**), fumaritriline (?), fumaritrine (**357**), fumarofine (**355**), fumaroline (?),
    fumarophycine (**370**), fumaroficinaline (?), *N*-methylhydrasteine (**339**), *N*-methylhydrastine
    (**336**), *N*-methyloxohydrasteine (**344**), *N*-methylsinactine (**248**), *O*-methylfumarophycine
    (**371**), parfumidine (**365**), protopine (**309**), sanguinarine (**296**), (+)- and (±)-scoulerine
    (**212**), (−)-sinactine (**215a**), (±)-stylopine (**217a**) (*3,208*)

F. parviflora Lam.
  Adlumidiceine (**340**), (−)-adlumine (**326**), (−)-adlumidine (**325**), adlumine methohydroxide
    (**335**), (+)-bicuculline (**325**), (−)-cheilanthifoline (**210a**), coclaurine (**13**), coptisine (**217b**),
    (−)-corlumine (**326**), cryptopine (**311**), dihydrofumariline (**368**), dihydroparfumine (**369**),
    dihydrosanguinarine (**280**), fumaramine (**347**), fumaramidine (**348**), (+)-fumaricine (**362**),

TABLE LIV   *(Continued)*

fumaridine (**349**), fumariflorine (**10**), (+)-fumariline (**367**), fumariflorine ethyl ester (**11**), fumaritine (**363**), (+)-α-hydrastine (**331**), (+)-isoboldine (**147**), izmirine (**313**), lahoramine (**354**), lahorine (**355**), *N*-methylhydrasteine (**339**), 8-oxocoptisine (**254**), oxysanguinarine (**281**), norjuziphine (**30**), noroxyhydrastinine (**7**), parfumidine (**365**), parfumine (**366**), parviflorine (**372**), protopine (**309**), sanguinarine (**296**), scoulerine (**212**), (−)-stylopine (**217a**) (*3,209–213*)

F. rostelata Knaf
(+)-Adlumine (**326**), cryptopine (**311**), fumariline (**367**), fumaritridine (**356**), fumaritrine (**357**), fumarostelline (**377**), parfumine (**366**), protopine (**309**), sinactine (**215a**), stylopine (**217a**) (*3*)

F. schleicheri Soyer-Willemet
Cryptopine (**311**), fumaramine (**347**), fumaridine (= hydrastine imide) (**349**), fumarinine (?), fumaritine (**363**), fumschleicherine (**351**), *N*-methylhydrasteine (**344**), *N*-methylhydrastine (**336**), (+)-parfumine (**366**), protopine (**309**), (±)-sinactine (**215a**) (*214–217*)

F. schrammii (Ascherson) Velen.
Adlumiceine (**341**), adlumiceine enol lactone (**338**), adlumidiceine (**340**), adlumidiceine enol lactone (**337**, *trans*), (−)-adlumine (**326**), bicucullinidine (**346**), bicucullinine (**345**), chelerythrine (**295**), coptisine (**217b**), cryptopine (**311**), dihydrofumariline (**368**), fumaricine (**362**), fumariline (**367**), parfumine (**366**), protopine (**309**), sanguinarine (**296**), (±)-sinactine (**215a**), (−)-stylopine (**217a**) (*218,219*)

F. vaillantii Loisl.
(±)-8-Acetonyldihydrosanguinarine (**285**), adlumidiceine (**340**), (−)-adlumidine (**325**), (−)-adlumilantine (?), (−)-adlumine (**326**), (+)-bicuculline (**325**), cheilanthifoline (**210a**), coclaurine (**13**), (−)-corledine (**329**), cryptopine (**311**), dihydrosanguinarine (**280**), enganine (**334**), fumaramine (**347**), fumaridine (**349**), fumariline (**367**), fumaramidine (**348**), fumaritine (**363**), fumvailine (**326**), (+)-α-hydrastine (**331**), isoboldine (**147**), (+)-isocorydine (**132**), (+)-juziphine (**29**), ledecorine (**31**), (±)-8-methoxydihydrosanguinarine (**282**), (−)-*N*-methyladlumine (**335**), *N*-methylcorydaldine (**5**), *N*-methylhydrasteine (**339**), *N*-methylhydrastine (**336**), norsanguinarine (**292**), oxysanguinarine (**281**), parfumidine (**365**), parfumine (**366**), reticuline (**20**), sanguinarine (**296**), (−)-scoulerine (**212**), sinactine (**215a**), (−)- and (+)-stylopine (**217a**), (−)-stylopine methohydroxide (**249**), vaillantine (= 2,3-didemethylmuramine) (**315**) (*3,220–225*)

---

greatly. Hegi (*196*) mentions 40 for Europe, Tutin *et al.* (*11*) 44, *Index Kewensis* (*141*) 215. The high number reported in *Index Kewensis* is also due to the listing of many taxons under several synonyms. To date, 12 species have been studied for their alkaloid content (Tables LIV and LV).

Fumaria chemically resembles *Corydalis*. Similarly, the genus *Fumaria* also contains protoberberines, protopines, and phthalideisoquinolines, and more frequently, secophthalideisoquinolines and spirobenzylisoquinolines. They also yield indenobenzazepines, which cannot be demonstrated in the genus *Corydalis* (Table LVI), but no aporphines and cularines, which have been isolated from the plants of the genus *Corydalis*.

TABLE LV
Alkaloids Found in Plants of the Genus *Fumaria*

Simple isoquinolines and benzylisoquinolines
  Fumariflorine, fumariflorine ethylester, *N*-methylcorydaldine, noroxyhydrastinine; coclaurine,
    (+)-juziphine, ledecorine, norjuziphine, reticuline

Aporphines
  (+)-Isoboldine, (+)-isocorydine

Protoberberines
  (−)-Canadine, (−)-cheilanthifoline, *N*-methylsinactine, (−)-, (+)-, and (±)-scoulerine, (−)-
    and (±)-sinactine, (−)-, (+)-, and (±)-stylopine, (−)-stylopine methohydroxide; berberine,
    coptisine, dehydrocheilanthifoline, 8-oxocoptisine, palmatine

Benzophenanthridines
  (±)-8-Acetonyldihydrosanguinarine, chelerythrine, dihydrosanguinarine, (±)-8-
    methoxydihydrosanguinarine, norsanguinarine, oxysanguinarine, sanguinarine

Protopines
  Cryptocavine (= cryptopine), cryptopine, izmirine, protopine, vaillantine

Phthalideisoquinolines
  (−)-Adlumidine, (−)-adlumine, (+)-bicuculline, (−)-corledine, (−)-corlumine, enganine,
    fumvailine [= (−)-adlumine], (+)-α-hydrastine, (−)-*N*-methyladlumine [= (−)-adlumine
    methohydroxide]

Secophthalideisoquinolines
  Adlumiceine, adlumiceine enol lactone, adlumidiceine, adlumidiceine enol lactone,
    bicucullinidine, bicucullinine, fumaramidine, fumaramine, fumaridine (= hydrastine imide),
    fumschleicherine, *N*-methylhydrasteine, *N*-methylhydrastine, narceine imide, narceine imine,
    narlumidine, *N*-methyloxohydrasteine

Indenobenzazepines
  Fumaritridine, fumaritrine, fumarofine, lahoramine, lahorine

Spirobenzylisoquinolines
  Densiflorine, dihydrofumariline, dihydroparfumine, fumaricine, (+)-fumariline, fumaritine,
    fumaritine *N*-oxide, fumarophycine, fumarostelline, *O*-methylfumarophycine, parfumidine,
    parfumine, parviflorine

Alkaloids of unknown structure
  Adlumilantine (*F. vailantii*), fumarilicine (*F. indica, F. officinalis*), fumarinine (*F.
    schleicheri*), fumaroficinaline (*F. officinalis*), fumaroline (*F. officinalis*), fumaritriline (*F.
    officinalis*)

C. The Genus *Dicentra*

Seven of the 17 species classified in the genus *Dicentra* Bernh. have been studied. The genus is characterized by the presence of cularine alkaloids (Tables LVII and LVIII). Chemically, this genus shows greater similarity to the genus *Corydalis* than to the genus *Fumaria*.

TABLE LVI
Alkaloid Types Found in the Genus *Fumaria*

	Isoquinolines	Benzylisoquinolines	Aporphines	Protoberberines	Benzophenanthridines	Protopines	Phthalideisoquinolines	Secophthalideisoquinolines	Indenobenzazepines	Spirobenzylisoquinolines
Fumaria agraria Lag.	−	−	−	−	−	+	−	−	−	−
F. capreolata	−	−	−	+	+	+	−	−	−	−
F. densiflora DC.	−	−	−	+	−	+	−	+	−	+
F. indica Pugsley	−	−	−	+	+	+	+	+	−	+
F. judaica Boiss.	−	−	−	+	−	+	+	−	−	+
F. kralikii Jord.	−	−	−	+	−	+	+	+	+	+
F. macrocarpa Parl.	−	−	−	−	−	−	+	−	−	+
F. micrantha Lag.	−	−	−	−	+	+	−	+	−	−
F. microcarpa (Hausskn.) Pugsley	−	−	−	−	−	−	−	−	+	−
F. muralis Green.	−	−	−	+	−	+	−	−	−	−
F. officinalis L.	+	+	+	+	+	+	−	+	+	+
F. parviflora Lam.	−	−	−	+	+	+	+	+	+	+
F. rostelata Knaf	+	+	−	+	−	+	+	−	+	+
F. schleicheri Soyer-Willemet	−	−	−	+	−	+	−	+	−	+
F. schrammii (Ascherson) Velen	−	−	−	+	+	+	+	+	−	+
F. vaillantii Loisl.	+	+	+	+	+	+	+	+	−	+

TABLE LVII
Plants of the Genus *Dicentra* and Their Alkaloids

Dicentra canadensis (Goldie) Walp.
Cancentrine (**94**), dehydrocancentrine (**95, 96**) (*3*)

D. cucullaria (L.) Bernh.
Corlumine (**326**), cularidine (**72**), cularine (**74**), ochotensine (**358**) (*3*)

D. eximia Torr.
Corydine (**136**), cularimine (**75**), cularine (**74**), dicentrine (**140**), glaucine (**145**), norprotosinomenine (**26**) (*3*)

D. formosa Walp.
Cularine (**74**) (*3*)

D. oregana Eastn.
Cularine (**74**) (*3*)

D. peregrina Rudolph.
Allocryptopine (**310**), bicuculline (**325**), chelerythrine (**295**), corydine (**136**), dicentrine (**140**), dihydrosanguinarine (**280**), isoboldine (**147**), isocorydine (**132**), lederine (**385**), predicentrine (**151**), protopine (**309**), reticuline (**20**), sanguinarine (**296**) (*226*)

D. spectabilis Lam.
Cheilanthifoline (**210a**), chelerythrine (**295**), chelilutine (**301**), chelirubine (**300**), coptisine (**217b**), corydine (**136**), dihydrosanguinarine (**280**), protopine (**309**), sanguinarine (**296**), scoulerine (**212**) (*3,226*)

TABLE LVIII
Alkaloids Found in Plants of the Genus *Dicentra*

Benzylisoquinolines
Norprotosinomenine, reticuline

Cularines
Cancentrine, cularidine, cularimine, cularine, dehydrocancentrine

Aporphines
Corydine, dicentrine, glaucine, isoboldine, isocorydine, predicentrine

Protoberberines
Cheilanthifoline, scoulerine; coptisine

Benzophenanthridines
Chelerythrine, chelilutine, chelirubine, dihydrosanguinarine, sanguinarine

Protopines
Allocryptopine, protopine

Phthalideisoquinolines
Bicuculline, corlumine

Spirobenzylisoquinolines
Lederine, ochotensine

TABLE LIX

Plants of the Genus *Sarcocapnos* and Their Alkaloids

Sarcocapnos crassifolia DC.
  Claviculine (**78**), culacorine (**68**), oxosarcocapnidine (**87**), ribasine (**305**), sarcocapnidine (**79**) (*227*)

S. enneaphylla DC.
  Aristoyagonine (**92**), celtyne (**71**), celtysine (**70**), 4-hydroxysarcocapnine (**82**), oxocompostelline (**85**), oxosarcocapnine (**80**), oxosarcophylline (**86**), sarcocapnine (**80**), secocularine (**90**), yagonine (**83**) (*158,228,229*)

TABLE LX

Alkaloids Found in the Genus *Sarcocapnos*

Cularines
  Aristoyagonine, celtyne, celtysine, claviculine, culacorine, 4-hydroxysarcocapnine, oxocompostelline, oxosarcocapnidine, oxosarcocapnine, oxosarcophylline, sarcocapnidine, sarcocapnine, secocularine, yagonine
Benzophenathridines
  Ribasine

D. THE GENUS *Sarcocapnos*

The genus *Sarcocapnos* DC. is also characterized by the presence of cularine alkaloids (Tables LIX and LX).

E. THE GENUS *Adlumia*

Only in one species of the genus *Adlumia* Rafin. have protopines and phthalideisoquinolines been found (Tables LXI and LXII).

TABLE LXI

Plants of the Genus *Adlumia* and Their Alkaloids

Adlumia cirrhosa Rafin. (syn. *A. fungosa* Greene)
  Adlumidine (**325**), adlumine (**326**), bicuculline (**325**), α-allocryptopine (**310**), protopine (**309**) (*230*)

TABLE LXII

Alkaloids Found in Plants of the Genus *Adlumia*

Protopines
  α-Allocryptopine, protopine
Phthalideisoquinolines
  Adlumidine, adlumine, bicuculline

IV. Alphabetic Listing of the Papaveraceae and Fumariaceae Alkaloids and Their Formulas

Simple Isoquinolines

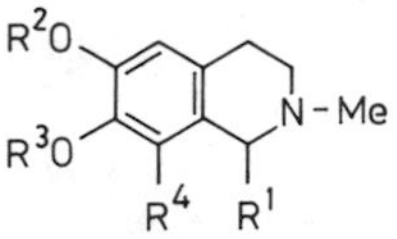

$$1 \quad R^1 = R^3 = R^4 = H; \; R^2 = Me$$
$$2 \quad R^1 = R^4 = H; \; R^2 = R^3 = Me$$
$$3 \quad R^1 = R^2 = Me; \; R^3 = R^4 = H$$
$$4 \quad R^1 = H; \; R^2 + R^3 = CH_2; \; R^4 = OMe$$

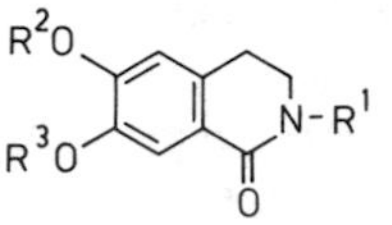

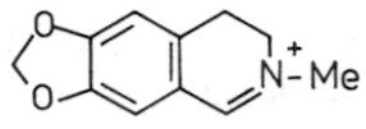

$$5 \quad R^1 = R^2 = R^3 = Me$$
$$6 \quad R^1 = Me; \; R^2 + R^3 = CH_2$$
$$7 \quad R^1 = H; \; R^2 + R^3 = CH_2$$

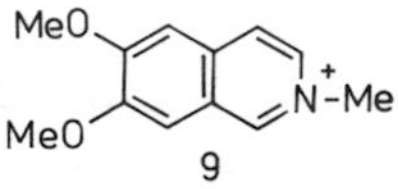

8

9

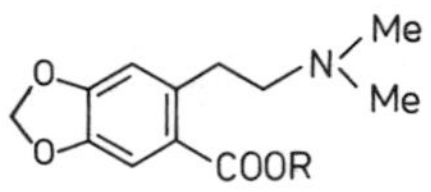

$$10 \quad R = H$$
$$11 \quad R = Et$$

Corypalline (**1**)
Dehydrocorypalline (**9**)
Fumariflorine (**10**)
Fumariflorine ethyl ester (**11**)
Hydrastinine (**8**)
Hydrocotarnine (**4**)

N-Methylcorydaldine (**5**)
1-Methylcorypalline (**3**)
N-Methyl-6,7-dimethoxytetrahydro-
    isoquinolone-1 (= N-methylcorydaldine) (**5**)
Noroxyhydrastinine (**7**)
Oxyhydrastinine (**6**)
O-Methylcorypalline (**2**)

Benzylisoquinolines

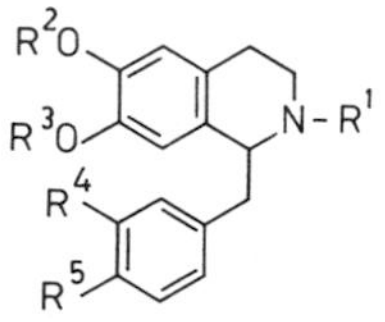

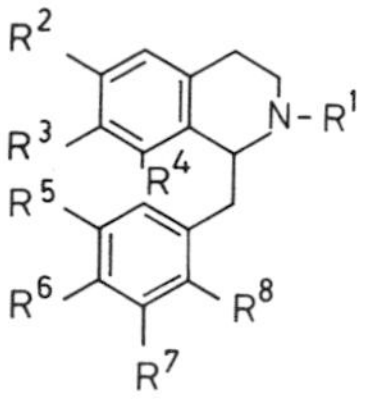

12 $R^1 = R^2 = R^3 = Me; R^4 = H; R^5 = OH$
13 $R^1 = R^3 = R^4 = H; R^2 = Me; R^5 = OH$
14 $R^1 = R^2 = Me; R^3 = H; R^4 = R^5 = OMe$
15 $R^1 = R^2 = Me; R^3 = D$-xylose$; R^4 = H; R^5 = OH$
16 $R^1 = R^2 = R^3 = Me; R^4 = OH; R^5 = OMe$
17 $R^1 = R^2 = R^3 = Me; R^4 = R^5 = OMe$
18 $R^1 = R^2 = Me; R^3 = R^4 = H; R^5 = OH$
19 $R^1 = R^2 = R^3 = Me; R^4 = H; R^5 = OMe$
20 $R^1 = R^2 = Me; R^3 = H; R^4 = OH; R^5 = OMe$
21 $R^1 = H; R^2 = R^3 = Me; R^4 = R^5 = OMe$
22 $R^1 = (Me)_2; R^2 + R^3 = CH_2; R^4 = R^5 = OMe$
23 $R^1 = H; R^2 + R^3 = CH_2; R^4 = R^5 = OMe$
24 $R^1 = Me; R^2 + R^3 = CH_2; R^4 = R^5 = OMe$
25 $R^1 = R^2 = R^4 = Me; R^3 = R^5 = H$
26 $R^1 = R^2 = R^4 = H; R^3 = Me; R^5 = OMe$

27 $R^1 = Me; R^2 = R^7 = R^8 = H; R^3 = R^6 = OMe; R^4 = R^5 = OH$
28 $R^1 = Me; R^2 = R^5 = R^7 = R^8 = H; R^3 = R^6 = OMe; R^4 = OH$
29 $R^1 = Me; R^2 = R^5 = R^7 = R^8 = H; R^3 = OMe; R^4 = R^6 = OH$
30 $R^1 = R^2 = R^5 = R^7 = R^8 = H; R^3 = OMe; R^4 = R^6 = OH$
31 $R^1 = Me; R^2 + R^3 = R^6 + R^7 = OCH_2O; R^4 = R^5 = H; R^8 = OH$
32 $R^1 = Me; R^2 + R^3 = R^6 + R^7 = OCH_2O; R^4 = R^5 = H; R^8 = OMe$

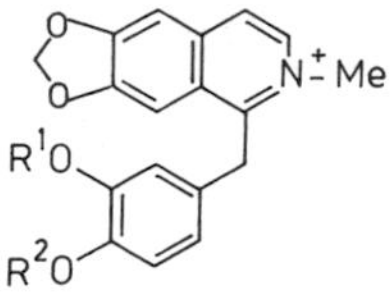

33 $R^1 + R^2 = CH_2$
34 $R^1 = Me; R^2 = H$

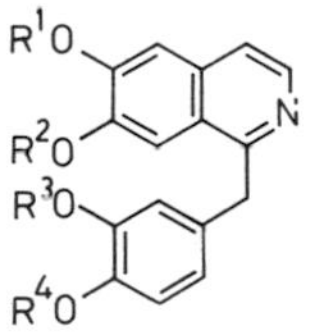

35 $R^1 = R^2 = R^3 = R^4 = Me$
36 $R^1 = R^2 = R^4 = Me; R^3 = H$
37 $R^1 = Me; R^2 = H; R^3 + R^4 = CH_2$
38 $R^1 = Me; R^2 = D$-glucose$; R^3 + R^4 = CH_2$

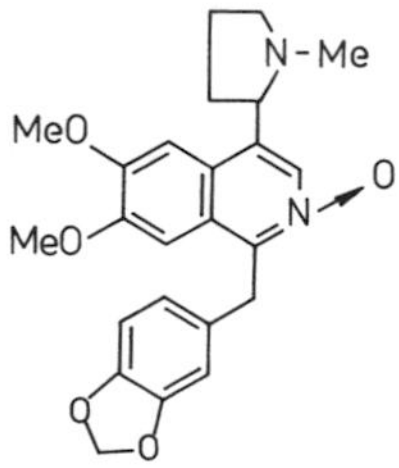

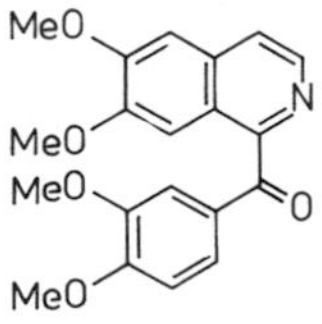

39

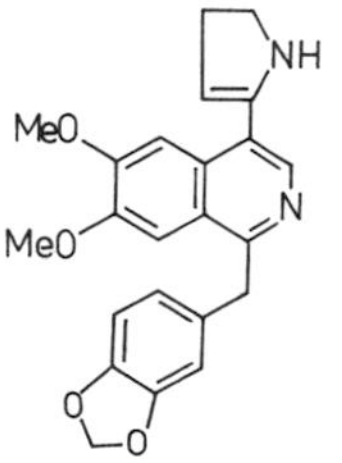

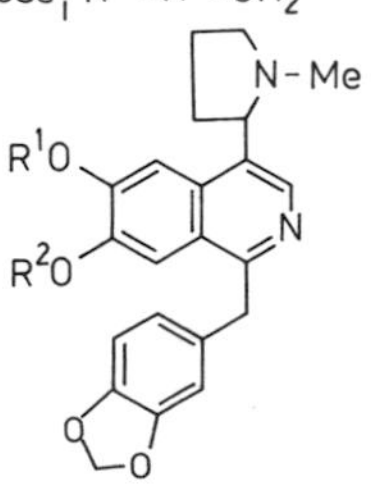

40 $R^1 = R^2 = Me$
41 $R^1 = Me; R^2 = H$

42

43

Arenine (**41**)
Armepavine (**12**)
Coclaurine (**13**)
Codamine (**14**)
Crassifoline (**27**)
Dehydronormacrostomine (**43**)
Escholamidine (**34**)
Escholamine (**33**)
Escholinine (**22**)
Glycomarine (**38**)
Gortschakoine (**28**)
Juziphine (**29**)
Latericine (**15**)
Laudanidine [= (±)-laudanine] (**16**)
Laudanosine (**17**)
Ledecorine (**31**)
Macrostomine (**40**)

Macrostomine *N*-oxide (**42**)
Marschaline (**32**)
N-Methylcoclaurine (**18**)
O-Methylarmepavine (**19**)
Norjuziphine (**30**)
Norprotosinomenine (**26**)
Norromneine (**23**)
Orientaline (**25**)
Palaudine (**36**)
Papaveraldine (**39**)
Papaverine (**35**)
Reticuline (**20**)
Romneine (**24**)
Sevanine (**37**)
Tetrahydropapaverine (**21**)
Papaveramine (structure unknown)

N-Benzylisoquinolines

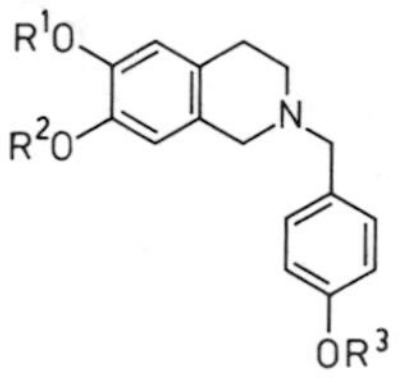

44   $R^1 = Me$; $R^2 = R^3 = H$
45   $R^1 = R^3 = Me$; $R^2 = H$
46   $R^1 + R^2 = CH_2$; $R^3 = Me$

Corgoine (**44**)
Sendaverine (**45**)

Viguine (**46**)

Pavines and Isopavines

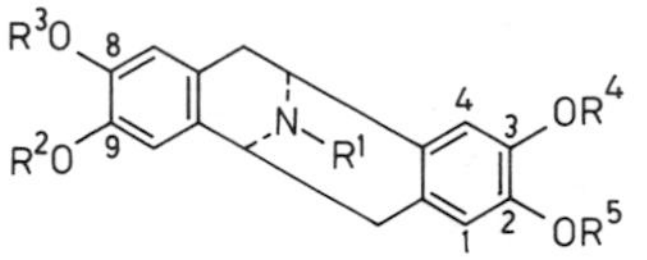

47 $R^1 = R^2 = R^3 = R^4 = R^5 = Me$
48 $R^1 = (Me)_2; R^2 = R^3 = R^4 = R^5 = Me$
49 $R^1 = MeO; R^2 = R^3 = R^4 = R^5 = Me$
50 $R^1 = R^2 = R^5 = Me; R^3 = R^4 = H$
51 $R^1 = (Me)_2; R^2 = R^3 = Me; R^4 + R^5 = CH_2$
52 $R^1 = R^4 = R^5 = Me; R^2 + R^3 = CH_2$
53 $R^1 = Me; R^2 + R^3 = R^4 + R^5 = CH_2$
54 $R^1 = R^2 = R^3 = R^4 = Me; R^5 = H$
55 $R^1 = R^3 = R^4 = R^5 = Me; R^2 = H$

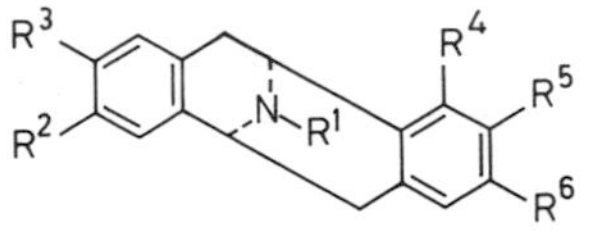

56 $R^1 = Me; R^2 = R^6 = OMe; R^3 = R^4 = H; R^5 = OH$
57 $R^1 = Me; R^2 = R^4 = OH; R^3 = R^5 = OMe; R^6 = H$
58 $R^1 = Me; R^2 = R^3 = R^5 = OMe; R^4 = OH; R^6 = H$
59 $R^1 = (Me)_2; R^2 = R^3 = R^5 = OMe; R^4 = OH; R^6 = H$

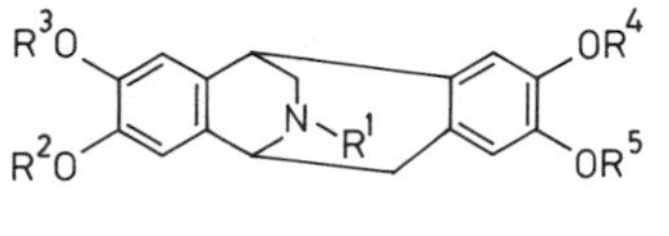

60 $R^1 = R^4 = Me; R^2 + R^3 = CH_2; R^5 = H$
61 $R^1 = R^4 = R^5 = Me; R^2 + R^3 = CH_2$
62 $R^1 = (Me)_2; R^2, R^3 = Cl_2; R^4 = R^5 = Me$
63 $R^1 = R^2 = R^3 = R^4 = R^5 = Me$
64 $R^1 = Me; R^2 + R^3 = R^4 + R^5 = CH_2$
65 $R^1 = R^2 = R^3 = Me; R^4 + R^5 = CH_2$
66 $R^1 = (Me)_2; R^2 = R^3 = Me; R^4 + R^5 = CH_2$
67 $R^1 = R^2 = Me; R^3 = H; R^4 + R^5 = CH_2$

Argemonine (**47**)
Argemonine methohydroxide (**48**)
Argemonine N-oxide (**49**)
Bisnorargemonine (**50**)
Californidine (**51**)
2,9-Dimethoxy-3-hydroxypavine (**56**)
Eschscholtzidine (**52**)
Eschscholtzine (= californine) (**53**)
Isonorargemonine (**54**)
Munitagine (**57**)
Norargemonine (**55**)

Platycerine (**58**)
Platycerine methohydroxide (**59**)

Amurensine (**60**)
Amurensinine (**61**)
Amurensinine methohydroxide (**62**)
O-Methylthalisopavine (**63**)
Reframidine (**64**)
Reframine (**65**)
Reframoline (**67**)
Remrefine (= reframine methohydroxide) (**66**)

Cularines

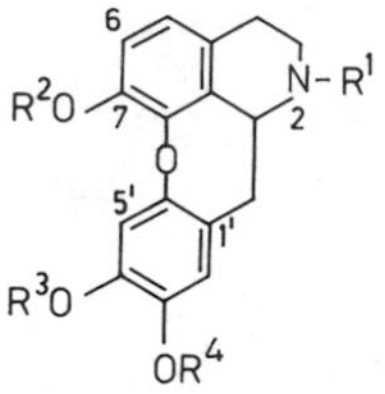

68 $R^1 = R^3 = Me; R^2 = R^4 = H$
69 $R^1 = R^2 = R^4 = H; R^3 = Me$
70 $R^1 = R^4 = Me; R^2 = R^3 = H$
71 $R^1 = R^2 = R^4 = Me; R^3 = H$
72 $R^1 = R^3 = R^4 = Me; R^2 = H$
73 $R^1 = R^2 = H; R^3 = R^4 = Me$
74 $R^1 = R^2 = R^3 = R^4 = Me$
75 $R^1 = H; R^2 = R^3 = R^4 = Me$
76 $R^1 = Me; R^2 = H; R^3 + R^4 = CH_2$
77 $R^1 = R^2 = H; R^3 + R^4 = CH_2$

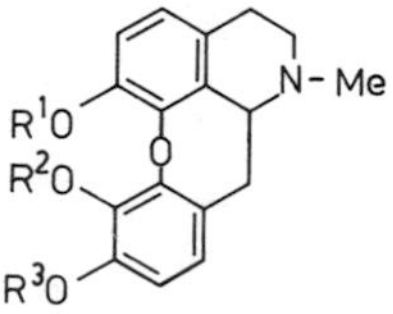

78 $R^1 = R^2 = H; R^3 = Me$
79 $R^1 = R^3 = Me; R^2 = H$
80 $R^1 = R^2 = R^3 = Me$

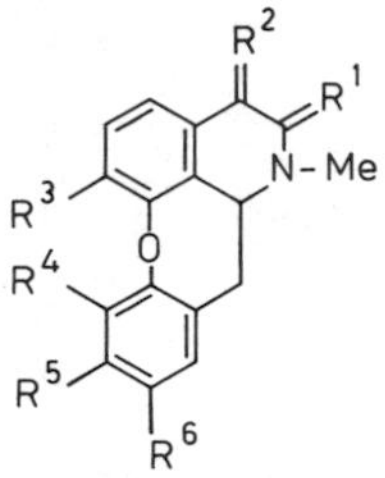

81 $R^1 = H_2; R^2 = H, OH; R^3 = OH; R^4 = H; R^5 = R^6 = OMe$
82 $R^1 = H_2; R^2 = H, OH; R^3 = R^4 = R^5 = OMe; R^6 = H$
83 $R^1 = R^2 = O; R^3 = R^4 = R^5 = OMe; R^6 = H$

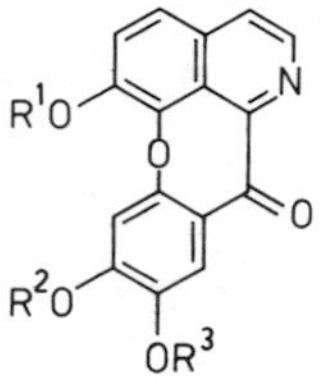

84 $R^1 = R^2 = R^3 = Me$
85 $R^1 = Me; R^2 + R^3 = CH_2$

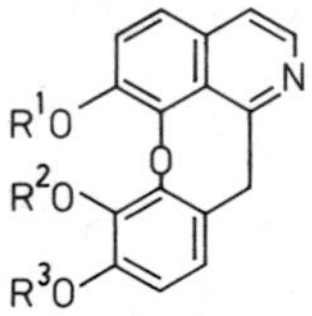

86 $R^1 = H; R^2 = R^3 = Me$
87 $R^1 = R^3 = Me; R^2 = H$
88 $R^1 = R^2 = R^3 = Me$

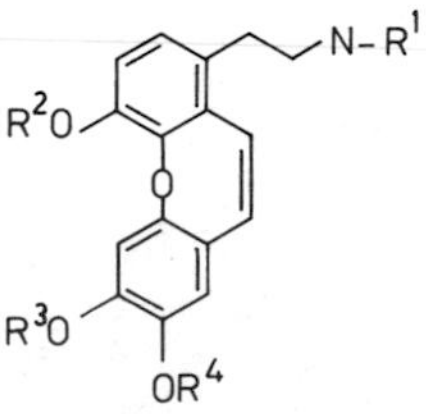

89 $R^1 = (Me)_2; R^2 = H; R^3 = R^4 = Me$
90 $R^1 = (Me)_2; R^2 = R^3 = R^4 = Me$
91 $R^1 = Me, H; R^2 = R^3 = R^4 = Me$

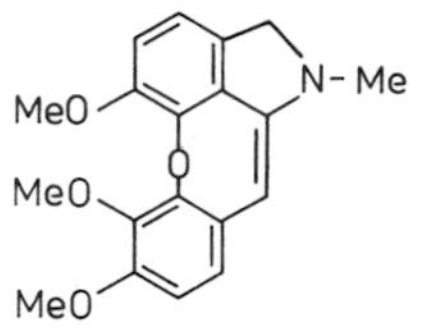

92

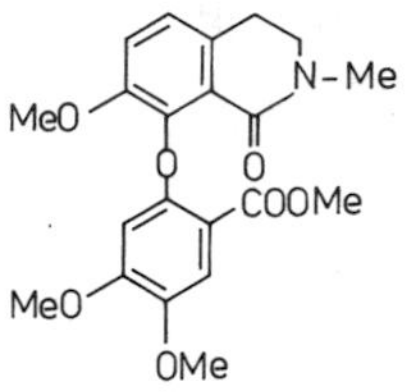

93

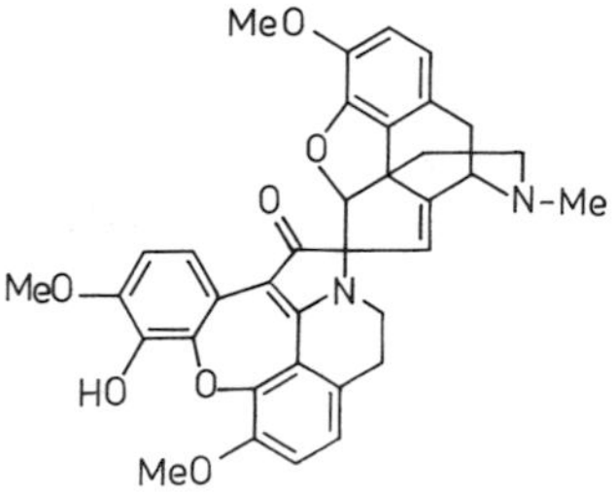

95

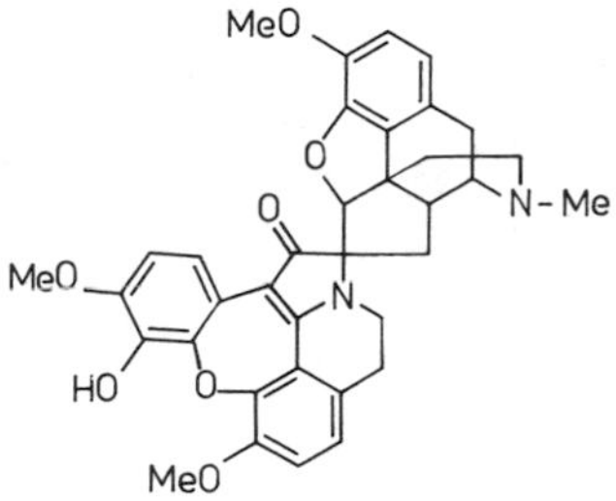

94

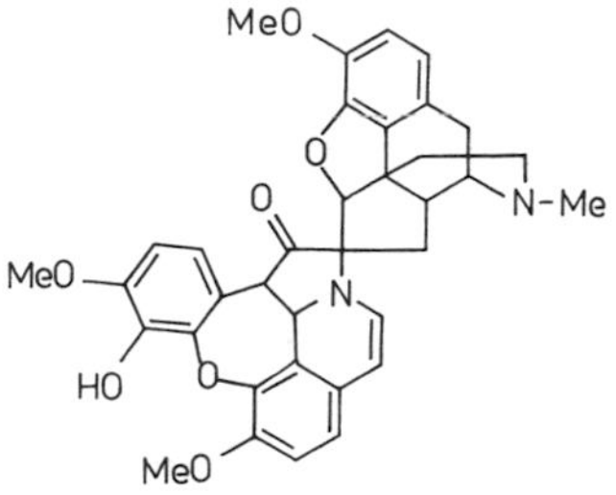

96

Aristoyagonine (92)
Cancentrine (94)
Celtyne (71)
Celtysine (70)
Claviculine (78)
Culacorine (68)
Cularicine (76)
Cularidine (72)
Cularimine (75)
Cularine (74)
Dehydrocancentrine (95, 96)
4-Hydroxysarcocapnine (82)
Limousamine (81)
Norculacorine (69)

Norcularicine (77)
Norcularidine (73)
Norsecocularine (91)
Noyaine (93)
Oxocompostelline (85)
Oxocularine (84)
Oxosarcocapnidine (87)
Oxosarcocapnine (88)
Oxosarcophylline (86)
Sarcocapnidine (79)
Sarcocapnine (80)
Secocularidine (89)
Secocularine (90)
Yagonine (83)

Proaporphines

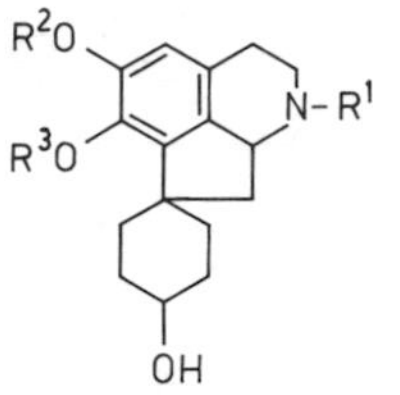

97 $R^1 = R^2 = R^3 = Me$
98 $R^1 = Me; R^2 + R^3 = CH_2$
99 $R^1 = R^2 = H; R^3 = Me$
100 $R^1 = R^2 = Me; R^3 = H$
101 $R^1 = H; R^2 = R^3 = Me$
102 $R^1 = R^3 = H; R^2 = Me$

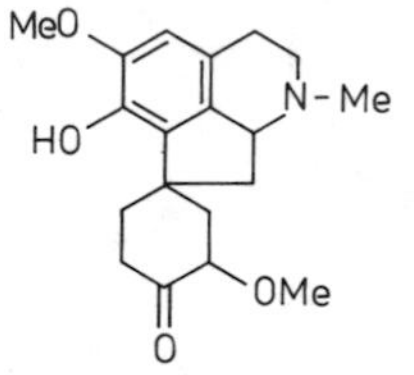

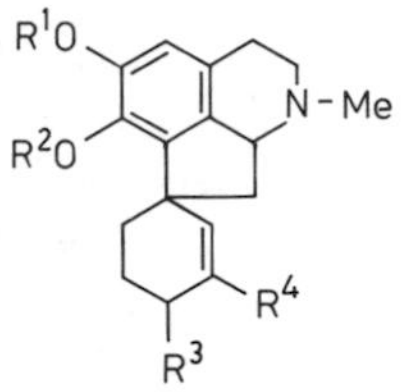

104 $R^1 = R^2 = Me; R^3 = H, OH; R^4 = H$
105 $R^1 + R^2 = CH_2; R^3 = H, OH; R^4 = H$
106 $R^1 = R^2 = Me; R^3 = O; R^4 = H$
107 $R^1 + R^2 = CH_2; R^3 = O; R^4 = H$
108 $R^1 = Me; R^2 = H; R^3 = O; R^4 = OMe$

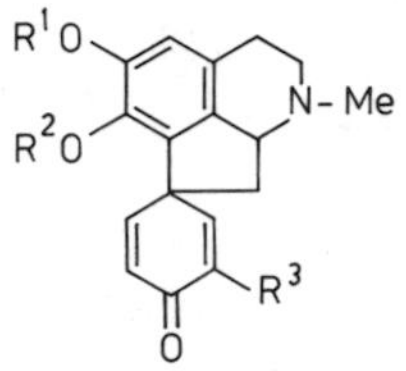

103

109 $R^1 = Me; R^2 = R^3 = H$
110 $R^1 + R^2 = CH_2; R^3 = H$
111 $R^1 = R^3 = H; R^2 = Me$
112 $R^1 = R^2 = Me; R^3 = H$
113 $R^1 = Me; R^2 = H; R^3 = OMe$

Amuroline (104)
Amuronine (106)
Dihydroorientalinone (108)
O,N-Dimethyloridine (97)
Glaziovine (109)
Hexahydromecambrine B (98)
Isooridine (99)
Mecambrine (110)
N-Methylcrotonosine (111)

N-Methyloridine (100)
O-Methyloridine (101)
Oridine (102)
Orientalinone (= bracteine) (113)
Pronuciferine (112)
Roehybrine (103)
Roemeramine (105)
Roemeronine (107)

Aporphines

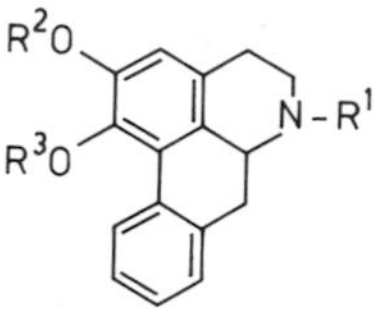

114 $R^1 = H$; $R^2 + R^3 = CH_2$
115 $R^1 = Me$; $R^2 + R^3 = CH_2$
116 $R^1 = Me_2$; $R^2 + R^3 = CH_2$
117 $R^1 = R^2 = R^3 = Me$
118 $R^1 = R^2 = Me$; $R^3 = H$
119 $R^1 = R^3 = Me$; $R^2 = H$
120 $R^1 = R^3 = Me$; $R^2 = rhamnose$

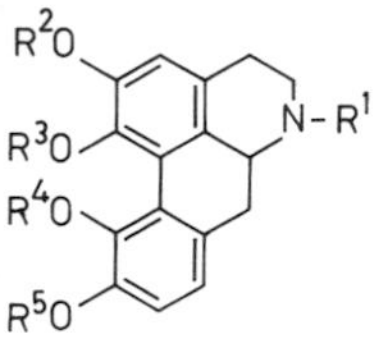

129 $R^1 = R^5 = Me$; $R^2 + R^3 = CH_2$; $R^4 = H$
130 $R^1 = Me_2$; $R^2 + R^3 = CH_2$; $R^4 = H$; $R^5 = Me$
131 $R^1 = R^4 = H$; $R^2 = R^3 = R^5 = Me$
132 $R^1 = R^2 = R^3 = R^5 = Me$; $R^4 = H$
133 $R^1 = Me_2$; $R^2 = R^3 = R^5 = Me$; $R^4 = H$
134 $R^1 = R^2 = R^4 = Me$; $R^3 = R^5 = H$
135 $R^1 = R^3 = H$; $R^2 = R^4 = R^5 = Me$
136 $R^1 = R^2 = R^4 = R^5 = Me$; $R^3 = H$
137 $R^1 = R^2 = R^5 = Me$; $R^3 = R^4 = H$
138 $R^1 = Me_2$; $R^2 = R^5 = Me$; $R^3 = R^4 = H$
139 $R^1 = R^3 = Me$; $R^2 = R^4 = R^5 = H$

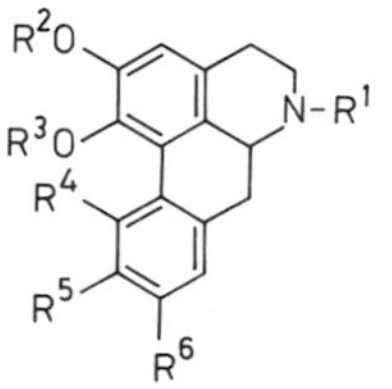

121 $R^1 = R^2 = Me$; $R^3 = R^5 = R^6 = H$; $R^4 = OH$
122 $R^1 = R^2 = Me$; $R^3 = R^5 = R^6 = H$; $R^4 = OMe$
123 $R^1 = Me_2$; $R^2 = Me$; $R^3 = R^5 = R^6 = H$; $R^4 = OMe$
124 $R^1 = Me$; $R^2 + R^3 = CH_2$; $R^4 = R^6 = H$; $R^5 = OH$
125 $R^1 = R^2 = R^3 = Me$; $R^4 = R^6 = H$; $R^5 = OH$
126 $R^1 = R^2 = R^3 = Me$; $R^4 = R^5 = H$; $R^6 = OMe$
127 $R^1 = Me$; $R^2 + R^3 = CH_2$; $R^4 = R^5 = H$; $R^6 = OH$
128 $R^1 = R^2 = Me$; $R^3 = R^4 = R^5 = H$; $R^6 = OMe$

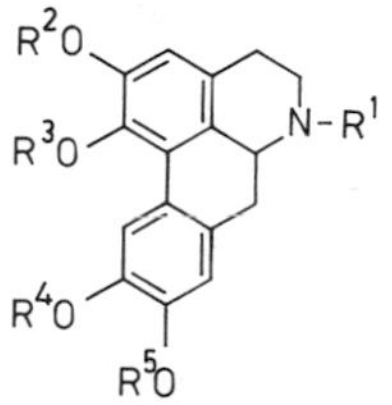

140 $R^1 = R^4 = R^5 = Me$; $R^2 + R^3 = CH_2$
141 $R^1 = R^3 = R^4 = H$; $R^2 = R^5 = Me$
142 $R^1 = R^2 = R^5 = Me$; $R^3 = R^4 = H$
143 $R^1 = R^2 = Me$; $R^3 = H$; $R^4 + R^5 = CH_2$
144 $R^1 = R^2 = R^3 = Me$; $R^4 + R^5 = CH_2$
145 $R^1 = R^2 = R^3 = R^4 = R^5 = Me$
146 $R^1 = R^2 = R^4 = R^5 = Me$; $R^3 = H$
147 $R^1 = R^2 = R^4 = Me$; $R^3 = R^5 = H$
148 $R^1 = R^5 = H$; $R^2 = R^3 = R^4 = Me$
149 $R^1 = R^2 = R^3 = R^4 = Me$; $R^5 = H$
150 $R^1 = Me$; $R^2 + R^3 = R^4 + R^5 = CH_2$
151 $R^1 = R^3 = R^4 = R^5 = Me$; $R^2 = H$
152 $R^1 = R^3 = H$; $R^2 = R^4 = R^5 = Me$
153 $R^1 = R^2 = R^4 = R^5 = Me$; $R^3 = H$

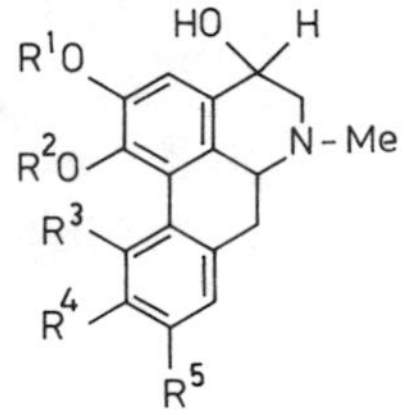

154   $R^1 + R^2 = CH_2$; $R^3 = OH$; $R^4 = OMe$ $R^5 = H$
155   $R^1 = Me$; $R^2 = R^5 = H$; $R^3 = R^4 = OMe$
156   $R^1 = R^2 = Me$; $R^3 = H$; $R^4 = R^5 = OMe$

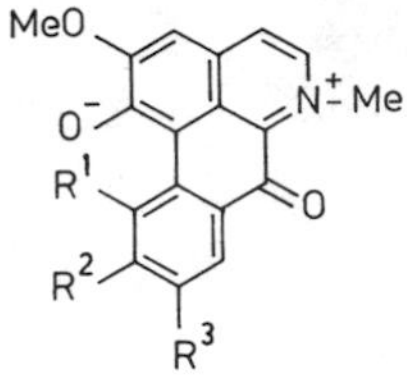

162   $R^1 = OMe$; $R^2 = R^3 = H$
163   $R^1 = R^2 = OMe$; $R^3 = H$
164   $R^1 = OH$; $R^2 = OMe$; $R^3 = H$
165   $R^1 = H$; $R^2 = R^3 = OMe$
166   $R^1 = H$; $R^2 + R^3 = OCH_2O$

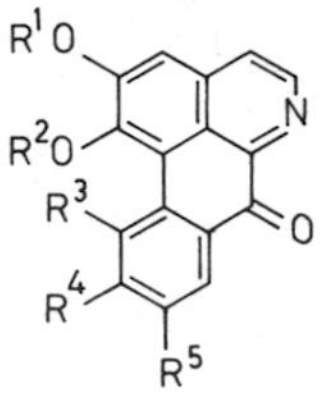

157   $R^1 = R^2 = Me$; $R^3 = R^4 = R^5 = H$
158   $R^1 = R^2 = Me$; $R^3 = OH$; $R^4 = OMe$; $R^5 = H$
159   $R^1 + R^2 = CH_2$; $R^3 = H$; $R^4 = R^5 = OMe$
160   $R^1 = R^2 = Me$; $R^3 = H$; $R^4 = R^5 = OMe$
161   $R^1 = R^2 = Me$; $R^3 = H$; $R^4 + R^5 = OCH_2O$

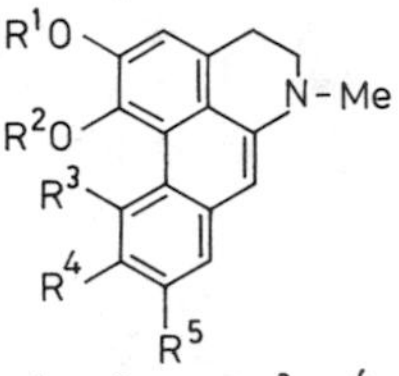

167   $R^1 + R^2 = CH_2$; $R^3 = R^4 = R^5 = H$
168   $R^1 = R^2 = Me$; $R^3 = OMe$; $R^4 = R^5 = H$
169   $R^1 = Me$; $R^2 = R^4 = R^5 = H$; $R^3 = OMe$
170   $R^1 = Me$; $R^2 = R^5 = H$; $R^3 = R^4 = OMe$
171   $R^1 + R^2 = CH_2$; $R^3 = H$; $R^4 = R^5 = OMe$
172   $R^1 = R^2 = Me$; $R^3 = H$; $R^4 = R^5 = OMe$
173   $R^1 = R^2 = Me$; $R^3 = H$; $R^4 + R^5 = CH_2$

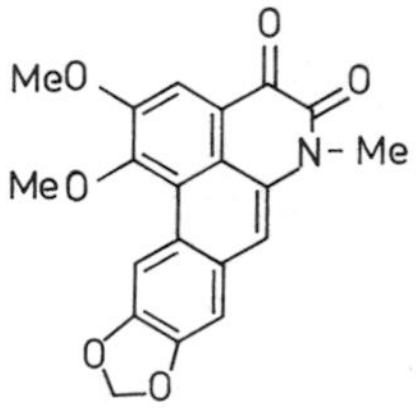

174 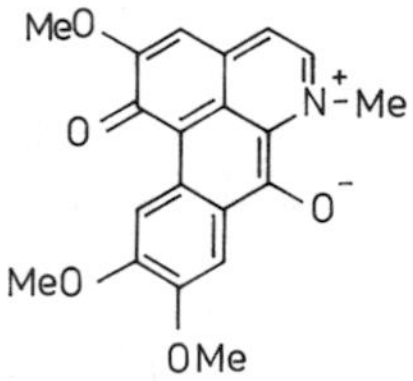   175 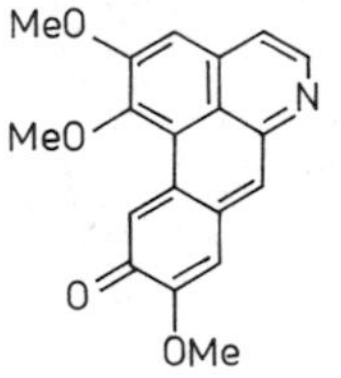   178 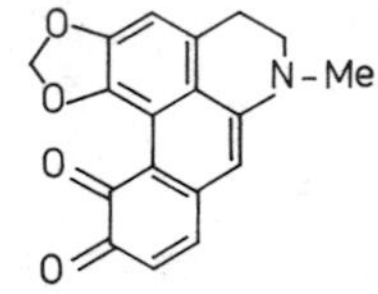   179

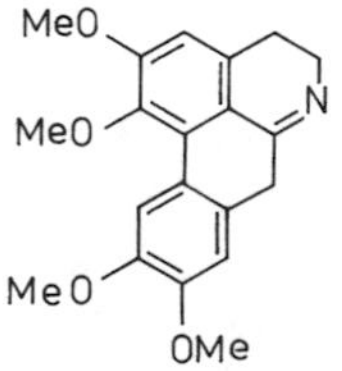

176 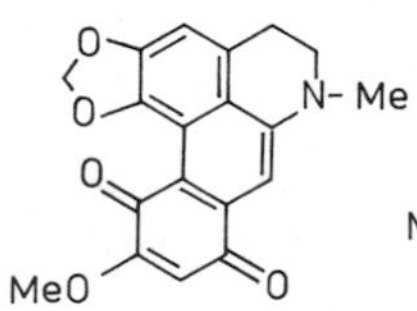   177

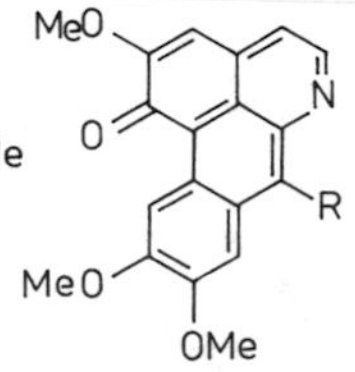

180   R = H
181   R = NH_2

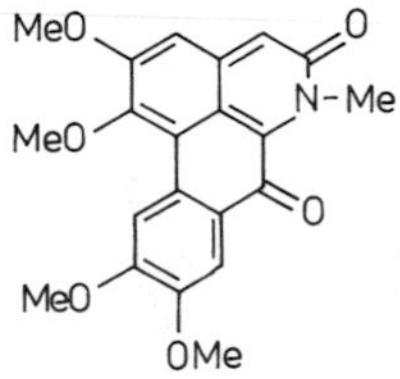

182

Alkaloid PO-3 (**162**)
Anonaine (**114**)
Aporheine (**115**)
Aporheine methohydroxide (**116**)
Arosine (= glaunidine) (**163**)
Arosinine (**164**)
Bracteoline (**142**)
Bulbocapnine (**129**)
Bulbocapnine methohydroxide (**130**)
Bulbodione (**177**)
Cataline (**156**)
Corunnine (**175**)
Corydine (**136**)
Corydione (**174**)
Corytuberine (**137**)
Corytuberine methohydroxide (**138**)
Dehydrocorydine (**170**)
Dehydrodicentrine (**171**)
Dehydroglaucine (**172**)
6a,7-Dehydroisothebaine (**169**)
Dehydronantenine (**173**)
Dehydronorglaucine (**176**)
Dehydroroemerine (**167**)
O-Demethylnuciferine (= lirinidine) (**118**)
Dicentrine (**140**)
Dicentrinone (**159**)
Domesticine (**143**)
Domestine (**144**)
Escholine (= magnoflorine) (**138**)
Floripavidine (**120**)
Glaucine (**145**)
Glaucinone (**165**)
Glaufine (**139**)
Glaufidine (**155**)
Glaunine (**158**)
Glaunidine (= arosine) (**163**)
Glauvine (**178**)
4-Hydroxybulbocapnine (**154**)
1-Hydroxy-2,9,10-trimethoxyaporphine (**146**)
Isoboldine (**147**)
Isocorydine (**132**)
Isocorytuberine (**134**)
Isothebaidine (**121**)

Isothebaine (**122**)
Isothebaine methohydroxide (**123**)
Laurotetanine (**148**)
Lauroscholtzine (= laurotetanine metho-
   hydroxide) (**149**)
Lirinidine (= O-demethylnuciferine) (**118**)
Liriodenine (**157**)
Magnoflorine (= escholine) (= corytuberine
   methohydroxide) (**138**)
Mecambroline (**124**)
Menisperine (= isocorydine methohydroxide)
   (**133**)
N-Methylasimilobine (**119**)
O-Methylatheroline (= oxoglaucine) (**160**)
O-Methyl-6a,7-dehydroisothebaine (**168**)
N-Methyllaurotetanine (= lauroscholtzine)
   (**149**)
N-Methyllindcarpine (**139**)
O-Methylorientine (**126**)
1,2-Methylenedioxy-6a,7-dehydro-
   aporphine-10,11-quinone (**179**)
Nandazurine (**166**)
Nantenine (= domestine) (**144**)
Neolitsine (**150**)
Norbracteoline (**141**)
Norcorydine (**135**)
Norisocorydine (**131**)
Nuciferine (**117**)
Nuciferoline (**125**)
Orientine (**128**)
Oxonantenine (**161**)
Pancoridine (**180**)
Pancorinine (**181**)
Pantevedrine (**182**)
Predicentrine (**151**)
Remrefidine (= aporheine methohydroxide)
   (**116**)
Roemerine (= aporheine) (**115**)
Roemerine methohydroxide (**116**)
Roemeroline (**127**)
Thalicmidine (= thaliporphine) (**153**)
Thaliporphine (= O-methylisoboldine) (**153**)
Wilsonirine (**152**)

Promorphinanes

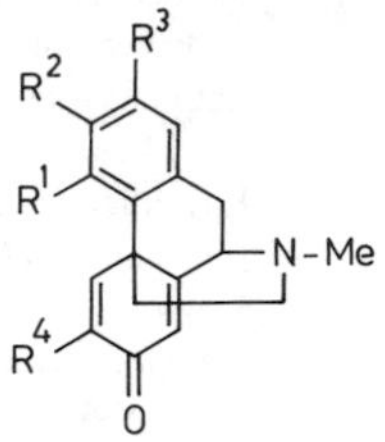

183   $R^1 = H_i R^2 + R^3 = OCH_2O_i R^4 = OMe$
184   $R^1 = H_i R^2 = R^3 = R^4 = OMe$
185   $R^1 = H_i R^2 = OH_i R^3 = R^4 = OMe$
186   $R^1 = H_i R^2 = R^4 = OMe_i R^3 = OH$
187   $R^1 = R^2 = R^4 = OMe_i R^3 = H$
188   $R^1 = OH_i R^2 = R^4 = OMe_i R^3 = H$

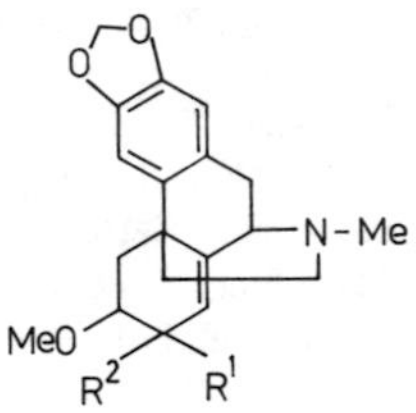

189   $R^1 = OH_i R^2 = H$
190   $R^1_i R^2 = O$

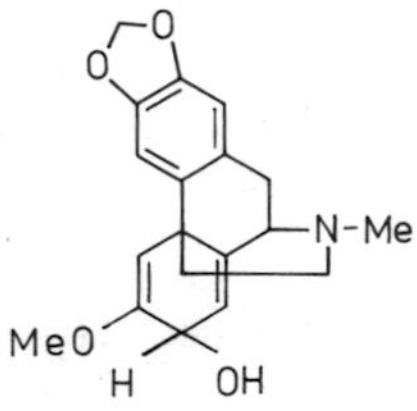

191

Amurine (**183**)
Amurinine (**190**)
Dihydronudaurine (**189**)
Epiamurinine (**190**)
Flavinantine (**185**)
O-Methylflavinantine (**184**)

O-Methylsalutaridine (**187**)
Nudaurine (**191**)
Pallidine (= isosalutaridine) (**186**)
Salutaridine (**188**)
Sinoacutine (**188**)

Morphinanes

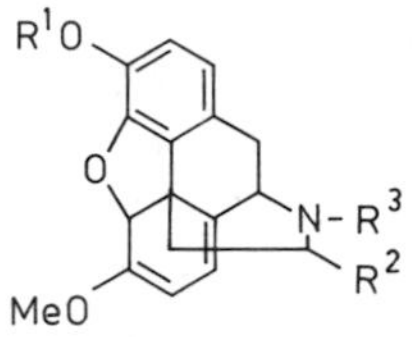

192   $R^1 = R^3 = Me ; R^2 = H$
193   $R^1 = Me ; R^2 = H ; R^3 = (Me)_2$
194   $R^1 = Me ; R^2 = H ; R^3 = Me_1 O$
195   $R^1 = R^2 = H ; R^3 = Me$
196   $R^1 = R^2 = R^3 = H$
197   $R^1 = R^3 = Me ; R^2 = OH$

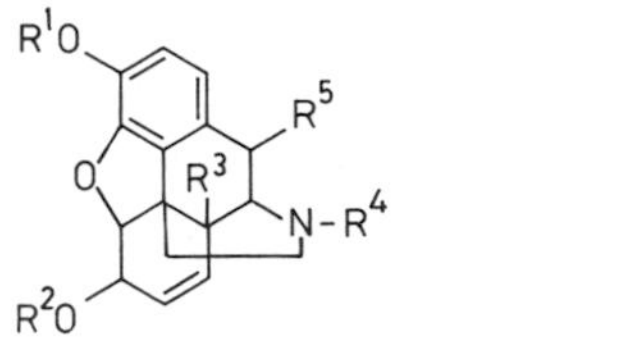

198   $R^1 = R^4 = Me ; R^2 = R^3 = R^5 = H$
199   $R^1 = Me ; R^2 = R^3 = R^5 = H ; R^4 = Me_1 O$
200   $R^1 = R^4 = Me ; R^2 = R^3 = H ; R^5 = OH$
201   $R^1 = R^4 = Me ; R^2 = R^5 = H ; R^3 = OH$
202   $R^1 = R^2 = R^4 = Me ; R^3 = R^5 = H$
203   $R^1 = R^2 = R^3 = R^4 = R^5 = H$
204   $R^1 = R^2 = R^3 = R^5 = H ; R^4 = Me$
205   $R^1 = R^2 = R^3 = R^5 = H ; R^4 = Me_1 O$

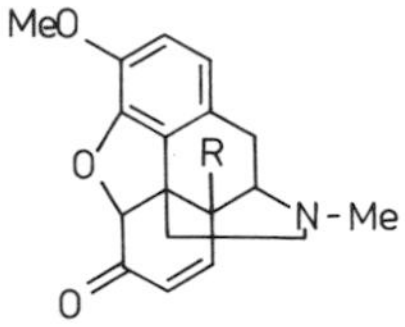

206   $R = H$
207   $R = OH$

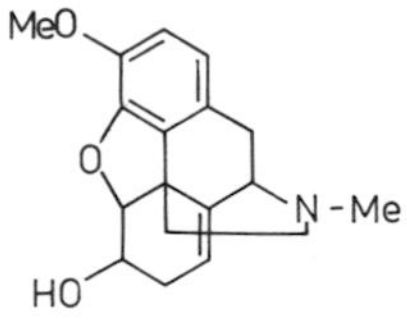

208

Codeine (**198**)
Codeine *N*-oxide (**199**)
Codeinone (**206**)
10-Hydroxycodeine (**200**)
14-Hydroxycodeine (**201**)
14-Hydroxycodeinone (**207**)
16-Hydroxythebaine (**197**)
6-Methylcodeine (**202**)
Morphine (**204**)

Morphine *N*-oxide (**205**)
Neopine (**208**)
Normorphine (**203**)
Oripavidine (**196**)
Oripavine (**195**)
Pseudomorphine (**204**)
Thebaine (**192**)
Thebaine methohydroxide (**193**)
Thebaine *N*-oxide (**194**)

Protoberberines

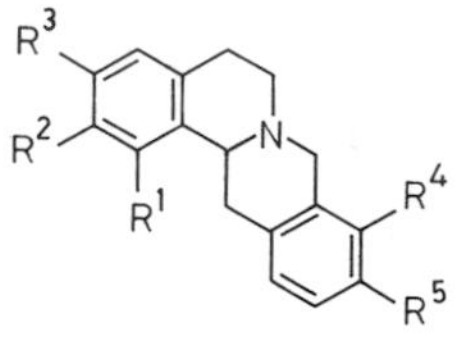

a 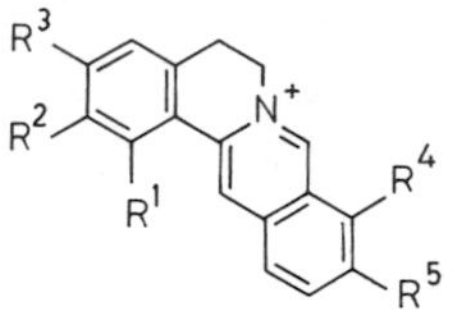   b

209 a,b R^1 = H ; R^2 + R^3 = OCH$_2$O ; R^4= R^5= OMe
210 a,b R^1 = H ; R^2= OH ; R^3= OMe ; R^4+ R^5= OCH$_2$O
211 a,b R^1 = H ; R^2= R^4= R^5= OMe ; R^3= OH
212 a R^1 = H ; R^2= R^4= OH ; R^3= R^5= OMe
213 a,b R^1 = H ; R^2= R^4= OMe ; R^3= R^5= OH
214 a,b R^1 = H ; R^2= OH ; R^3= R^4= R^5= OMe
215 a R^1 = H ; R^2= R^3= OMe ; R^4+ R^5 = OCH$_2$O
216 a,b R^1 = H ; R^2= R^5= OH ; R^3= R^4= OMe
217 a,b R^1 = H ; R^2+ R^3= R^4+ R^5= OCH$_2$O
218 a,b R^1 = H ; R^2= R^3= R^4= R^5= OMe
219 a,b R^1 = OH ; R^2= R^3= R^4= R^5= OMe
220 a,b R^1 = R^5= OH ; R^2= R^3= R^4= OMe

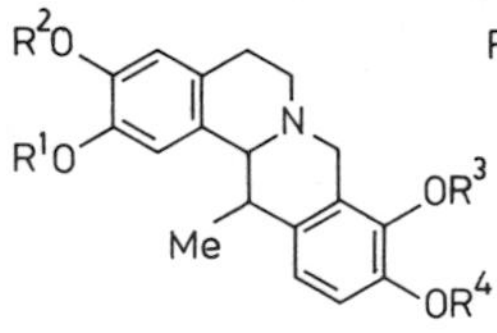

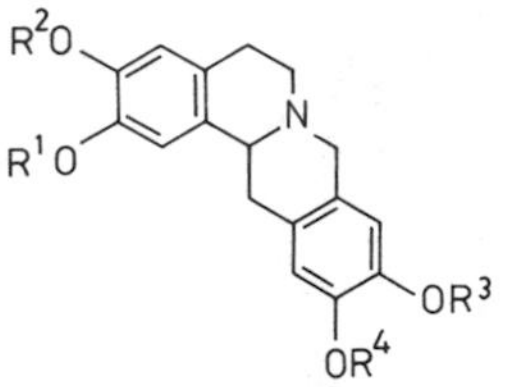

221   R^1 = R^4= H ; R^2 = R^3 = Me
222   R^1 = R^2 = Me ; R^3 + R^4 = CH$_2$
223   R^1 = R^2 = R^3 = Me ; R^4 = H
224   R^1 = R^3 = H ; R^2 = R^4 = Me
225   R^1 = H ; R^2= R^3= R^4 = Me

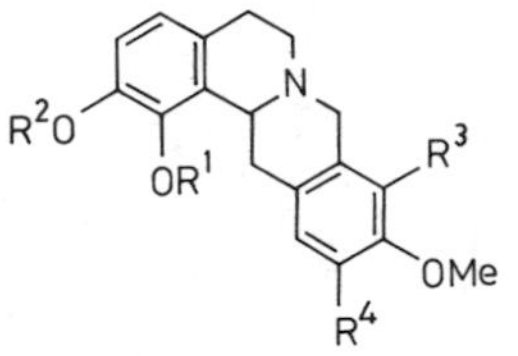

226   R^1 = R^3 = H ; R^2 = Me ; R^4 = OMe
227   R^1 = Me ; R^2= R^3= H ; R^4= OH
228   R^1 = R^4 = H ; R^2= Me ; R^3 = OMe

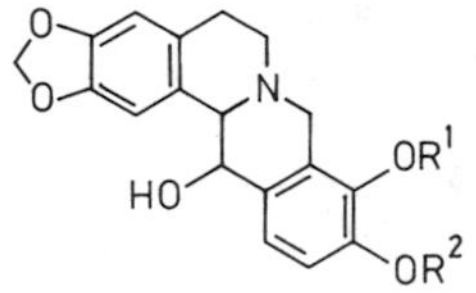

a 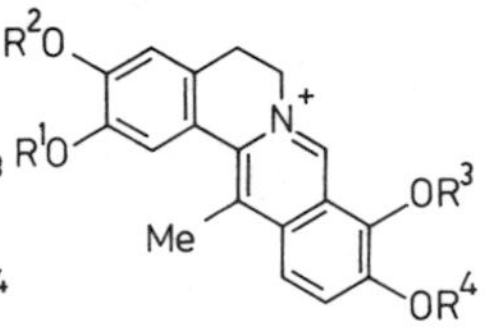   b

229 a,b R^1 = H ; R^2= Me ; R^3 + R^4 = CH$_2$
230 a,b R^1 = R^3= R^4 = Me ; R^2 = H
231 a R^1 = R^3 = Me ; R^2 = R^4 = H
232 a,b R^1 = R^2 = R^3 = R^4= Me
233 a,b R^1 = R^2 = Me ; R^3 + R^4= CH$_2$
234 a,b R^1 + R^2 = R^3+ R^4 = CH$_2$
235 a,b R^1 + R^2 = CH$_2$; R^3 = R^4 = Me
236 a R^1 = H ; R^2= R^3 = R^4 = Me

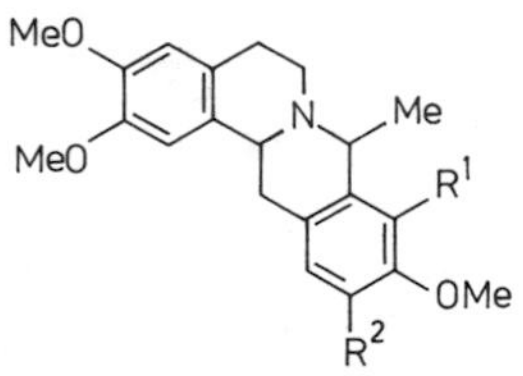

239   R^1 + R^2 = CH$_2$
240   R^1 = R^2 = Me

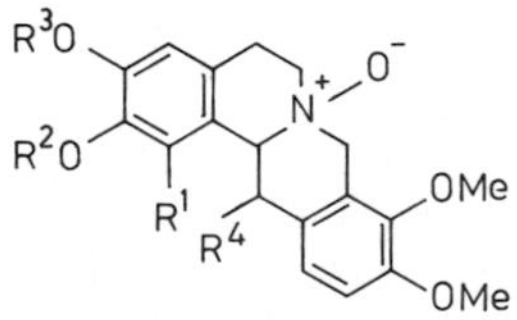

237   R^1 = OH ; R^2= H
238   R^1 = H ; R^2= OH

241   R^1 = R^4= H ; R^2= R^3 = Me
242   R^1 = H ; R^2+ R^3= CH$_2$; R^4= OH
243   R^1= OH ; R^2= R^3= Me ; R^4= H

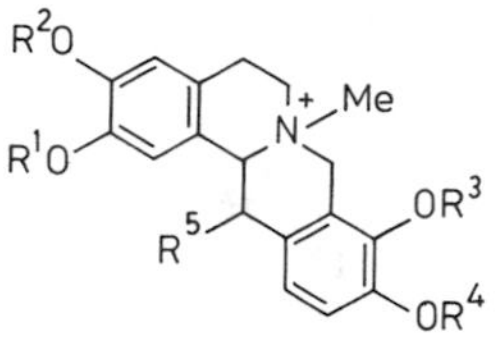

244   $R^1 + R^2 = CH_2$; $R^3 = R^4 = Me$; $R^5 = H$
245   $R^1 = R^3 = H$; $R^2 = R^4 = Me$; $R^5 = H$
246   $R^1 + R^2 = CH_2$; $R^3 = Me$; $R^4 = OH$; $R^5 = H$
247   $R^1 = H$; $R^2 = R^3 = R^4 = Me$; $R^5 = H$
248   $R^1 = R^2 = Me$; $R^3 + R^4 = CH_2$; $R^5 = H$
249   $R^1 + R^2 = R^3 + R^4 = CH_2$; $R^5 = H$
250   $R^1 = R^2 = R^3 = R^4 = Me$; $R^5 = H$
251   $R^1 + R^2 = R^3 + R^4 = CH_2$; $R^5 = Me$

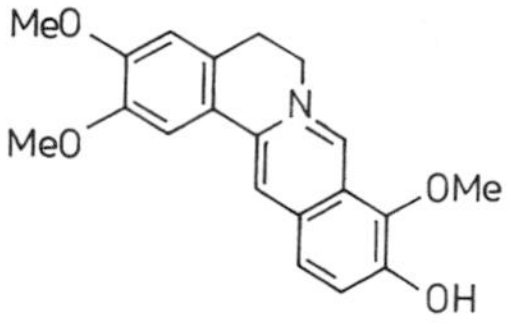

253

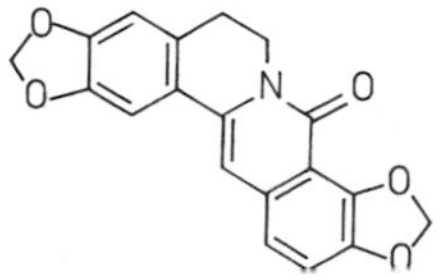

254

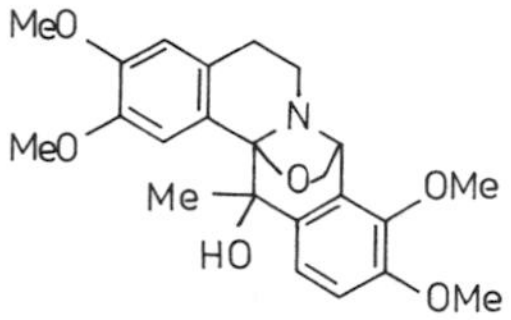

252

Apocavidine (**229a**)
Canadine (**209a**)
Canadine methohydroxide (**244**)
Capauridine [= (±)-capaurine] (**219a**)
Capaurimine (**220a**)
Capaurine (**219a**)
Caseadine (**226**)
Caseamine (**227**)
Caseanadine (**228**)
Cavidine (= thalictrifoline) (**233a**)
Cheilanthifoline (**210a**)
Coreximine (**221**)
Corybulbine (= corydalmine) (**230a**)
Corydalidzine (**231a**)
Corydaline (**232a**)
Corydalmine (= kikemanine) (= corybulbine) (**230a**)
Corygovanine (**222**)
Corypalmine (= tetrahydrojatrorrhizine) (**211a**)
trans-Corynoxidine (**241**)
Corytenchine (**223**)
Corytenchirine (**238**)
Cyclanoline [= (−)-α-scoulerine metho-hydroxide] (**245**)

Ophiocarpine N-oxide (**242**)
Scoulerine (**212**)
Scoulerine methohydroxide (= cyclanoline) (**245**)
Sinactine (**215a**)
Solidaline (**252**)
Stepholidine (**216a**)
Stylopine (**217a**)
Stylopine methohydroxide (**249**)
Tetrahydrocorysamine (**234a**)
Tetrahydrocorysamine methohydroxide (**251**)
Tetrahydropalmatine (**218a**)
Tetrahydropalmatine methohydroxide (**250**)
Thalictricavine (**235a**)
Thalictrifoline (= cavidine) (**233a**)

Berberine (**209b**)
Columbamine (**214b**)
Coptisine (**217b**)
Corysamine (**234b**)
Dehydroapocavidine (**229b**)
Dehydrocapaurine (**219b**)
Dehydrocapaurimine (**220b**)
Dehydrocavidine (**233b**)

Discretamine (**213a**)
cis-Epicorynoxidine (**241**)
Escholidine (= tetrahydrothalifendine metho-
  hydroxide) (**246**)
Govadine (**224**)
Govanine (**225**)
13-Hydroxystylopine (**239**)
Isocorybulbine (**236a**)
Isocorypalmine (= tetrahydrocolumbamine)
  (**214a**)
Isocorypalmine methohydroxide (**247**)
Lienkonine (**237**)
N-Methylsinactine (**248**)
13-Methyltetrahydroprotoberberine
  (= thalictricavine) (**235a**)
Nokoensine (**243**)
Ophiocarpine (**240**)

Dehydrocheilanthifoline (**210b**)
Dehydrocorybulbine (**230b**)
Dehydrocorydaline (**232b**)
Dehydrocorydalmine (**253**)
Dehydrodiscretamine (**213b**)
Dehydroisocorypalmine (**214b**)
Dehydrothalictrifoline (**233b**)
Dehydrothalictricavine (**235b**)
Didehydrocheilanthifoline (**210b**)
Jatrorrhizine (**211b**)
8-Oxocoptisine (**254**)
Palmatine (**218b**)
Stepharanine (**216b**)

Retroprotoberberines

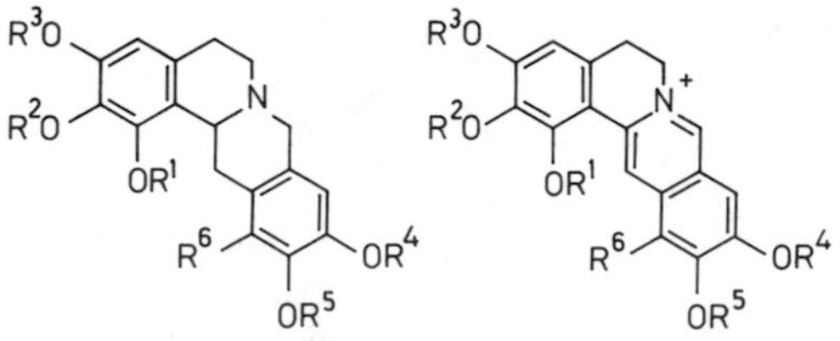

a   b

255 a $R^1 = R^4 = Me ; R^2 + R^3 = CH_2 ; R^5 = H ; R^6 = CH_2OH$
256 a,b $R^1 = R^4 = R^5 = Me ; R^2 + R^3 = CH_2 ; R^6 = CH_2OH$
257 a,b $R^1 = R^4 = Me ; R^2 + R^3 = CH_2 ; R^5 + R^6 = CH_2-O-CH_2$

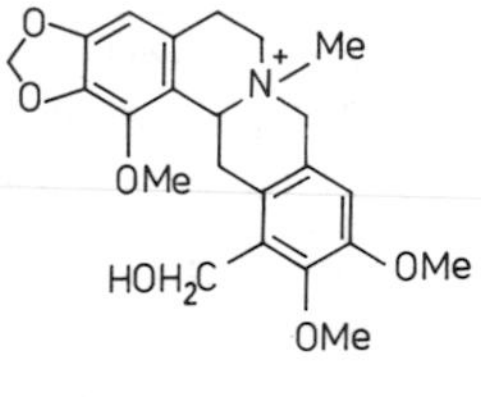

258

Aryapavine (**255a**)
Mecambridine (**256a**)
Mecambridine methohydroxide (**258**)

Orientalidine (**257a**)
Alkaloid PO-4 (**257b**)
Alkaloid PO-5 (= alborine) (**256b**)

Secoberbines

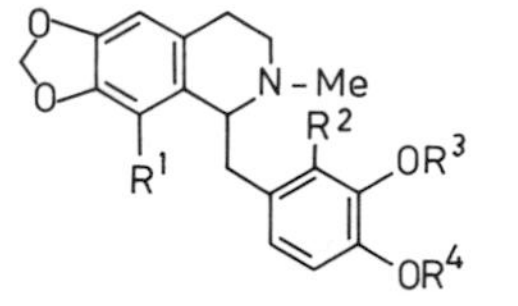

259 $R^1 = H ; R^2 = CHO ; R^3 + R^4 = CH_2$
260 $R^1 = H ; R^2 = CH_2OH ; R^3 + R^4 = CH_2$
261 $R^1 = OMe ; R^2 = CH_2OH ; R^3 = R^4 = Me$
262 $R^1 = OMe ; R^2 = COOH ; R^3 = R^4 = Me$

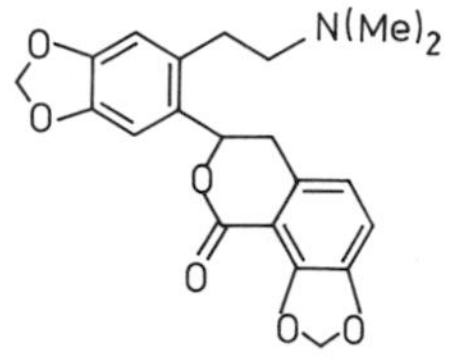

265

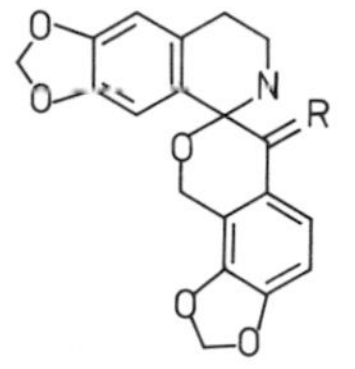

263 R = O
264 R = H_2

Aobamine (259)
Corydalisol (260)
Corydalispiron (= hypecorinine) (263)
Hypecorine (264)

Hypecorinine (= corydalispiron) (263)
Macrantaline (261)
Macrantoridine (262)
Peshawarıne (265)

Benzophenanthridines

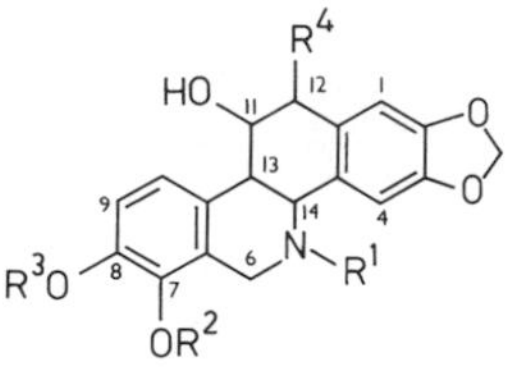

266 $R^1 = Me ; R^2 + R^3 = CH_2 ; R^4 = H$
267 $R^1 = R^4 = H ; R^2 + R^3 = CH_2$
268 $R^1 = Me ; R^2 + R^3 = CH_2 ; R^4 = OH$
269 $R^1 = R^2 = R^3 = Me ; R^4 = H$

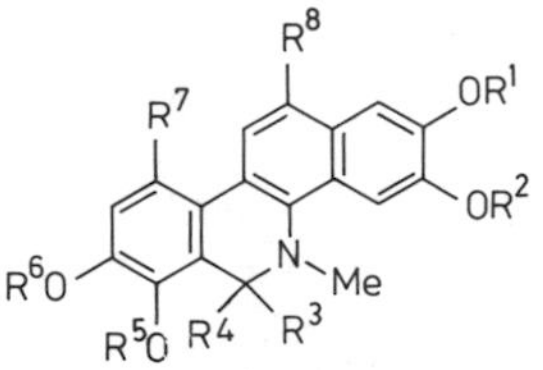

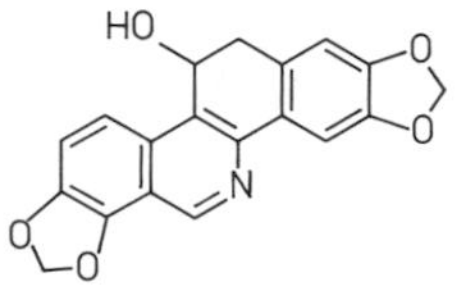

270

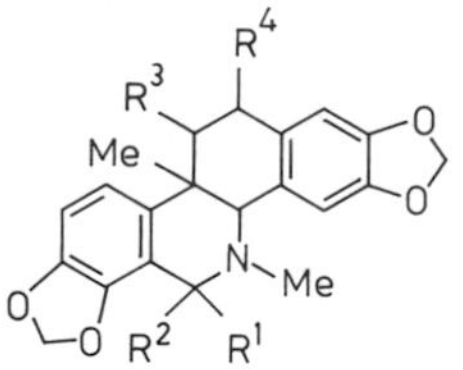

271 $R^1 = R^2 = R^4 = H ; R^3 = OH$
272 $R^1 = R^2 = R^4 = H ; R^3 = OCOCH_3$
273 $R^1 = R^2 = R^4 = H ; R^3 = SO_3H$
274 $R^1 + R^2 = O ; R^3 = OH ; R^4 = H$
275 $R^1 = R^2 = H ; R^3 = R^4 = OH$
276 $R^1 = CH_2COCH_3 ; R^2 = R^4 = H ; R^3 = OCOCH_3$
277 $R^1 = R^2 = R^4 = H ; R^3 = OH$
277a $R^1 = CH_2OH ; R^2 = R^4 = H ; R^3 = OH$

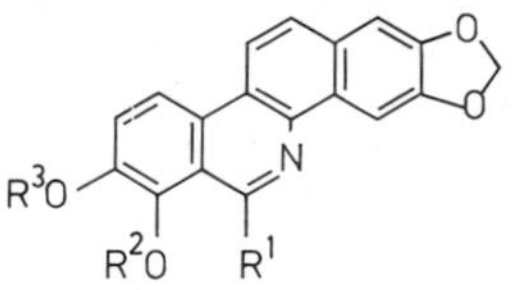

278 $R^1 + R^2 = CH_2 ; R^3 = R^4 = R^7 = R^8 = H ; R^5 = R^6 = Me$
279 $R^1 + R^2 = CH_2 ; R^3 = CH_2OH ; R^4 = R^7 = R^8 = H ; R^5 = R^6 = Me$
280 $R^1 + R^2 = R^5 + R^6 = CH_2 ; R^3 = R^4 = R^7 = R^8 = H$
281 $R^1 + R^2 = R^5 + R^6 = CH_2 ; R^3 + R^4 = O ; R^7 = R^8 = H$
282 $R^1 + R^2 = R^5 + R^6 = CH_2 ; R^3 = OMe ; R^4 = R^7 = R^8 = H$
283 $R^1 + R^2 = R^5 + R^6 = CH_2 ; R^3 + R^4 = NH ; R^7 = R^8 = H$
284 $R^1 + R^2 = R^5 + R^6 = CH_2 ; R^3 = CH_2OH ; R^4 = R^7 = R^8 = H$
285 $R^1 + R^2 = R^5 + R^6 = CH_2 ; R^3 = CH_2COCH_3 ; R^4 = R^7 = R^8 = H$
286 $R^1 + R^2 = CH_2 ; R^3 = R^4 = R^8 = H ; R^5 = R^6 = R^7 = Me$
287 $R^1 = R^2 = R^5 = R^6 = Me ; R^3 = R^4 = R^8 = H ; R^7 = OMe$
288 $R^1 + R^2 = R^5 + R^6 = CH_2 ; R^3 = R^4 = R^8 = H ; R^7 = OMe$
289 $R^1 + R^2 = R^5 + R^6 = CH_2 ; R^3 = R^4 = H ; R^7 = R^8 = OMe$
290 $R^1 + R^2 = CH_2 ; R^3 = OMe ; R^4 = R^7 = R^8 = H ; R^5 = R^6 = Me$

291 $R^1 = H ; R^2 = R^3 = Me$
292 $R^1 = H ; R^2 + R^3 = CH_2$
293 $R^1 = OMe ; R^2 + R^3 = CH_2$

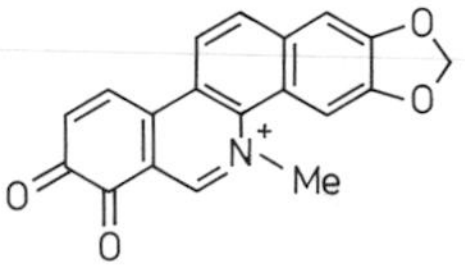

294

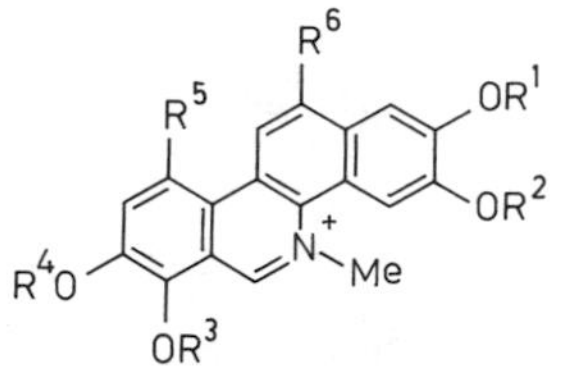

295 $R^1 + R^2 = CH_2$; $R^3 = R^4 = Me$; $R^5 = R^6 = H$
296 $R^1 + R^2 = R^3 + R^4 = CH_2$; $R^5 = R^6 = H$
297 $R^1 + R^2 = CH_2$; $R^3 = R^4 = R^5 = R^6 = H$
298 $R^1 = R^2 = Me$; $R^3 + R^4 = CH_2$; $R^5 = OMe$; $R^6 = H$
299 $R^1 = R^2 = R^3 = R^4 = Me$; $R^5 = OMe$; $R^6 = H$
300 $R^1 + R^2 = R^3 + R^4 = CH_2$; $R^5 = OMe$; $R^6 = H$
301 $R^1 + R^2 = CH_2$; $R^3 = R^4 = Me$; $R^5 = OMe$; $R^6 = H$
302 $R^1 + R^2 = R^3 + R^4 = CH_2$; $R^5 = R^6 = OMe$

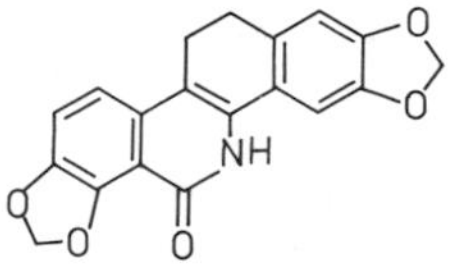

303

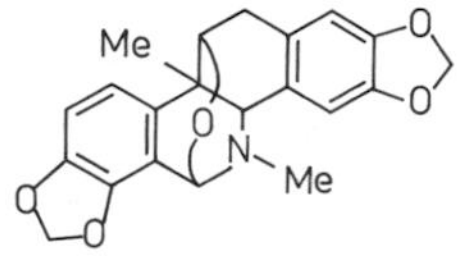

304

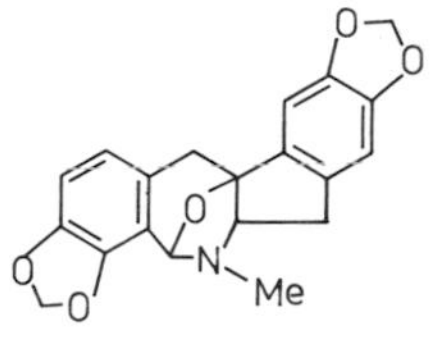

305

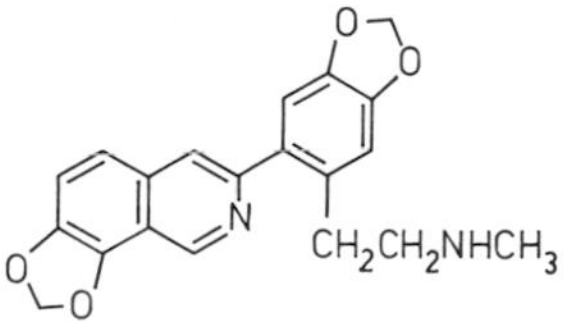

306

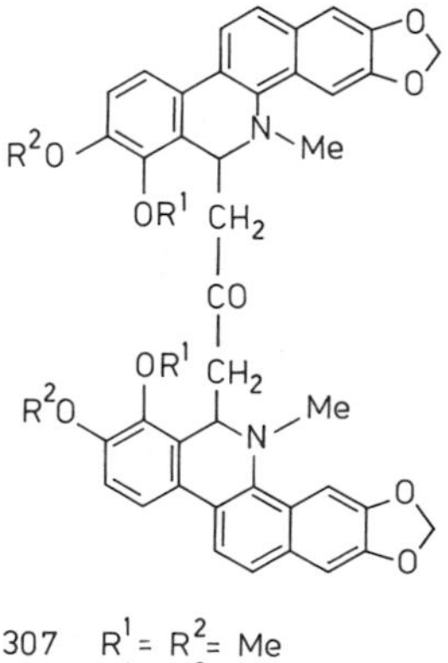

307 $R^1 = R^2 = Me$
308 $R^1 + R^2 = CH_2$

6-Acetonyldihydrosanguinarine (**285**)
8-Acetonyldihydrosanguinarine (**285**)
Acetylcorynoline (**272**)
Acetylisocorynoline (**272**)
1,3-Bis(6-hydroxychelerythrinyl)acetone
  (= chelerythrindimerine) (**307**)
Bocconine (= chelirubine) (**300**)
Bocconoline (**279**)
Chelamidine (= 11-hydroxyhomochelidonine)
  (**269**)
Chelamine (= 11-hydroxychelidonine) (**268**)
Chelerythrine (**295**)
Chelerythrindimerine (**307**)
Chelidonine (**266**)
Chelilutine (**301**)
Chelirubine (**300**)
Consperine (**276**)
Corydamine (**306**)
Corynolamine (**277a**)
Corynoline (**271**)
Corynoline 11-*O*-sulfate (**273**)
Corynoloxine (**304**)
9,10-Demethylene-9,10-dihydrosanguinarine
  (**294**)
9,10-Demethylenesanguinarine (**297**)
N-Demethyl-9,10-dihydrooxysanguinarine
  (**303**)

Dihydrochelerythrine (**278**)
Dihydrochelilutine (**286**)
Dihydrochelirubine (**288**)
Dihydromacarpine (**289**)
Dihydrosanguilutine (**287**)
Dihydrosanguinarine (**280**)
11-Epicorynoline (**277**)
12-Hydroxycorynoline (**275**)
6-Hydroxymethyldihydrosanguinarine (**284**)
6-Iminosanguinarine (**283**)
Isocorynoline (= 14-epicorynoline) (**277**)
Luguine (**270**)
Macarpine (**302**)
6-*O*-Methyldihydrochelerythrine (= angoline)
  (**290**)
8-Methoxydihydrosanguinarine (**282**)
Norchelerythrine (**291**)
Norchelidonine (**267**)
Norsanguinarine (**292**)
6-Oxocorynoline (**274**)
Oxysanguinarine (**281**)
Pancorine (**293**)
Ribasine (**305**)
Sanguidimerine (**308**)
Sanguilutine (**299**)
Sanguinarine (**296**)
Sanguirubine (**298**)

Protopines

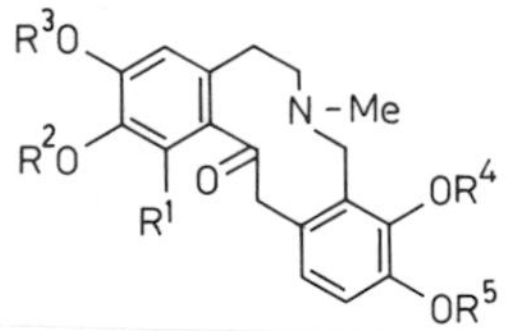

309	$R^1 = H; R^2 + R^3 = R^4 + R^5 = CH_2$
310	$R^1 = H; R^2 + R^3 = CH_2; R^4 = R^5 = Me$
311	$R^1 = H; R^2 = R^3 = Me; R^4 + R^5 = CH_2$
312	$R^1 = R^4 = H; R^2 + R^3 = CH_2; R^5 = Me$
313	$R^1 = R^3 = H; R^2 = Me; R^4 + R^5 = CH_2$
314	$R^1 = H; R^2 = R^3 = R^4 = R^5 = Me$
315	$R^1 = R^2 = R^3 = H; R^4 = R^5 = Me$
316	$R^1 = OMe; R^2 + R^3 = CH_2; R^4 = R^5 = Me$

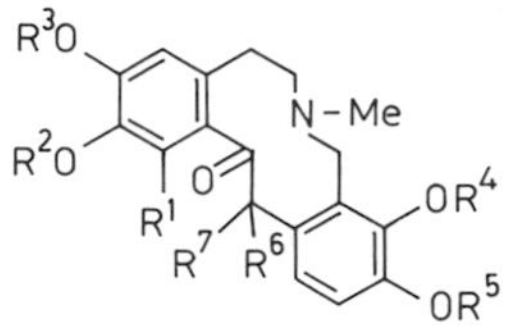

317 $R^1 = H$; $R^2 = R^3 = Me$; $R^4 + R^5 = CH_2$; $R^6, R^7 = O$
318 $R^1 = H$; $R^2 = R^3 = R^4 = R^5 = Me$; $R^6, R^7 = O$
319 $R^1 = H$; $R^2 + R^3 = R^4 + R^5 = CH_2$; $R^6 R^7 = O$
320 $R^1 = OMe$; $R^2 + R^3 = CH_2$; $R^4 = R^5 = Me$; $R^6_1 R^7 = O$
321 $R^1 = R^6 = H$; $R^2 + R^3 = R^4 + R^5 = CH_2$; $R^7 = Me$
322 $R^1 = R^6 = H$; $R^2 + R^3 = CH_2$; $R^4 = R^5 = Me$; $R^7 = Me$

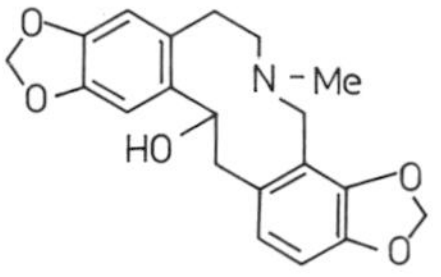

323 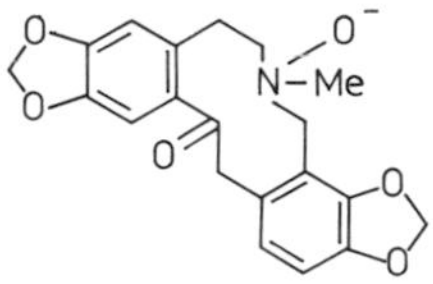   324 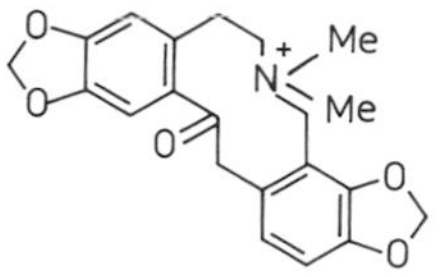   324 a

Allocryptopine (**310**)	1-Methoxyallocryptopine (= coulteropine)
Corycavamine (**321**)	(**316**)
Corycavidine (**322**)	1-Methoxy-13-oxoallocryptopine (**320**)
Corycavine [= (±)-corycavamine] (**322**)	Muramine (**314**)
Coulteropine (**310**)	13-Oxocryptopine (**317**)
Cryptocavine (= cryptopine) (**311**)	13-Oxomuramine (**318**)
Cryptopine (**311**)	13-Oxoprotopine (**319**)
Dihydroprotopine (**323**)	Protopine (**309**)
Hunnemannine (**312**)	Protopine *N*-oxide (**324**)
Izmirine (**313**)	Protopine methohydroxide (**324a**)
	Vaillantine (**315**)

Phthalideisoquinolines

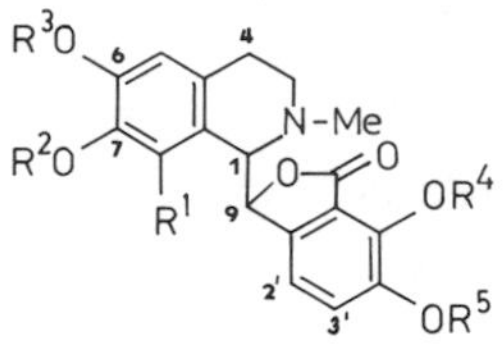

325 $R^1 = H$; $R^2 + R^3 = R^4 + R^5 = CH_2$
326 $R^1 = H$; $R^2 = R^3 = Me$; $R^4 + R^5 = CH_2$
327 $R^1 = H$; $R^2 = R^3 = R^4 = R^5 = Me$
328 $R^1 = R^5 = H$; $R^2 + R^3 = CH_2$; $R^4 = Me$
329 $R^1 = R^3 = H$; $R^2 = Me$; $R^4 + R^5 = CH_2$
330 $R^1 = R^2 = H$; $R^3 = Me$; $R^4 + R^5 = CH_2$
331 $R^1 = H$; $R^2 + R^3 = CH_2$; $R^4 = R^5 = Me$
332 $R^1 = OMe$; $R^2 + R^3 = CH_2$; $R^4 = R^5 = Me$
333 $R^1 = OH$; $R^2 + R^3 = CH_2$; $R^4 = R^5 = Me$

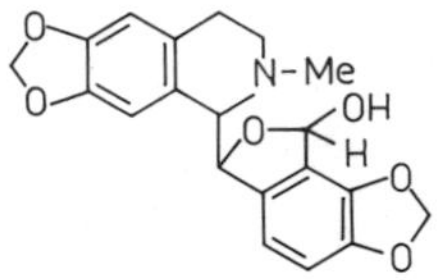

334

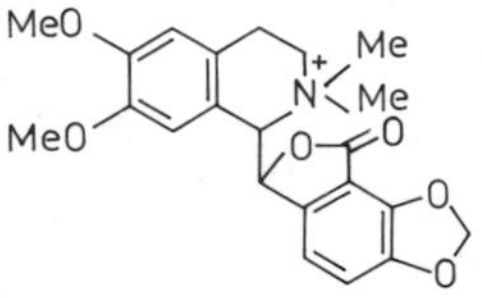

335

Adlumidine (**325**)
Adlumine (**326**)
Bicuculline (**325**)
Capnoidine (**325**)
Cordrastine (**327**)
Corftaline (**328**)
Corledine (**329**)
Corlumidine (**330**)
Corlumine (**326**)

Enganine (**334**)
Fumvailine [= (−)-adlumine] (**326**)
Gnoscopine [= (±)-α-narcotine] (**332**)
Hydrastine (**331**)
N-Methyladlumine (**335**)
Narcotine (**332**)
Narcotoline (**333**)
Stylophylline [= (−)-α-hydrastine] (**331**)

Secophthalideisoquinolines

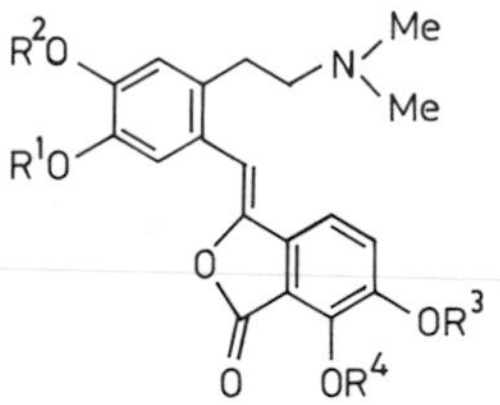

336 $R^1 + R^2 = CH_2$; $R^3 = R^4 = Me$
337 $R^1 + R^2 = R^3 + R^4 = CH_2$
338 $R^1 = R^2 = Me$; $R^3 + R^4 = CH_2$

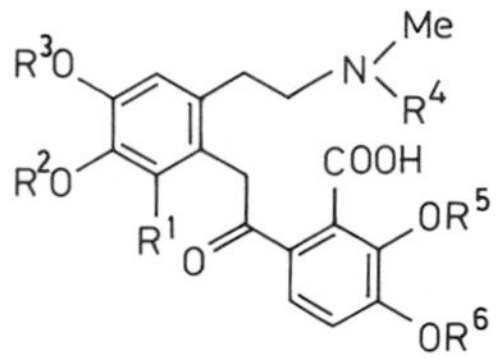

339   $R^1 = H$; $R^2 + R^3 = CH_2$; $R^4 = R^5 = R^6 = Me$
340   $R^1 = H$; $R^2 + R^3 = R^5 + R^6 = CH_2$; $R^4 = Me$
341   $R^1 = H$; $R^2 = R^3 = R^4 = Me$; $R^5 + R^6 = CH_2$
342   $R^1 = OMe$; $R^2 + R^3 = CH_2$; $R^4 = H$; $R^5 = R^6 = Me$
343   $R^1 = OMe$; $R^2 + R^3 = CH_2$; $R^4 = R^5 = R^6 = Me$

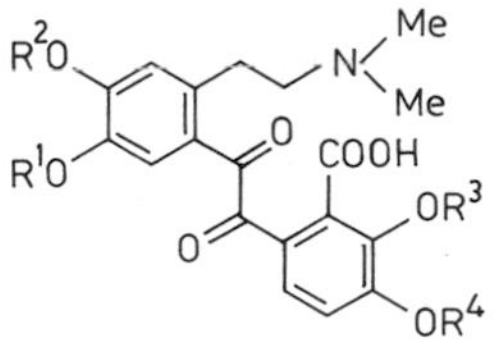

344   $R^1 + R^2 = CH_2$; $R^3 = R^4 = Me$
345   $R^1 + R^2 = R^3 + R^4 = CH_2$
346   $R^1 = R^2 = Me$; $R^3 + R^4 = CH_2$

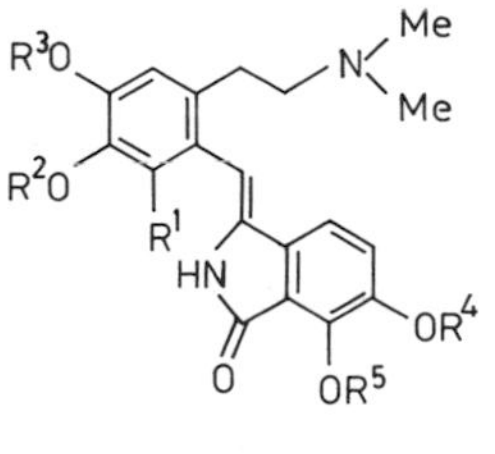

347   $R^1 = H$; $R^2 + R^3 = R^4 + R^5 = CH_2$
348   $R^1 = H$; $R^2 = R^3 = Me$; $R^4 + R^5 = CH_2$
349   $R^1 = H$; $R^2 + R^3 = CH_2$; $R^4 = R^5 = Me$
350   $R^1 = OMe$; $R^2 + R^3 = CH_2$; $R^4 = R^5 = Me$

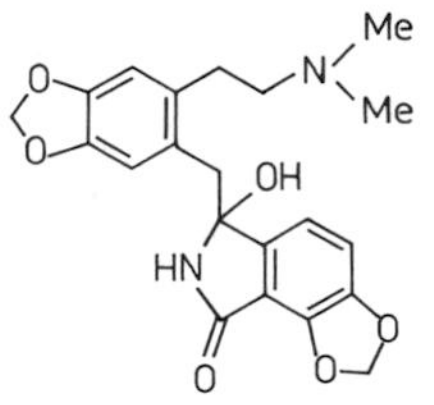

351 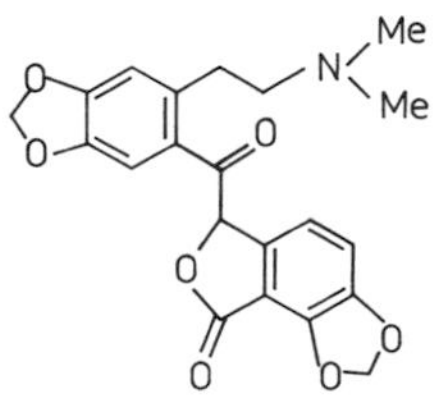   352

Adlumiceine (**341**)
Adlumiceine enol lactone (**338**)
Adlumidiceine (**340**)
Adlumidiceine enol lactone (**337**)
Aobamidine (= adlumidiceine enol lactone)
  (**337**, *cis*)
Bicucullinidine (**346**)
Bicucullinine (**345**)
Fumaramidine (**348**)
Fumaramine (**347**)

Fumaridine (= hydrastine imide) (**349**)
Fumschleicherine (**351**)
N-Methylhydrasteine (**339**)
N-Methylhydrastine (**336**)
N-Methyloxohydrasteine (**344**)
Narceine (**343**)
Narceine imide (**350**)
Narceine imine (= bicucullinine) (**345**)
Narlumidine (**352**)
Nornarceine (**342**)

Indenobenzazepines

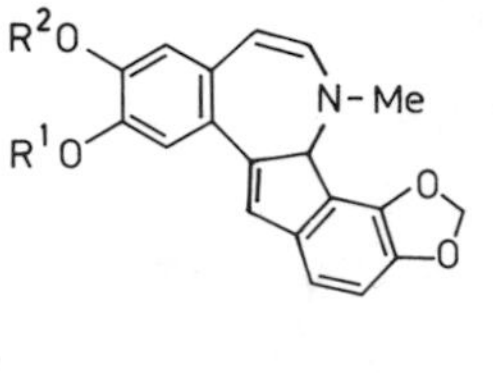

353 $R^1 + R^2 = CH_2$
354 $R^1 = R^2 = Me$

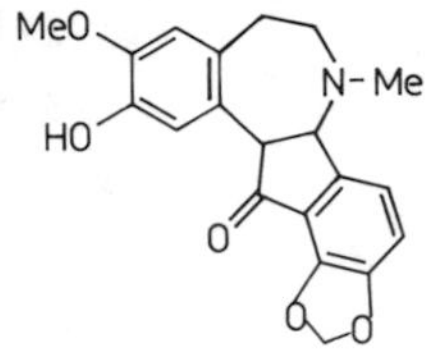

355

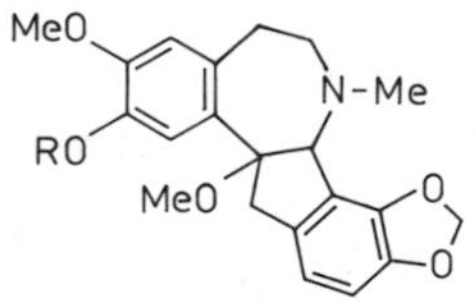

356 R = H
357 R = Me

Fumaritridine (**356**)
Fumaritrine (**357**)
Fumarofine (**355**)

Lahoramine (**354**)
Lahorine (**353**)

Spirobenzylisoquinolines

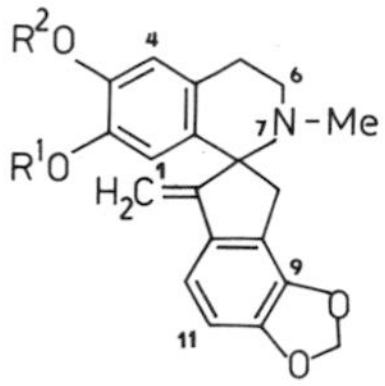

358 $R^1 = Me$; $R^2 = H$
359 $R^1 = R^2 = Me$

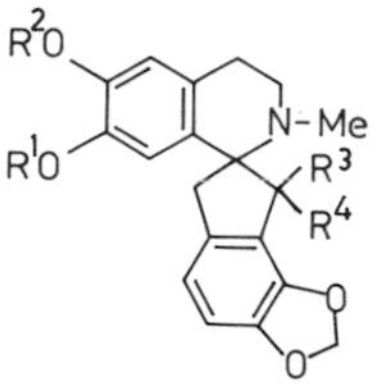

360 $R^1 = R^4 = H$; $R^2 = Me$; $R^3 = OH$
361 $R^1 = R^2 = Me$; $R^3 = OH$; $R^4 = H$
362 $R^1 = R^2 = Me$; $R^3 = H$; $R^4 = OH$
363 $R^1 = R^3 = H$; $R^2 = Me$; $R^4 = OH$
364 $R^1 = R^2 = R^3 = Me$; $R^4 = OH$
365 $R^1 = R^2 = Me$; $R^3 + R^4 = O$

366 $R^1 = Me$; $R^2 = H$; $R^3 + R^4 = O$
367 $R^1 + R^2 = CH_2$; $R^3 + R^4 = O$
368 $R^1 + R^2 = CH_2$; $R^3 = H$; $R^4 = OH$
369 $R^1 = R^3 = H$; $R^2 = Me$; $R^4 = OH$
370 $R^1 = R^3 = H$; $R^2 = Me$; $R^4 = Acetyl$
371 $R^1 = R^2 = Me$; $R^3 = H$; $R^4 = Acetyl$
372 $R^1 = D-glucose$; $R^2 = Me$; $R^3 + R^4 = O$

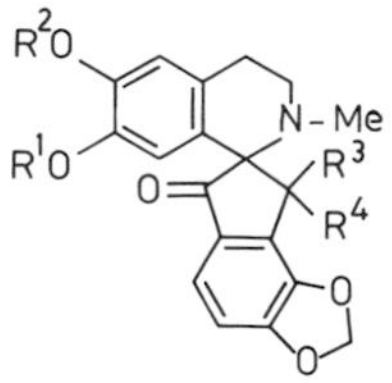

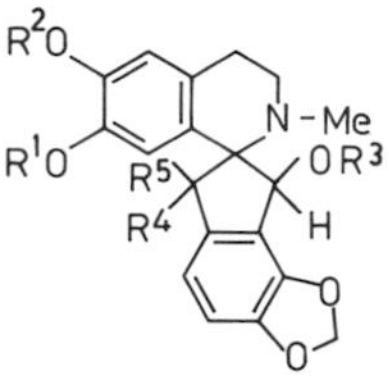

379 $R^1 + R^2 = CH_2$; $R^3 = R^4 = H$; $R^5 = OH$
380 $R^1 = R^2 = Me$; $R^3 = R^4 = H$; $R^5 = OH$
381 $R^1 = R^2 = Me$; $R^3 = R^5 = H$; $R^4 = OH$
382 $R^1 = R^2 = Me$; $R^3 = Ac$; $R^4 = OH$; $R^5 = H$
383 $R^1 = R^3 = R^5 = H$; $R^2 = Me$; $R^4 = OH$
384 $R^1 + R^2 = CH_2$; $R^3 = R^4 = H$; $R^5 = OH$
385 $R^1 + R^2 = CH_2$; $R^3 = Ac$; $R^4 = OH$; $R^5 = H$

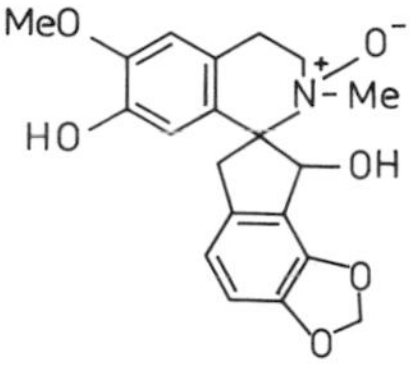

386

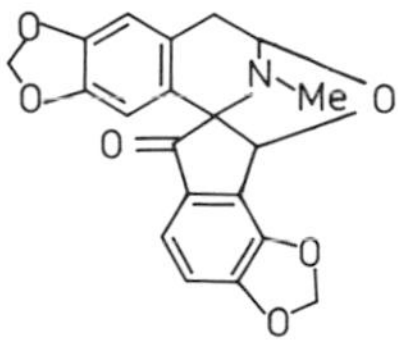

387

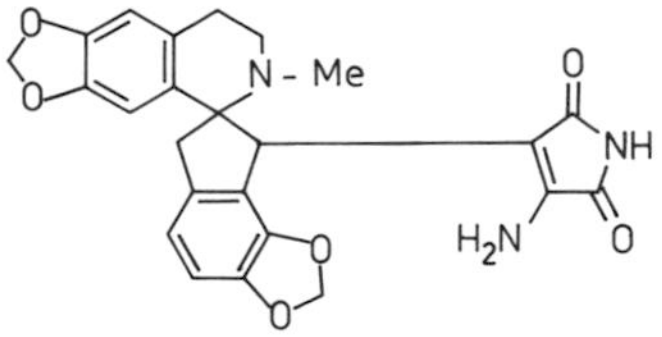

388

373 $R^1 + R^2 = CH_2$; $R^3 = OH$; $R^4 = H$
374 $R^1 + R^2 = CH_2$; $R^3 = H$; $R^4 = OH$
375 $R^1 = R^2 = Me$; $R^3 = OH$; $R^4 = H$
376 $R^1 = R^2 = Me$; $R^3 = H$; $R^4 = OH$
377 $R^1 = H$; $R^2 = Me$; $R^3 + R^4 = H + OH$
378 $R^1 = Me$; $R^2 = R^3 = H$; $R^4 = OH$

Corpaine (**360**)
Corydaine (**374**)
Densiflorine (**387**)
Dihydrofumariline (**368**)
Dihydroparfumine (**369**)
Fumaricine (**362**)
Fumariline (**367**)
Fumaritine (**363**)
Fumaritine *N*-oxide (**386**)
Fumarophycine (**370**)
Fumarostelline (**377**)
Hyperectine (**388**)
Ledeboridine (**383**)
Ledeborine (**378**)
Lederine (**385**)
O-Methylcorpaine (**361**)

O-Methylfumarophycine (**371**)
Ochotensimine (**359**)
Ochotensine (**358**)
Ochrobirine (**379**)
Parfumidine (**365**)
Parfumine (**366**)
Parviflorine (**372**)
Raddeanamine (**364**)
Raddeanidine (**382**)
Raddeanine (**381**)
Raddeanone (**376**)
Severcinine (**384**)
Sibiricine (**373**)
Yenhusomidine (**375**)
Yenhusomine (**380**)

Rhoeadines

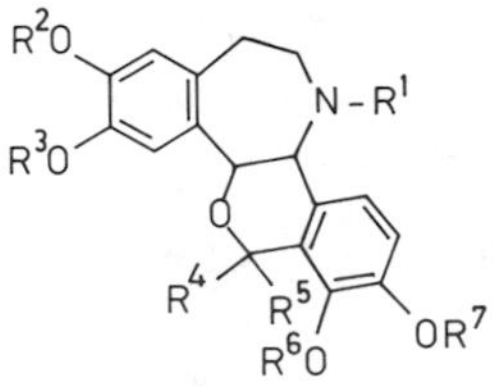

389 R^1 = Me ; R^2 + R^3 = R^6 + R^7 = CH_2 ; R^4 = H ; R^5 = OMe
390 R^1 = Me ; R^2 + R^3 = R^6 + R^7 = CH_2 ; R^4 = α-glu ; R^5 = H
391 R^1 = Me ; R^2 + R^3 = R^6 + R^7 = CH_2 ; R^4 = OH ; R^5 = H
392 R^1 = R^2 = R^3 = Me ; R^4 = H ; R^5 = OMe ; R^6 + R^7 = CH_2
393 R^1 = R^2 = R^3 = Me ; R^4 = OH ; R^5 = H ; R^6 + R^7 = CH_2
394 R^1 = R^2 = R^3 = R^6 = R^7 = Me ; R^4 = H ; R^5 = OMe
395 R^1 = R^2 = R^3 = R^6 = R^7 = Me ; R^4 = OMe ; R^5 = H
396 R^1 = R^4 = H ; R^2 + R^3 = R^6 + R^7 = CH_2 ; R^5 = OMe
397 R^1 = R^4 = H ; R^2 = R^3 = Me ; R^5 = OMe ; R^6 + R^7 = CH_2
398 R^1 = R^3 = R^4 = H ; R^2 = Me ; R^5 = OMe ; R^6 + R^7 = CH_2
399 R^1 = R^4 = H ; R^2 = R^3 = R^6 = R^7 = Me ; R^5 = OMe
400 R^1 = R^3 = R^5 = H ; R^2 = Me ; R^4 = OMe ; R^6 + R^7 = CH_2
401 R^1 = R^2 = Me ; R^3 = R^5 = H ; R^4 = OH ; R^6 + R^7 = CH_2
402 R^1 = R^2 = R^3 = Me ; R^4 = OMe ; R^5 = H ; R^6 + R^7 = CH_2
403 R^1 = R^2 = R^3 = R^6 = R^7 = Me ; R^4 = OH ; R^5 = H
404 R^1 = Me ; R^2 + R^3 = R^6 + R^7 = CH_2 ; R^4 = H ; R^5 = OH
405 R^1 = R^2 = R^3 = Me ; R^4 = H ; R^5 = OH ; R^6 + R^7 = CH_2

Alpinigenine (**403**)
Alpinine (**394**)
Epialpinine (**395**)
Epiglaudine (**402**)
Glaucamine (**393**)
Glaudine (**392**)
Glaupavine (= epiglaudine?) (**402**)
Isorhoeadine (**389**)
Isorhoeagenine (**391**)
Isorhoeagenine glycoside (**390**)
N-Methyl-14-*O*-demethylepiporphyroxine (**401**)
Oreodine (**392**)
Oreogenine (**405**)
Papaverrubine A (**396**)
Papaverrubine B (**397**)
Papaverrubine C (**400**)
Papaverrubine D (**398**)
Papaverrubine E (**396**)
Papaverrubine F (**397**)
Papaverrubine G (**399**)
Rhoeadine (**389**)
Rhoeagenine (**404**)

Alkaloids of Types Other Than Isoquinolines

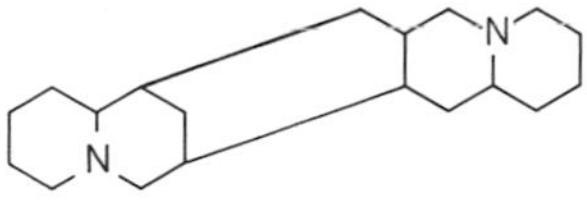

407

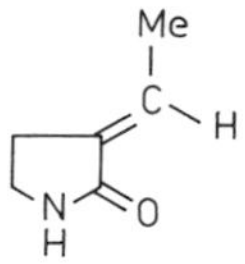

411

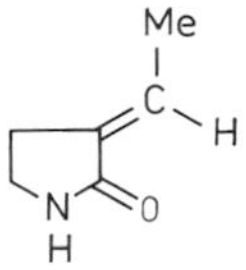

408

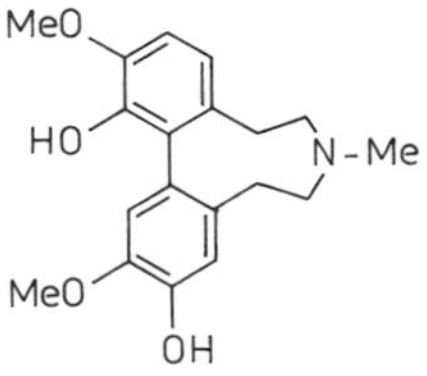

414

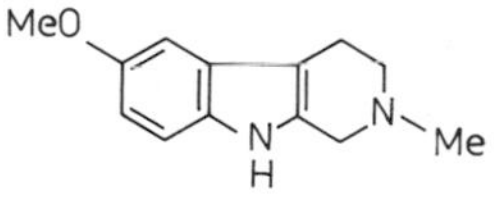

412

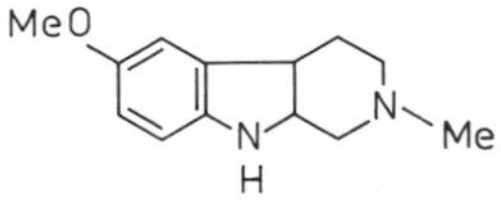

409

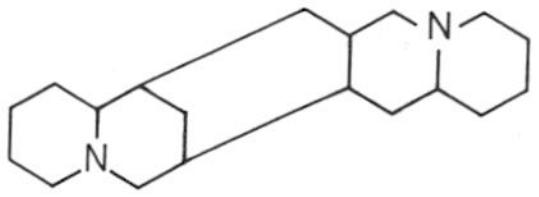

410

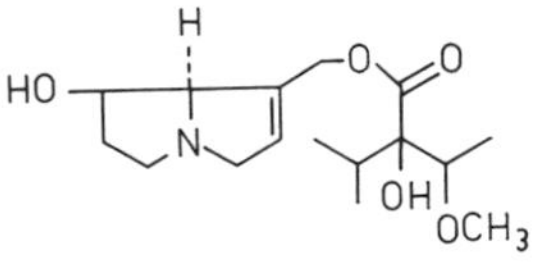

413

Sparteine (*Chelidonium majus*) (**407**)
trans-3-Ethylideno-2-pyrrolidone (= corydalactame) (*Corydalis pallida* var. *tenuis*) (**408**)
N-Methyltetrahydronorharmane (6-methoxy-2-methyl-1,2,3,4-tetrahydro-β-carboline) (*Meconopsis rudis*) (**409**)
Heliotrine (pyrrolizidines) (*Glaucium corniculatum*) (**410**)
Crassifolazonine (dibenzazonines) (*Corydalis claviculata*) (**411**)

REFERENCES

1. V. Preininger, in "The Alkaloids" (R. H. F. Manske, ed.), Vol. XV, p. 207. Academic Press, New York, 1975.
2. Y. Kondo, *Heterocycles* **4**, 197 (1976).
3. F. Šantavý, in "The Alkaloids" (R. H. F. Manske, ed.), Vol. XII, p. 333. Academic Press, New York, 1970; F. Šantavý, in "The Alkaloids" Vol. XVII, p. 385. Academic Press, New York, 1979.
3a. L. Kühn and S. Pfeiffer, *Pharmazie* **18**, 819 (1963).
4. M. Shamma, "The Isoquinoline Alkaloids." Academic Press, New York, 1972; M. Shamma and J. L. Moniot, "Isoquinoline Alkaloids Research." Plenum, New York, 1978.
5. P. Tétényi, *Herba Hungarica* **23**, 19 (1984).
6. T. Kametani, "The Chemistry of the Isoquinoline Alkaloids," Vol. II. Kinkodo Publ. Co., Sendai, Japan, 1975.
7. T. Kametani, *Heterocycles, Supplement* **1** (1983).
8. J. E. Saxton, ed., "The Alkaloids," Vols. 1–5; M. E. Grundon, ed., Vols. 6–13. The Chemical Society, London, 1971–1983.
9. R. Hegnauer, "Chemotaxonomie der Pflanzen." Birkhäuser Verlag, Basel, Stuttgart, 1962–1973.
10. F. Fedde, *In* "Das Pflanzenreich" (A. Engler, ed.), Part IV, No. 40, p. 288. Engelmann, Leipzig, 1909.
11. T. G. Tutin, V. H. Heywood, N. A. Burges, D. H. Valentine, S. M. Walters, and D. A. Webb, "Flora Europea," Vol. I. Cambridge Univ. Press, London and New York, 1964.
12. H. Melchior, in "Syllabus der Pflanzenfamilien" (A. Engler, ed.), 12th Ed., Part II, p. 179. Borntraeger, Berlin, 1964.
13. J. Hutchinson, "Families of Flowering Plants," 2nd ed., Vol. 2. Oxford Univ. Press, London and New York, 1959.
14. R. Hegnauer, *Planta Med.* **9**, 37 (1961).
15. A. Takhtajan, "Die Evolution der Angiospermen." Fischer, Jena, 1959.
16. J. Slavík, *Acta Univ. Olomuc. (Olomouc), Fac. Med.* **93**, 5 (1980).
17. J. W. Fairbairn and E. Williamson, *Phytochemistry* **17**, 2087 (1978).
18. M. M. Nijland, *Pharm. Weekbl.* **99**, 1165 (1964); **100**, 99 (1965).
19. C. Tani and K. Tagahara, *Chem. Pharm. Bull.* **22**, 2457 (1974).
20. N. Takao, K. Iwasa, M. Kamigauchi, and M. Sugiura, *Chem. Pharm. Bull.* **24**, 2859 (1976).
21. C. Tani and K. Tagahara, *Yakugaku Zasshi* **97**, 87 (1977).
22. L. Slavíková and J. Slavík, *Collect. Czech. Chem. Commun.* **36**, 2385 (1971).
23. J. Slavík and L. Slavíková, *Collect. Czech. Chem. Commun.* **42**, 2686 (1977).
24. V. Preininger, *Acta Univ. Olomuc. (Olomouc), Fac. Med.* **106**, 9 (1984).
25. J. J. Bernhardi, *Linnaea* **8**, 401 (1833).
26. L. Elkan, "Tentamen monographiae generis *Papaver*." Regimentii Borussorum, Königsberg, 1839.
27. K. Prantl and L. Künding, in "Die natürlichen Pflanzenfamilien" (A. Engler and K. Prantl, eds.), Part III/2, p. 130. Engelmann, Leipzig, 1889.
28. K. F. Günther, *Flora (Jena)* **164**(4), 393 (1975).
29. J. Novák and V. Preininger, *Preslia* **52**, 97 (1980).
30. V. Preininger, J. Novák, and F. Šantavý, *Planta Med.* **41**, 119 (1981).
31. J. Slavík, L. Slavíková, and L. Dolejš, *Collect. Czech. Chem. Commun.* **46**, 2587 (1981).
32. I. A. Israilov, M. A. Manushakyan, V. M. Mnatsakanyan, M. S. Yunusov, and S. Yu. Yunusov, *Khim. Prir. Soedin.*, 852 (1980); 76 (1984).
33. V. A. Mnatsakanyan, M. A. Manushakyan, and N. E. Mesropjan, *Khim. Prir. Soedin.*, 417 (1978).

34. J. Slavík, K. Picka, L. Slavíková, E. Táborská, and F. Věžník, *Collect. Czech. Chem. Commun.* **45**, 914 (1980).
35. S. F. Hussain, S. Nakkady, L. Khan, and M. Shamma, *Phytochemistry* **22**, 319 (1983).
36. G. Sariyar and J. D. Phillipson, *J. Nat. Prod.* **44**, 239 (1981).
37. S. El-Masny, M. G. El-Ghazooly, A. A. Omar, S. M. Khafagy, and J. D. Phillipson, *Planta Med.* **41**, 61 (1981).
38. W. Egels, *Planta Med.* **39**, 76 (1959).
39. J. Slavík, *Collect. Czech. Chem. Commun.* **29**, 1314 (1964).
40. V. Preininger, P. Vácha, B. Šula, and F. Šantavý, *Planta Med.* **10**, 124 (1962).
41. A. Němečková, A. D. Cross, and F. Šantavý, *Naturwiss.* **54**, 45 (1967).
42. K. Kubát, *in* "Flora of Czechoslovakia" (S. Hejný, ed.) (in press).
43. K. Mothes and H. R. Schütte, "Biosynthese der Alkaloide," VEB Deutscher Verlag Wissenschaften, Berlin, 1969.
44. F. Šantavý, M. Maturová, A. Němečková, H. B. Schröter, H. Potěšilová, and V. Preininger, *Planta Med.* **8**, 167 (1960).
45. O. Gašić, V. Preininger, H. Potěšilová, B. Belia, and F. Šantavý, *Bull. Soc. Chim. Beograd.* **39**, 499 (1974).
46. J. Cullen, *in* "Flora of Turkey and the East Aegean Islands" (P. H. Davis, ed.), Vol. I. Edinburgh, 1965.
47. S. Pfeifer and S. K. Bannerjee, *Pharmazie* **19**, 286 (1964).
48. V. A. Mnatsakanyan, V. Preininger, V. Šimánek, J. Juřina, A. Klásek, L. Dolejš, and F. Šantavý, *Collect. Czech. Chem. Commun.* **42**, 1421 (1977).
49. V. Preininger, V. Tošnarová, and F. Šantavý, *Planta Med.* **20**, 70 (1971).
50. B. Nielson, J. Röe, and E. Brochmann-Hanssen, *Planta Med.* **48**, 205 (1983).
51. J. D. Phillipson, A. I. Gray, A. A. R. Askari, and A. A. Khalil, *J. Nat. Prod.* **44**, 296 (1981).
51a. G. Sariyar, T. Baytop, and J. D. Phillipson, *Abstracts of the Intern. Symp. on the Chemistry and Biology of Isoquinoline Alkaloids,* 16th–18th April, London, 1984.
52. J. Slavík, *Collect. Czech. Chem. Commun.* **45**, 2706 (1980).
53. F. R. Stermitz, *Recent Adv. Phytochem.* **1**, 161 (1968).
54. J. Novák and V. Preininger, "Taxonomické a fytochemické hodnocení rodu *Papaver* (Papaveraceae)." Vysoká škola zemědělská, Praha, 1981.
55. V. Preininger and F. Šantavý, *Pharmazie* **25**, 356 (1970).
56. J. D. Phillipson, O. O. Thomas, A. I. Gray, and G. Sariyar, *Planta Med.* **41**, 105 (1981).
57. G. Sariyar, *Planta Med.* **39**, 216 (1980).
58. G. Sariyar and J. D. Phillipson, *Phytochemistry* **19**, 2189 (1980).
59. G. Sariyar, *Planta Med.* **49**, 43 (1983).
60. M. A. Manushakyan, I. A. Israilov, V. A. Mnatsakanyan, M. S. Yunusov, and S. Yu. Yunusov, *Khim. Prir. Soedin.,* 849 (1980).
61. G. Sariyar, Dissertation. University of Istanbul, Istanbul, 1975.
62. V. Preininger, V. Appelt, L. Slavíková, and J. Slavík, *Collect. Czech. Chem. Commun.* **32**, 2682 (1967).
63. J. D. Phillipson, *Phytochemistry* **21**, 2441 (1982).
64. J. D. Phillipson, *Planta Med.* **48**, 187 (1983).
65. F. Věžník, P. Sedmera, V. Preininger, V. Šimánek, and J. Slavík, *Phytochemistry* **20**, 347 (1981).
66. F. Věžník, E. Táborská, and J. Slavík, *Collect. Czech. Chem. Commun.* **46**, 926 (1981).
67. R. Hocquemiller, A. Oztekin, F. Roblot, M. Hutin, and A. Cavé, *J. Nat. Prod.* **47**, 342 (1984).
68. L. Slavíková and J. Slavík, *Collect. Czech. Chem. Commun.* **45**, 761 (1980).
69. H. Böhm and K. F. Günther, *Pharmazie* **27**, 125 (1972).
70. J. W. Fairbairn and F. Hakim, *J. Pharm. Pharmac.* **25**, 353 (1973).

71. U. Nyman and J. G. Bruhn, *Planta Med.* **35,** 97 (1979).
72. J. J. Kettenes—Van den Bosch, C. A. Salemink, and I. Khan, *J. Ethnopharmacol.* **3,** 21 (1981).
73. H. Böhm, *Pharmazie* **36,** 660 (1981).
74. K. W. Bentley and D. G. Hardy, *J. Am. Chem. Soc.* **89,** 3267 (1967).
75. G. B. Lockwood, *Phytochemistry* **20,** 1463 (1981).
76. D. Corrigan and E. M. Martyn, *Planta Med.* **42,** 45 (1981).
77. S. W. Zito and E. J. Staba, *Planta Med.* **45,** 53 (1982).
78. M. Bastart-Malsot and M. Paris, *Planta Med.* **45,** 35 (1982).
79. J. K. Wold, B. S. Paulsen, O. F. Ellingsen, and A. Nordal, *Nor. Pharm. Acta* **45,** 103 (1983).
80. P. Goldblatt, *Ann. Mo. Bot. Gard.* **61,** 264 (1974).
81. J. Novák, *Preslia* **51,** 341 (1979).
82. J. D. Phillipson, A. Scutt, A. Baytop, N. Özhatay, and G. Sariyar, *Planta Med.* **43,** 261 (1981).
83. H. Böhm, *Planta Med.* **15,** 215 (1967).
83a. I. Lalezari, P. Nasseri, and R. Asgharian, *Phytochemistry* **63,** 1331 (1974).
84. P. Cheng, M.Sc. Thesis, Univ. of Mississippi, 1972.
85. H. Rönsch and W. Schade, *Phytochemistry* **18,** 1089 (1979).
86. H. Meshulam and D. Lavie, *Phytochemistry* **19,** 2633 (1980).
87. H. G. Theuns, E. J. Vlietstra, and C. A. Salemink, *Phytochemistry* **22,** 247 (1983).
88. I. A. Israilov, O. N. Denisenko, M. S. Yunusov, D. A. Muraveva, and S. Yu. Yunusov, *Khim, Prir. Soedin.,* 474 (1978).
89. I. A. Israilov, M. A. Manushakyan, V. A. Mnatsakanyan, M. S. Yunusov, and S. Yu. Yunusov, *Khim, Prir. Soedin.,* 81 (1984).
90. I. A. Israilov, M. A. Manushakyan, V. A. Mnatsakanyan, M. S. Yunusov, and S. Yu. Yunusov, *Khim, Prir. Soedin.,* 258 (1984).
91. J. W. Fairbairn and M. J. Steele, *Phytochemistry* **19,** 2317 (1980).
92. A. Shafiee, I. Lalezari, P. Nasseri-Nouri, and R. Asgharian, *J. Pharm. Sci.* **64,** 1570 (1975).
93. A. Shafiee, I. Lalezari, F. Assadi, and F. Khalafi, *J. Pharm. Sci.* **66,** 1050 (1977).
93a. E. Brochmann-Hanssen and C. Y. Cheng, *J. Nat. Prod.* **45,** 434 (1982).
94. G. Sariyar and J. D. Phillipson, *Phytochemistry* **16,** 2009 (1977).
95. H. Böhm, *Planta Med.* **43,** 660 (1981).
96. J. Novák, "Taxonomické hodnocení sekce *Scapiflora*." Vysoká škola zemědělská, Praha, 1978.
96a. H. Böhm, L. Dolejš, V. Preininger, F. Šantavý, and V. Šimánek, *Planta Med.* **28,** 210 (1975).
97. B. A. Celombitko, V. A. Mnatsakanyan, and L. V. Salnikova, *Khim. Prir. Soedin.,* 270 (1978).
98. V. V. Melik-Guseinov, D. A. Muraveva, and V. A. Mnatsakanyan, *Khim. Prir. Soedin.,* 239 (1979).
99. I. A. Israilov, S. U. Karimova, O. N. Denisenko, M. S. Yunusov, D. A. Muraveva, and S. Yu. Yunusov, *Khim. Prir. Soedin.,* 751 (1984).
100. A. Ghanbarpour, A. Shafiee, and M. Parchami, *Lloydia* **41,** 472 (1978).
101. S. U. Karimova, I. A. Israilov, and S. Yu. Yunusov, *Khim. Prir. Soedin.,* 814 (1978); 104 (1979); 224 (1980).
102. I. A. Israilov, S. U. Karimova, M. S. Yunusov, and S. Yu. Yunusov, *Khim. Prir. Soedin.,* 415 (1979).
103. L. Castedo and D. Dominguez, *Tetrahedron Lett.,* 2923 (1978).
104. L. Castedo, D. Dominguez, M. Pereira, J. Saá, and R. Suau, *Heterocycles* **16,** 535 (1981).
105. L. Castedo, D. Dominguez, J. M. Saá, and R. Suau, *Tetrahedron Lett.,* 4589 (1979).

106. T. Goezler, *Planta Med.* **46**, 179 (1982).
107. S. U. Karimova and I. A. Israilov, *Khim. Prir. Soedin.*, 259 (1984).
108. J. Slavík, L. Slavíková, and L. Dolejš, *Collect. Czech. Chem. Commun.* **49**, 1318 (1984).
109. A. Shafiee, A. Ghanbarpour, I. Lalezari, and S. Lajevardi, *J. Nat. Prod.* **42**, 174 (1979).
110. T. F. Platonova, P. S. Massagatev, A. D. Kuzuvkov, and L. M. Utkin, *Zh. Obshch. Khim.* **26**, 173 (1956).
111. J. Slavík, L. Dolejš, and L. Slavíková, *Collect. Czech. Chem. Commun.* **39**, 888 (1974).
112. R. Konovalova, S. Yunusov, and A. Orechov, *Bull. Soc. Chim. France* **6**, 1479 (1939).
113. M. S. Yunusov, S. T. Akramov, and S. Yu. Yunusov, *Dokl. Akad. Nauk Uz. SSR* **23**, 38 (1966).
114. J. Slavík, L. Slavíková, and L. Dolejš, *Collect. Czech. Chem. Commun.* **33**, 4066 (1968).
115. J. Slavík and L. Slavíková, *Collect. Czech. Chem. Commun.* **42**, 132 (1977).
116. S. R. Hemingway, J. D. Phillipson, and R. Verpoorte, *J. Nat. Prod.* **44**, 67 (1981).
117. G. B. Ownbey, *Memoirs Torrey Botanical Club* **21**, No. 1 (1958); *Brittonia* **13**, 91 (1960).
118. F. R. Stermitz, *Recent Adv. Phytochem.* **1**, 161 (1968).
119. S. F. Hussain, S. Nakkady, L. Khan, and M. Shamma, *Phytochemistry* **22**, 319 (1983).
120. J. Slavík and L. Slavíková, *Collect. Czech. Chem. Commun.* **28**, 1728 (1963).
121. S. Parfeinikov and D. A. Muraveva, *Khim. Prir. Soedin.* 242 (1983).
122. J. Berlin, E. Forche, and V. Wray, *Z. Naturforsch.* **38B**, 346 (1983).
123. J. Slavík, V. Novák, and L. Slavíková, *Collect. Czech. Chem. Commun.* **41**, 2429 (1976).
124. E. Táborská, F. Věžník, L. Slavíková, and J. Slavík, *Collect. Czech. Chem. Commun.* **43**, 1108 (1978).
125. J. Cullen, *Baileya* **15**, 112 (1967).
126. W. Vent and B. Mory, *Gleditschia* **1**, 33 (1973).
127. H. Itokawa, A. Ikuta, N. Tsutsui, and I. Ishigura, *Phytochemistry* **17**, 839 (1973).
128. J. Slavík and L. Slavíková, *Collect. Czech. Chem. Commun.* **42**, 2686 (1977).
129. J. Slavík, *Collect. Czech. Chem. Commun.* **20**, 198 (1955).
130. O. E. Lasskaya and O. N. Tolkachev, *Khim. Prir. Soedin.*, 764 (1978).
131. K. Iwasa, M. Okada, and N. Takao, *Phytochemistry* **22**, 627 (1983).
132. H. Ishii, T. Ishikawa, K. Hosoya, and N. Takao, *Chem. Pharm. Bull.* **26**, 166 (1978).
133. J. Slavík and L. Slavíková, *Collect. Czech. Chem. Commun.* **40**, 3206 (1975).
134. M. E. Perelson, G. G. Alexandrov, L. D. Yakhontova, O. N. Tolkachev, D. A. Fesenko, M. N. Komarova, and S. E. Esipov, *Khim. Prir. Soedin.*, 628 (1984).
135. L. D. Yakhontova, O. N. Tolkachev, M. N. Komarova, and A. I. Shreter, *Khim. Prir. Soedin.*, 673 (1984).
136. T. Gözler, M. A. Önür, R. Minard, and M. Shamma, *J. Nat. Prod.* **46**, 414 (1983).
137. J. Slavík and L. Slavíková, *Collect. Czech. Chem. Commun.* **49**, 704 (1984).
138. M. A. El-Shanawany, A. M. El-Fishaway, D. J. Slatkin, and P. L. Schiff, Jr., *J. Nat. Prod.* **46**, 753 (1983).
139. J. Slavík, K. Picka, L. Slavíková, E. Táborská, and F. Věžník, *Collect. Czech. Chem. Commun.* **45**, 914 (1980).
140. J. Slavík, *Collect. Czech. Chem. Commun.* **32**, 4431 (1967).
141. B. D. Jackson, "Index Kewensis," Part I. Oxford Univ. Press (Clarendon), London and New York, 1893.
142. M. G. Popov, *in* "Flora SSSR" (V. L. Komarov and B. K. Šiškin, eds.), Vol. 7, p. 573. Izd. Akad. Nauk SSSR, 1937.
143. M. Ryberg, *Acta Horti Bergiani* **17**, 115 (1955).
144. H. Taguchi and I. Imaseki, *J. Pharm. Soc. Jpn.* **83**, 578 (1963); **84**, 773 (1964).
145. H. G. Boit and H. Emke, *Naturwiss.* **46**, 427 (1959).
146. H. G. Kiryakov, E. Iskrenova, B. Kuzmanov, and L. Evstatieva, *Planta Med.* **43**, 51 (1981).

147. H. G. Kiryakov and E. S. Iskrenova, *Planta Med.* **50**, 136 (1984).
148. V. Preininger, V. S. Thakur, and F. Šantavý, *J. Pharm. Sci.* **65**, 294 (1976).
149. J. Slavík and L. Slavíková, *Collect. Czech. Chem. Commun.* **44**, 2261 (1979).
150. R. H. F. Manske, *Can. J. Res.* **18B**, 100 (**1940**); **20B**, 57 (1942).
151. P. Oroszlán, L. Dolejš, V. Šimánek, and V. Preininger, *Planta Med.* **286** (1985).
152. L. M. Boento, L. Castedo, R. Cuadros, J. M. Saá, R. Suau, A. Perales, M. Martinez-Ripoll, and J. Fayos, *Tetrahedron Lett.* **24**, 2020 (1983).
153. D. P. Allais and H. Guinaudeau, *Heterocycles* **20**, 2055 (1983).
154. H. Guinaudeau and D. P. Allais, *Heterocycles* **22**, 107 (1984).
155. D. P. Allais and H. Guinaudeau, *J. Nat. Prod.* **46**, 881 (1983).
156. J. M. Boento, L. Castedo, and D. Dominguez, *Int. Symp. London, 1984*, 26.
157. J. M. Boento, L. Castedo, A. Rodriguez de Lera, J. M. Saá, R. Suau, and M. C. Vidal, *Tetrahedron Lett.* **24**, 2295 (1983).
158. J. M. Boento, L. Castedo, D. Dominguez, A. Farina, A. Rodriguez de Lera, and M. C. Villaverde, *Tetrahedron Lett.* **25**, 889 (1984).
159. J. M. Boento, L. Castedo, R. Cuadros, A. Rodriguez de Lera, R. Suau, J. M. Saá, and M. C. Vidal, *Tetrahedron Lett.* **24**, 2303 (1983).
160. F. Qicheng, L. Mao, and W. Qingmei, *Planta Med.* **50**, 25 (1984).
161. R. H. F. Manske, *Can. J. Res.* **24B**, 66 (1946).
162. R. H. F. Manske, *Can. J. Res.* **17B**, 57 (1939).
163. N. N. Margvelašvili, O. N. Tolkachev, N. P. Prisjažnjuk, and A. T. Kirjanova, *Khim. Prir. Soedin.*, 592 (1978).
164. T. Irgashev and I. A. Israilov, *Khim. Prir. Soedin.* 260 (1984).
165. K. Iwasa, N. Takao, G. Nonaka, and I. Nishioka, *Phytochemistry* **18**, 1725 (1979).
166. N. Takao and K. Iwasa, *Chem. Pharm. Bull.* **27**, 2194 (1979).
167. I. A. Israilov and S. Yu. Yunusov, *Khim. Prir. Soedin.*, 418 (1979).
168. I. A. Israilov, M. S. Yunusov, and S. Yu. Yunusov, *Khim. Prir. Soedin.*, 537 (1978).
169. I. A. Israilov, F. M. Melikov, M. S. Yunusov, D. A. Muraveva, and S. Yu. Yunusov, *Khim. Prir. Soedin.*, 540 (1980).
170. K. S. Khusainova and Yu. D. Sadykov, *Khim. Prir. Soedin.*, 670 (1981).
171. F. Qicheng, K. Mao, and W. Qingmei, *Iaixue Xuebao* **16**, 708 (1981); *Chem. Abstr.* **96**, 31687 (1982).
172. V. Preininger, J. Novák, V. Šimánek, and F. Šantavý, *Planta Med.* **33**, 396 (1978).
173. H. G. Kiryakov, E. Iskrenova, B. Kuzmanov, and L. Evstatieva, *Planta Med.* **41**, 298 (1981).
174. I. A. Israilov, O. N. Denisenko, D. A. Muraveva, and M. S. Yunusov, *Khim. Prir. Soedin.*, 672 (1984).
175. D. S. Bhakuni and R. Chaturvedi, *J. Nat. Prod.* **46**, 466 (1983).
176. R. H. F. Manske, *Can. J. Res.* **20B**, 49 (1942).
177. R. H. F. Manske, *Can. J. Res.* **18B**, 288 (1940).
178. C. Tani, K. Tagahara, and S. Aratani, *Yakugaku Zasshi* **96**, 527 (1976).
179. S. T. Lu and E. Ch. Wang, *Phytochemistry* **21**, 809 (1982).
180. V. Preininger, L. Dolejš, B. Smysl, and V. Šimánek, *Planta Med.* **36**, 213 (1979).
181. C. Tani, N. Nagakura, and Ch. Kuriyama, *Yakugaku Zasshi* **98**, 1658 (1978).
182. K. Iwasa and N. Takao, *Phytochemistry* **21**, 611 (1982).
183. M. Alimova, I. A. Israilov, M. S. Yunusov, N. D. Abdulaev, and S. Yu. Yunusov, *Khim. Prir. Soedin.*, 727 (1982).
184. I. A. Israilov, T. Irgashev, M. S. Yunusov, and S. Yu. Yunusov, *Khim. Prir. Soedin.*, 851 (1980).
185. R. H. F. Manske, *Can. J. Res.* **14B**, 347 (1936).
186. V. Preininger, J. Veselý, O. Gašić, V. Šimánek, and L. Dolejš, *Collect. Czech. Chem. Commun.* **40**, 699 (1975).

187. H. G. Kiryakov, E. Iskrenova, E. Daskalova, B. Kuzmanov, and L. Evstatieva, *Planta Med.* **44**, 168 (1982).
188. R. H. F. Manske, R. Rodrigo, H. L. Holland, D. W. Hughes, D. B. MacLean, and J. K. Saunders, *Can. J. Chem.* **56**, 383 (1978).
189. H. G. Kiryakov, E. Daskalova, A. Georgieva, B. Kuzmanov, and L. Evstatieva, *Folia Med. (Plovdiv)* **24**, 19 (1982).
190. S. Luo and S. Wu, *Yaoxue Xuebao* **17**, 699 (1982); *Chem. Abstr.* **98**, 31431u (1983).
191. C. Tani, N. Nagakura, S. Saeki, and M. T. Kao, *Planta Med.* **41**, 403 (1981).
192. J. Go, *J. Pharm. Soc. Japan* **50**, 933, 940 (1930).
193. V. Preininger, L. Dolejš, and V. Šimánek, *Planta Med.* (in press).
194. P. Forgacs, G. Buffard, A. Jehanno, J. Provost, R. Tiberghien, and A. Touche, *Plant. Med. Phytother.* **16**, 99 (1982).
195. J. Novák and V. Preininger, *Acta Univ. Olomuc. (Olomouc), Fac. Rer. Nat.* **78**, 39 (1983).
196. G. Hegi, "Illustrierte Flora von Mitteleuropa," Part IV/1, p. 49. Lehmanns Verlag, München, 1931.
197. J. Susplugas, M. Lalaurie, G. Privat, and P. Chicaya, *Trav. Soc. Pharm., Montpellier* **17**, 134 (1957).
198. J. Susplugas, M. Lalaurie, S. El Nouri, V. Massa, P. Susplugas, and J. C. Rossi, *Trav. Soc. Pharm., Montpellier* **36**, 293 (1976).
199. M. E. Popova, J. Novák, V. Šimánek, L. Dolejš, P. Sedmera, and V. Preininger, *Planta Med.* **48**, 272 (1983).
200. V. B. Pandey, A. B. Ray, and B. Dasgupta, *Phytochemistry* **18**, 695 (1979).
201. B. Dasgupta, K. K. Seth, V. B. Pandey, and A. B. Ray, *Planta Med.* **50**, 481 (1984).
202. A. H. A. Abou-Donia, S. El-Masry, M. R. I. Saleh, and J. D. Phillipson, *Planta Med.* **40**, 295 (1980).
203. H. Kiryakov, D. W. Hughes, B. C. Nalliah, and D. B. MacLean, *Can. J. Chem.* **57**, 53 (1979).
204. M. E. Popova, V. Šimánek, L. Dolejš, B. Smysl, and V. Preininger, *Planta Med.* **45**, 120 (1982).
205. A. Loukis and S. Philianos, *J. Nat. Prod.* **47**, 187 (1984).
206. S. A. E. Hakim, V. Mijovic, and J. Walker, *Nature (London)* **189**, 198 (1961).
207. G. Blaskó, N. Murugesan, S. F. Hussain, R. D. Minard, and M. Shamma, *Tetrahedron Lett.* **22**, 3135 (1981).
208. Z. H. Mardirossian, H. G. Kiryakov, J. P. Ruder, and D. B. MacLean, *Phytochemistry* **22**, 759 (1983).
209. S. F. Hussain, R. D. Minard, A. J. Freyer, and M. Shamma, *J. Nat. Prod.* **44**, 169 (1981).
210. G. Blaskó, S. F. Hussain, and M. Shamma, *J. Nat. Prod.* **44**, 475 (1981).
211. G. Blaskó, S. F. Hussain, A. J. Freyer, and M. Shamma, *Tetrahedron Lett.* **22**, 3127 (1981).
212. M. Alimova, I. A. Israilov, M. S. Yunusov, and S. Yu. Yunusov, *Khim. Prir. Soedin.*, 642 (1982).
213. H. Guinaudeau, T. Gözler, and M. Shamma, *J. Nat. Prod.* **46**, 934 (1983).
214. T. F. Platonova, P. S. Massagetov, A. D. Kuzovkov, and L. M. Utkin, *Zh. Obshch. Khim.* **26**, 173 (1956).
215. P. Forgacs, J. Provost, R. Tiberghien, J. F. Desconlois, G. Buffard, and M. Pesson, *C. R. Acad. Sci. Paris* **276D**, 105 (1973).
216. P. Forgacs, J. Provost, J. F. Desconlois, A. Jehanno, and M. Pesson, *C. R. Acad. Sci. Paris* **279D**, 885 (1974).
217. H. G. Kiryakov, Z. H. Mardirossian, and P. P. Panov, *C. R. Acad. Sci. Bulg.* **33**, 1377 (1980).
218. M. E. Popova, A. N. Boeva, L. Dolejš, V. Preininger, V. Šimánek, and F. Šantavý, *Planta Med.* **40**, 156 (1980).
219. H. G. Kiryakov, Z. H. Mardirossian, and P. P. Panov, *C. R. Acad. Sci. Bulg.* **34**, 43 (1981).

220. G. P. Sheveleva, D. Sargazakov, N. V. Plekhanova, S. T. Aktanova, and A. S. Aldasheva, *Fiziol. Akt. Soedin. Rast. Kirg.*, 41 (1970).
221. A. Radu, M. Tamas, and B. Olah, *Farmacia (Bucharest)* **27**, 1 (1979).
222. B. Sener, Doctoral Dissertation Thesis, Eczacilik Fakultesi, Ankara Universitesi, Ankara, 1981.
223. M. Alimova and I. A. Israilov, *Khim. Prir. Soedin.*, 602 (1981).
224. B. Sener, B. Gözler, R. D. Minard, and M. Shamma, *Phytochemistry* **22**, 2073 (1983).
225. B. Gözler, T. Gözler, and M. Shamma, *Tetrahedron* **39**, 577 (1983).
226. I. A. Israilov, F. M. Melikov, and D. A. Muraveva, *Khim. Prir. Soedin.*, 79 (1984).
227. J. M. Boente, L. Castedo, R. Cuadros, A. Rodrigez de Lera, J. M. Saá, R. Suau, and M. C. Vidal, *Tetrahedron Lett.* **24**, 2303 (1983).
228. M. J. Campello, L. Castedo, J. M. Saá, R. Suau, and M. C. Vidal, *Tetrahedron Lett.* **23**, 239 (1982).
229. J. M. Boente, L. Castedo, A. Rodrigez de Lera, J. M. Saá, R. Suau, and M. C. Vidal, *Tetrahedron Lett.* **24**, 2295 (1983).
230. R. H. F. Manske, *Can. J. Res.* **8**, 210 (1933).

—— CHAPTER 2 ——

QUINAZOLINE ALKALOIDS*

SIEGFRIED JOHNE

Institute of Plant Biochemistry
The Academy of Sciences of the German Democratic Republic
Halle (Saale), German Democratic Republic

I. Introduction

The sources, chemistry, and biochemistry of simple quinazoline alkaloids have been reviewed by Openshaw in Volumes 3 and 7 of this treatise (*1,2*), by Johne and Gröger (*3*), Johne (*4*), Price (*5*), and by Prakash and Ghosal (*6*). Reports of progress in the chemistry of the quinazoline alkaloids of *Peganum harmala* (*7*) and of advances in Vasaka alkaloids (*8*) have also been published. Since the mid 1960s approximately 60 new quinazoline alkaloids have been isolated from different families of the plant kingdom, microorganisms, and animals. Additional significant progress has included structural elucidation with modern spectroscopic methods and the total syntheses of several known alkaloids. This chapter deals with work reported since 1960; results of earlier investigations are mentioned briefly, only if necessary. Table III lists molecular formulas, melting points, optical rotations, and sources of each natural compound covered by this chapter (see p. 132).

II. Simple Substituted Quinazolin-4-ones

A. ARBORINE

Arborine (**1**) (see Fig. 1) has been isolated from the leaves of *Glycosmis arborea* (*9,10,13*) and *G. bilocularis* (*11*). Much controversy centred around the structure and source of this alkaloid. "Glycosine" isolated from *Glycosmis pentaphylla* (*12,13*) (later identified as *Glycosmis arborea*) was shown to be identical with arborine (*14*). *Glycosmis arborea* occurs widely throughout India, whereas *G. pentaphylla* occurs only in the Eastern and Western Ghats and in Orissa, although it is much more common in the Malayan peninsula (*14*).

* This review is dedicated to Prof. Klaus Schreiber, teacher and colleague, on the occasion of his sixtieth birthday, January 25, 1987.

THE ALKALOIDS, VOL. 29

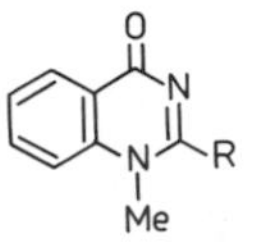

1 R = CH₂Ph Arborine
2 R = Me   Glomerin
3 R = H   Glycorine

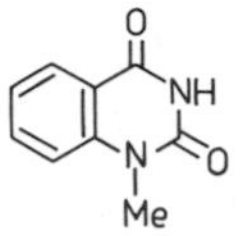

4 Glycosmicine

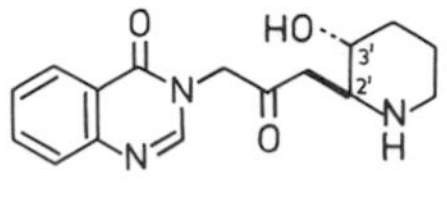

9 Febrifugine

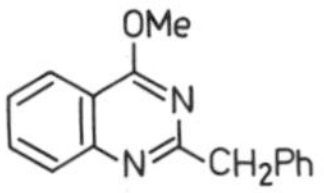

5 Glycophymoline

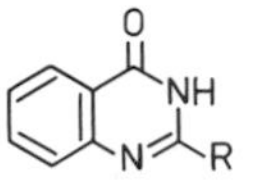

6 R = CH₂Ph Glycosminine
7 R = CH⟨Me/OH Chrysogine
8 R = (CH₂)₂CH₂OH Pegamine

FIG 1. Structures of simple substituted quinazolin-4-ones.

Compound **1** is a weak monoacidic tertiary base that, on hydrogenation, gives a dihydro derivative which is extremely stable to alkali but is hydrolyzed by hot dilute hydrochloric acid to phenylacetaldehyde, N-methylanthranilic acid, and its amide. Alkaline hydrolysis of arborine formed N-methylanthranilic acid, phenylacetic acid, ammonia, and arboricin (**10**). Oxidation with neutral permanganate yielded 1,2,3,4-tetrahydro-2,4-dioxoquinazoline. Treatment with periodic acid, ozone, or hydrogen peroxide gave benzaldehyde in extremely low yield. In distillation with soda lime, arborine yielded ammonia, toluene, and N-methylaniline. The hydrochloride of **1**, on vacuum sublimation above its melting point, yielded glycosminine (**6**) (Fig. 1). The remethylation of this alkaloid with alkaline methyl iodide gave a mixture of **1** and the 3-methyl derivative **11**. Oxidation of arborine with chromic oxide in acetic acid yielded glycorine (**3**) and glycosmicine (**4**) (*15*).

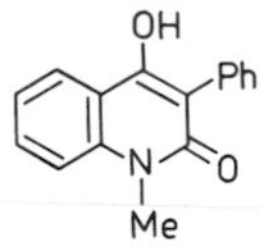

10 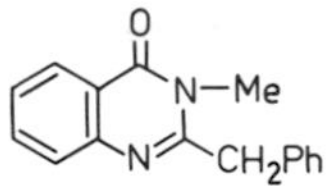   11

On the basis of a critical study of the UV and the IR absorption spectra of **1**, dihydroarborine, and some other quinazoline derivatives of unambigous structure, Chakravarti (*16*) has shown that arborine is 2-benzyl-1-methyl-quinazol-4-one (**1**) and not the tautomer, 2-benzylidene-1,2,3,4-tetrahydro-1-methyl-quinazolin-4-one, postulated by Chatterjee and Majumdar (*13*). Also, the absence of a vinyl proton in the ¹-H-NMR spectrum of arborine is unequivocal evidence against the benzylidene structure.

Mass spectral studies on **1** have been carried out by Pakrashi *et al.* (*17*). The formation of benzaldehyde by oxidative degradation can be explained by the formation of a small portion of the highly reactive tautomeric form (see above), which undergoes rapid oxidative fission; there are many well-known analogies (*14*). For further reports on the elucidation of the structure of **1**, see Refs. *16, 18,* and *19.* The structure was confirmed by a direct synthesis from N^2-methylanthranilamide and phenylacetyl chloride via N^2-methyl-N^2-phenylacetyl-anthranilamide (*9,20,21*).

Arboricine was shown to be 1,2-dihydro-4-hydroxy-1-methyl-2-oxo-3-phenylquinoline (**10**) or its diketo tautomer. It was synthesized by condensation of ethyl phenylmalonate with *N*-methylaniline (*9*). The formation of **10** during the alkaline hydrolysis of **1** probably takes place through the intermediate N^2-methyl-N^2-phenylacetylanthranilamide with loss of ammonia.

Chatterjee and Majumdar (*13*) synthesized **1** by refluxing *N*-methylanthranilamide with phenylacetic acid in xylene in the presence of an excess of P_2O_5. Pakrashi *et al.* (*22*) have realized the dehydrocyclization of the phenylacetyl anthranilamide and the corresponding *N*-methyl derivative to **1** and glycosminine (**6**), respectively, by passage through a column of alumina (acid-washed or neutral), by brief heating with a dilute solution of sodium hydrogen carbonate (pH 9.3), or by treatment with sodium hydroxide (pH 12.5) at room temperature. Some other new routes to **1** originate from *N*-methylisatoic anhydride and phenylacetic acid thioamide (*23*) or from the *N*-methylphenyl-acetanilide **12** by cyclization with ethylurethane in the presence of diphosphorus pentoxide (xylene as solvent) (*24*).

Möhrle and Seidel (*25*) realized the synthesis of **1** by the intramolecular Mannich reaction of *N*-methylanthranilamide with phenylacetaldehyde and dehydrogenation of the resulting tetrahydroquinazoline with mercuric EDTA.

Chakraborty *et al.* (*26*) synthesized **1** by cyclization of the *o*-acyl-aminobenzamide **13** with diphosphorus pentoxide. Kametani *et al.* have developed a one-step synthesis of quinazolinone derivatives by condensation of sulphinamide anhydrides generated from anthranilic acids and thionyl chloride with amides (*27,28*), imines (*29,30*), or thioamides (*31*). This reaction was applied to the synthesis of **1** (*28,31,32*), glycosminine (**6**) (*28,31*), glomerin (**2**) (*27,31*), homoglomerin (*27*), glycorine (**3**) (*27*), chrysogine (**7**) (*27*), and other quinazoline alkaloids (Scheme 1).

The application of this method to the synthesis of arborine from *N*-methylanthranilic acid gives this alkaloid in a yield of 54.5%. The treatment of the

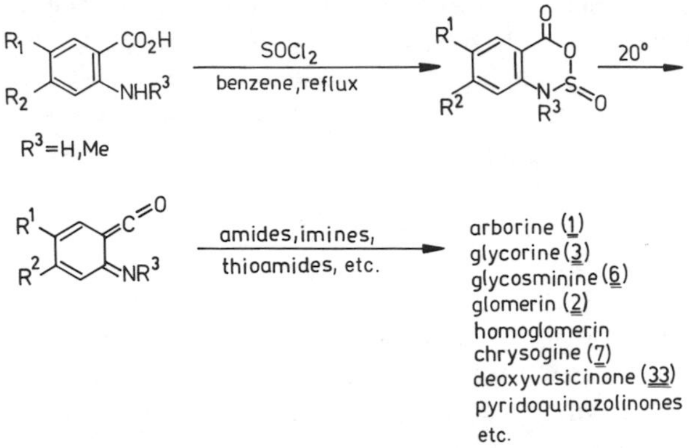

SCHEME 1. Synthesis of quinazoline alkaloids according to Kametani *et al.*

sulfinamide anhydride with an equimolar amount of thiophenylacetamide in dry benzene overnight at room temperature gave **1** in 66% yield (*31*).

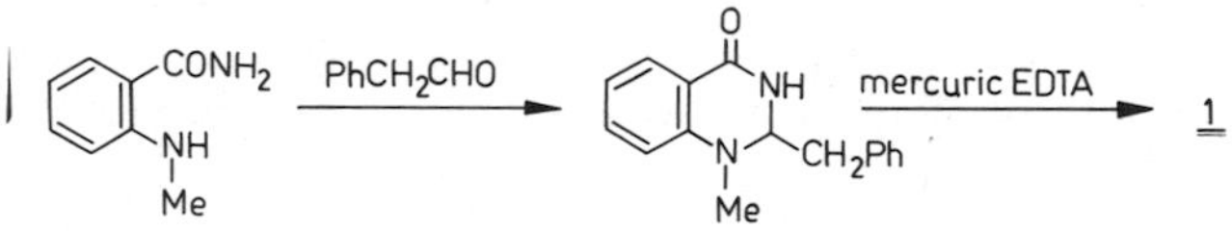

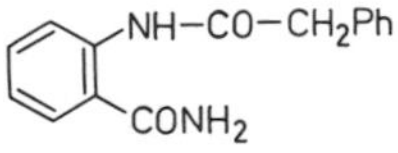

B. FEBRIFUGINE

Febrifugine (**9**) is perhaps one of the most interesting quinazoline alkaloids. Information on this alkaloid has been summarized in the earlier reviews published in this treatise (*1,2*); readers are also referred to the authoritative book by Armarego (*33*). The antimalarial activity of **9** especially provided a strong stimulus for the synthesis and biological screening of a vast number of quinazoline derivatives.

Hill and Edwards (*34*) have determined the absolute configuration of the C-2′ position as *S* by alkaline permanganate oxidation of (−)-*N*-benzoyl-β-furyl-β-alanine (**14**) to L-(+)-*N*-benzoylaspartic acid, using the conditions developed by Reichstein *et al.* (*35*) for the oxidation of the furan ring. Compound **14** is the starting material for the synthesis of the (+) enantiomer of febrifugine according to Baker *et al.*

Barringer *et al.* (*36,37*) have determined the stereochemistry of **9** by investi-

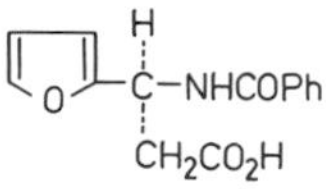

14

15   16

gating the thermal equilibration of cis-(3'-methoxy-2'-piperidyl)-2-propanone (**15**) and some analogous cis-(3'-substituted-2'-piperidyl)-2-propanones with their trans isomers (**16**). Compound **15** is a key intermediate in one synthesis of **9**. The authors have studied the effects of temperature, solvent, pH, and the steric requirement of the C-3 substituent on the equilibrium. Barringer et al. have synthesized cis and trans racemic febrifugines from the isomeric (3'-methoxy-2'-piperidyl)-2-propanones. Although trans-(3'-methoxy-2'-piperidyl)-2-propanone contained from 25–35% of the cis isomer, purification of the febrifugine salts by recrystallization removes the cis isomer and affords the trans compound in high purity. It could be shown that the cis isomer contained a small amount of the trans. The cis isomer underwent a rapid isomerization to the trans compound near its melting point. The 100-MHz NMR spectra of the diacid salts of the acetate esters of the natural product and the synthesized trans-**9** were compared and found to be identical. The resonance of the C-2' proton appears as a quartet centered at $\delta 3.98$ ($J = 7$ Hz) which collapses to a doublet ($J = 7$ Hz) upon irridiation of the side-chain methylene signal. The C-3' proton appears as a multiplet at $\delta 5.0$ ($J = 4-5$, 7, and 9–10 Hz). These couplings indicate a trans-diaxial hydrogen arrangement from which Barringer et al. have concluded that febrifugine was trans rather than cis as Baker et al. had proposed earlier. Attempts to repeat the old synthetic work in order to resolve the difference in stereochemical assignments was beset by difficulties (for details, see the original paper). While there are differences in the chemistry of the intermediates leading to febrifugine, Barringer et al. proposed the structure (2'S,3'R)-3-[3-(3-hydroxy-piperid-2-yl)-acetonyl]-4(3H)-quinazolinone for this alkaloid. The absolute configuration of the hydroxyl group is the same as that in δ-hydroxylysine, which is a plausible biogenetic precursor.

C. CHRYSOGINE

Hikino et al. (38) have isolated the chrysogine metabolite from a culture broth of Penicillium chrysogenum. The (−) isomer exhibits the following properties:

MS: m/z 190 (M$^+$), 175, 173, 161, 147, 145, 130, 119, 102, 90, 76, 63, 45; UV [λ_{max}, nm (log ϵ), ethanol]: 226 (4.34), 230 (4.32), 238 (4.11), 265 (3.86), 273 (3.82), 292 (3.46), 305 (3.59), 316 (3.51); UV [λ_{max}, nm (log ϵ), NaOH]: 226 (4.27), 230.5 (4.25), 238 (4.13), 274 (3.83), 284 (3.86), 312 (3.68), 324 (3.55); IR: 770, 1471, 1610, 1632, 1680, 3050, 3190 cm^{-1}; ^{1}H NMR [(CD$_3$)$_2$SO]: δ1.45 (d, 3H, J = 6 Hz, CH$_3$), 4.60 (1H, C-9 H), 5.59 (d, 1H, J = 6 Hz, OH), 7.35–7.85 (m, 3H, C-6 H, C-7 H and C-8 H), 8.08 (dd, 1H, J = 2,8 Hz, C-5 H). Oxidation of chrysogine with CrO$_3$ gave 2-acetyl-4(3H)-quinazolinone. Acetylation with acetic anhydride led to 2-(1-acetoxymethyl)-4(3H)-quinazolinone. Together, these results indicate the structure 7 [2-(1-hydroxyethyl)-4(3H)-quinazolinone] for chrysogine.

Fermentation of the fungus with anthranilic acid and lactic acid, or with anthranilic acid alone, increased the yield of chrysogine, while with lactic acid alone the yield decreased. Incubation of the fungus with anthranilic acid and analogs of lactic acid, such as propionic acid, glycolic acid, α-hydroxybutyric acid, pyruvic acid, and malic acid, however, gave no congeners but increased the yield of chrysogine. Kametani et al. (27) have reported the synthesis of 7 by reaction of the N-unsubstituted sulphinamide anhydride with O-benzyllactamide followed by debenzylation of the resulting 2-(1-benzyloxymethyl)quinazolin-4(3H)-one (see Scheme 1).

D. GLOMERIN AND HOMOGLOMERIN

Both glomerin and homoglomerin have been isolated as crystalline components from the defensive secretion of the glomerid millipede *Glomeris marginata*. ^{1}H-NMR spectra of both alkaloids show the presence of four aromatic protons, three as a complex multiplet centered at δ7.5 and one as a doublet (J = 9 Hz) of closely spaced doublets at δ8.2. Methyl groups appear as singlets at δ3.75 in both compounds, but glomerin shows an additional methyl singlet at δ2.65 and homoglomerin gives a typical ethyl group pattern at δ2.89 (quartet) and δ1.39 (triplet). Glomerin displays UV absorptions at 215 (log ϵ, 4.33), 228 (4.30), 266 (3.69), 275 (3.69), 304 (3.92), and 312 (3.80) nm. In the IR spectrum, bands at 1640 and 1530 cm^{-1} indicate the presence of C=O and C=N, respectively. From these data and high resolution mass spectra, the compounds were indentified as 1-methyl-2-ethylquinazolin-4-one (homoglomerin) and 1,2-dimethylquinazolin-4-one (glomerin) (2) by Meinwald et al. (39) and Schildknecht et al. (40,41) published approximately at the same time.

Both alkaloids have been synthesized by various methods. Chakravarti et al. (14) have prepared both 2 and homoglomerin from N^2-methylanthranilamide with acetic anhydride and propionic anhydride, respectively. Ziegler et al. (23) have described a one-step synthesis of 2 from N-methyl isatoic anhydride and thioacetamide. Another route to 2 and homoglomerin (Scheme 1) involves condensation of the N-methylsulphinamide anhydride with acetamide or pro-pionamide, respectively (27). The same group has reported a modified synthesis of glomerin in 57.8% yield by condensation of the sulfinamide anhydride with thioacetamide (31).

E. GLYCOPHYMOLINE

The optically inactive alkaloid glycophymoline was isolated from the flower heads of *Glycosmis pentaphylla* in very low yield (42). The mass spectrum has fragments at m/z 249 (M^+ − 1), 235, 119, and 91 indicating a quinazoline system. This is in accordance with the IR spectrum which shows bands at 1208 (OCH_3), 1545, 1610, and 1618 cm^{-1}. The UV spectrum [λ_{max}, nm, 230, 268, 276, 302, and 312 with log ϵ 4.35, 3.85, 3.78, 3.50, and 3.45] is very similar to that of a quinazoline alkaloid. The ^{1}H-NMR spectrum displays a signal for one aromatic OCH_3 group at $\delta3.7$. The signals at $\delta4.23$ ($2H$) and $\delta7.25$ ($5H$) corre-spond to a benzylic methylene group and a monosubstituted phenyl nucleus. The signal at $\delta8.1$ can be attributed to the deshielded aromatic proton in the peri position to a CO group of a quinazolinone system. The signals for the other aromatic protons are observed at $\delta7.5$–7.8. On this basis, glycophymoline has been assigned the structure **5,** which has been confirmed by O-methylation of glycosminine.

F. GLYCORINE

Glycorine was isolated in low yield from the leaves of *Glycosmis arborea* (43,44) and *G. bilocularis* (11). The difficulty of its identification was due to its hygroscopic nature. Glycorine (**3**) contains one N-methyl but no methoxy or active hydrogen. The UV spectrum shows maxima at λ_{max}, nm (log ϵ), at 269 (3.59), 278 (3.66), 306 (3.91), and 317 (3.84) identical to those of 1-methyl-quinazolin-4-one. In 0.01 N HCl, the maxima are shifted to 282 (3.68), 295 (3.73), and 304 (3.69) nm. The ^{1}H-NMR spectrum supports the presence of one N-CH$_3$ ($\delta4.08$); a triplet at $\delta7.73$ can be assigned to the proton which is meta to the proton peri to the carbonyl. The peaks at $\delta7.35$ and 7.49 are assignable to two other protons in the benzene ring. Bands in the IR spectrum at 1538, 1602, 1652, and 1704 cm^{-1} are consistent with the quinazolin-4-one system. Culbertson *et al.* (45) reported that the free base showed the acyl carbonyl band 69 cm^{-1} lower (1635 cm^{-1}) than that of the hydrochloride. This should be attributed to the conjugation of the carbonyl with the N=C function. This large shift has been explained by assuming that the base passes into two zwitterionic forms which are stabilized by salt formation.

The same authors have reported IR spectra of a number of quinazolinones. The ring system shows several characteristic absorptions in the region 1500–1700 cm^{-1}. Two bands at 1605 and 1484 cm^{-1} are typical of a quinazolin-2,4-dione system. Pakrashi *et al.* (*17*) have demonstrated that two strong bands at 1626–1676 cm^{-1} and 1593–1625 cm^{-1} may reasonably be regarded as characteristic of quinazolin-4-ones. Further details on IR spectra of this type can be found in Ref. *14*. ^{13}C-NMR analysis of some simple substituted quinazoline alkaloids and related compounds supported this supposition (*46*). The C-4 chemical shift of **3** hydrochloride (159.2 ppm) is very close to that of the free base. The N-1 CH$_3$ in **3** hydrochloride appears at a relatively low field (39.6 ppm). Structure **17** with significant contribution from **18** more accurately represents **3** hydrochloride.

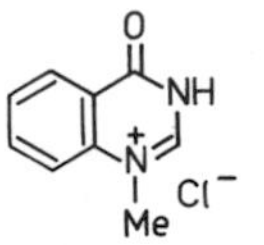

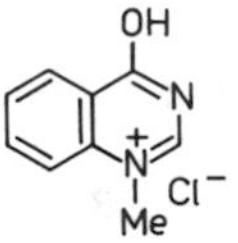

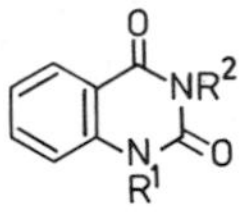

The hydrochloride of **3,** on vacuum sublimation above its melting point, yielded quinazolin-4-one (*15*). Based on its simple structure various syntheses of **3** have been reported. Pakrashi *et al.* (*17*) have published the synthesis from *N*-methylanthranilamide and ethyl orthoformate. Bhattacharyya *et al.* (*24*) have synthesized **3** from *N*-methylformanilide and ethylurethane. Compound **3** also was obtained by Kametani *et al.* (*27*) according to Scheme 1. Finally, **3** is obtained by oxidation of arborine (**1**) with chromic oxide in acetic acid (*15*).

G. GLYCOSMICINE

Glycosmicine (**4**), isolated as a minor alkaloid from *Glycosmis arborea* (*43,44*), contains one active hydrogen. The UV spectrum shows maxima (log ε) at 219 (4.70), 244 (3.97), and 311 (3.64) nm. The maxima remain unaltered in 0.01 *N* HCl, but in 0.01 *N* NaOH, the high intensity band at 244 nm disappears and the others shift to 223 (4.70) and 313 (3.75) nm, as is characteristic for a diketo compound. The IR spectrum shows strong absorptions at 1661 and 1701 cm^{-1} corresponding to two carbonyl functions, one at the 4 and the other at the 2 position, plus the two bands at 1484 and 1605 cm^{-1} typical of a quinazoline-dione system. The ^{1}H-NMR spectrum shows peaks at δ3.58 (*N*-CH$_3$), 8.22 (four

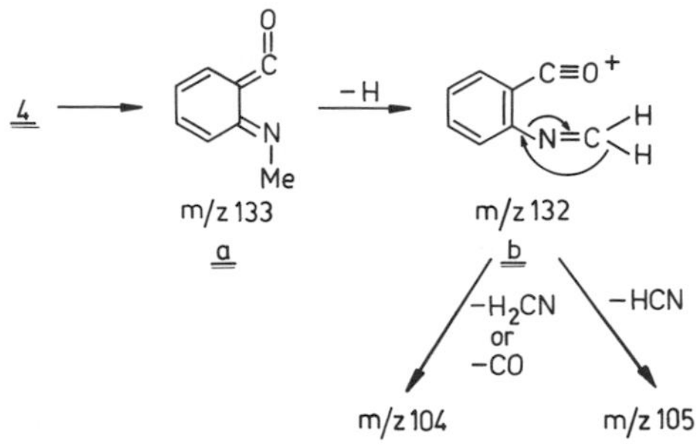

Scheme 2. Pathway of fragmentation of glycosmicine (4).

protons, one of which must be peri to a carbonyl), and 8.59 (*N*-H). The MS and IR characteristics of **4** and similar simple quinazolines have been discussed in detail by Pakrashi *et al.* (*17*). The extreme reluctance of **4** toward salt formation precluded its demethylation to **19**. Methylation of **19** with alkaline methyl iodide yield the 3-methyl derivative **20**, not the 1-methyl isomer **4** (glycosmicine), and then the 1,3-dimethyl derivative (**21**). It is also known that during the condensation of *N*-methylanthranilamide and urea in the preparation of glycosmicine, the methyl group migrates from position 1 to position 3 to give a mixture of **4** and **20** in 1:1.4 ratio.

Pakrashi's interpretation of the mass spectrometric fragmentation of **4** is shown in Scheme 2. The first fragmentation step seems to be the loss of atoms 2 and 3 along with their substituents by concerted cleavage. This transition could be verified by the presence of metastable ions in **4** at m/z 101 (calc. for 170 → 133:100.5). In a second step, further loss of one hydrogen may lead to structure **b**. The metastable ion at m/z 83.5 (calc. for 132 → 105:82.5) seems to indicate that the next fragmentation step starts from ion **b** and involves the loss of 27 units (HCN) and of 28 units (H_2CN or CO).

Glycosmicine was obtained by potassium permanganate, periodic acid (*12*), or chromic acid (*15*) oxidation of arborine (**1**). This fact and the spectral data indicate structure **4** for this alkaloid, which was confirmed by comparison with synthetic material prepared from *N*-methylanthranilamide and ethyl chloroformate.

H. GLYCOSMININE

Glycosminine (**6**) has been isolated from *Glycosmis arborea* (12,43) and *Ruta* species (*48*). The chemistry of this base and its syntheses have close parallels with the other simple substituted quinazolines. The structure was established by Pakrashi *et al.* (*17*). Physical data are as follows: UV [λ_{max}, nm (log ϵ)]: 225 (4.44), 265 (3.95), 303 (3.66), 312 (3.57); IR: 713, 748, 770, 1600, 1613,

1676, 3356, 3440 cm^{-1}; ^{1}H NMR: δ10.25 (N-H), 8.26 (d, protons on a benzene ring peri to the carbonyl), 4.08 (C-CH$_2$ group directly attached to a benzene ring and one other deshielding group), 7.5 (aromatic protons). These data suggest some tautomeric structural possibilities. Bhattacharyya and Pakrashi (46), in a ^{13}C-NMR study based on the C-4 and C-8a chemical shifts of some quinazolin-4-ones, have shown that this compound exists predominantly in the 3H,4 oxo form.

Hagiwara et al. (49) have investigated the intramolecular alkyl rearrangements and tautomerism of simple substituted quinazolinone derivatives. By means of IR, UV, and NMR spectral analysis, it could be shown that the 4(3H)-quinazolinone structure 22 is thermodynamically the most stable tautomer among three possible structures (22–24). Although the C=N double bond is located in the β,γ-nonconjugated position relative to the carbonyl group, the latter is stabilized by amide resonance. The greater stability of tautomer 22 is supported by the fact that the methyl group of, for example, 1-methyl-2-phenyl-4(1H)-quinazolinone rearranges intramolecularly to give 3-methyl-2-phenyl-4(3H)-quinazolinone at elevated temperatures.

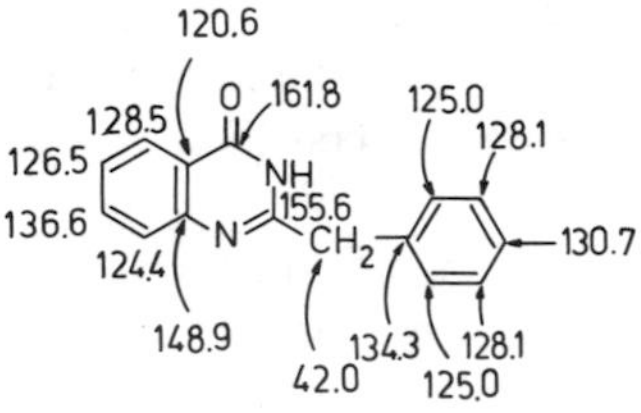

6

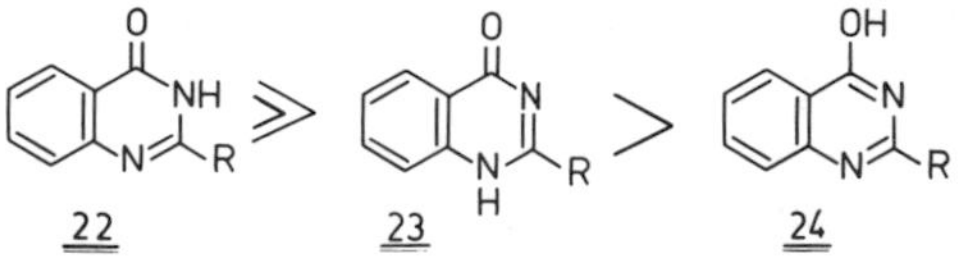

22   23   24

Glycosminine (6) was obtained by demethylation of arborine (1) (15). Methylation with alkaline methyl iodide gave a mixture of arborine (1) and the 3-methyl derivative 11. In contrast, methylation of quinazolin-4-one under the same conditions yielded almost exclusively 3-methylquinazolin-4-one (15). For the synthesis of 6 many methods have been reported to date:

From anthranilamide with phenylacetyl chloride, phenylacetic acid, or phenylacetamide and cyclization of the resulting N-phenylacetylanthranilamide (17,22,50)

By a multistep reaction starting with condensation of anthranilamide and phenylpyruvic acid (51)

By reaction of phenylacetanilide with ethyl carbonate in the presence of AlCl$_3$ to give o-phenylacetamidobenzamide and subsequent cyclodehydration with P$_2$O$_5$ (26)

By condensation of sulfinamide anhydride (prepared from anthranilic acid and thionyl chloride) with phenylacetamide or thiophenylacetamide (28,31) according to Scheme 1

By condensation of anthranilic acid with ethyl phenyliminoacetate (52)

"Glycophymine" isolated from the flowers of *Glycosmis pentaphylla* (53) is identical to glycosminine (54).

I. PEGAMINE

Pegamine was isolated from *Peganum harmala* in 0.005% of the total alkaloid yield by Khashimov *et al.* (55), and its structure was elucidated by the use of extensive spectral data. The UV absorptions [λ_{max} (log ϵ)] at 226 (4.27), 266 (3.74), 306 (3.35), and 318 (3.21) nm and the MS fragmentations [m/z (%)] at 204 (5), 187 (8), 185 (7), 174 (10), 173 (13), 160 (100), and 119 (10) (Scheme 3) were in agreement with the structure 8. The ^{1}H-NMR spectrum displays signals at δ7.5–8.3 for four aromatic protons. The doublet at δ8.15 (J = 7 Hz) is characteristic for the proton situated in the peri position to the carbonyl group. The triplet centered around δ7.47 is attributed to the C-7 proton. Signals for the protons of C-9, C-10, and C-11 appear at δ3.0 (t), 2.14 (m), and 4.2 (t). Acetylation of 8 led to an acetylderivative. Compound 8 may be an artifact formed during the isolation from deoxyvasicinone (33).

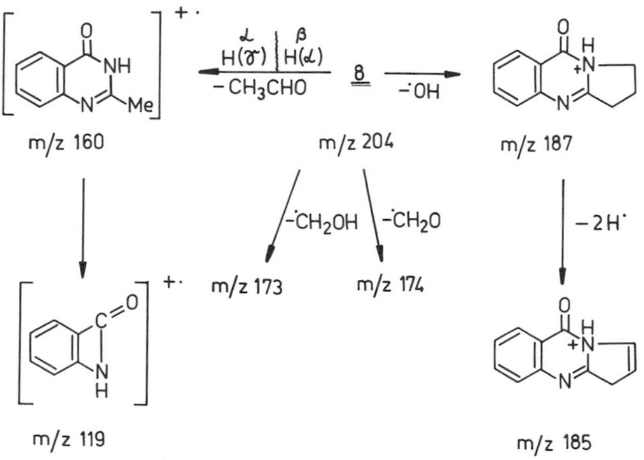

SCHEME 3. Mechanistic pathway of fragmentation of pegamine (8).

J. Quinazolines of *Zanthoxylum arborescens*

In 1980 Dreyer and Brenner (*56*) reported the isolation of two closely related quinazolinone alkaloids from seed husks of the Mexican wild citrus relative *Zanthoxylum arborescens*. Grina *et al.* (*57*) have found these alkaloids in the leaves, bark, and wood of the same species. The IR spectrum of the major alkaloid shows three intense bands in the carbonyl-aromatic region at 1620, 1655, and 1702 cm^{-1}. The ^{1}H-NMR spectrum indicates two two-proton multiplets at δ2.93 and 4.40 (consistent with the presence of a 2-phenylethyl system), an *O*-methyl or *N*-methyl three-proton singlet at δ3.55, and a complex pattern in the aromatic region representing nine protons. A one-proton downfield quartet (*J* = 7.1 Hz) at δ8.12 and a one-proton triplet at δ7.8 (*J* = 7 Hz) suggests the presence of four adjacent aromatic protons on an *o*-substituted benzene ring. The mass spectrum shows M$^+$ at *m/z* 280 and a base peak at *m/z* 176; other fragments are observed at *m/z* 105 and 133, supporting the presence of the *N*-methylanthranilic acid system. The position of the carbonyl resonance in the ^{13}C-NMR spectrum at 161.4 ppm indicates its amide character. Even with the aid of ^{13}C NMR is was not possible to discriminate between the two imaginable structures **25** and **26**. The distinction was made by synthesis of **25** from *N*-methylisatoic anhydride and 2-phenylethylamine in a straightforward fashion (Scheme 4).

$\underline{25}$ R = (CH$_2$)$_2$Ph $\underline{26}$

$\underline{27}$ R = (CH$_2$)$_2$PhOMe–p

The second alkaloid (**27**) is the methoxy analog of **25**. It shows IR absorptions at 1620, 1660, and 1700 cm^{-1}. The ^{1}H-NMR spectrum displays signals for two three-proton singlets at δ3.58 and 3.77, suggesting the presence of both *N*-methyl and methoxy groups. The signals in the aromatic region at δ6.90 and 7.20 can be recognized as the A$_2$B$_2$ system of an *N*-methylanthranilic acid moiety, while the multiplets at δ2.87 and 4.44 derive from the two adjacent methylenes. The structure was confirmed by synthesis (Scheme 4).

K. Quinazolines of *Pseudomonas aeruginosa*

In 1967 Mann (*58,59*) isolated some simple quinazoline derivatives from the purified chloroform extracts of culture material of *Pseudomonas aeruginosa,* other *Pseudomonas* species, *Sarcina* sp., and *Bacillus* sp. The following compounds have been identified by comparison with synthetic material: 4-methyl-

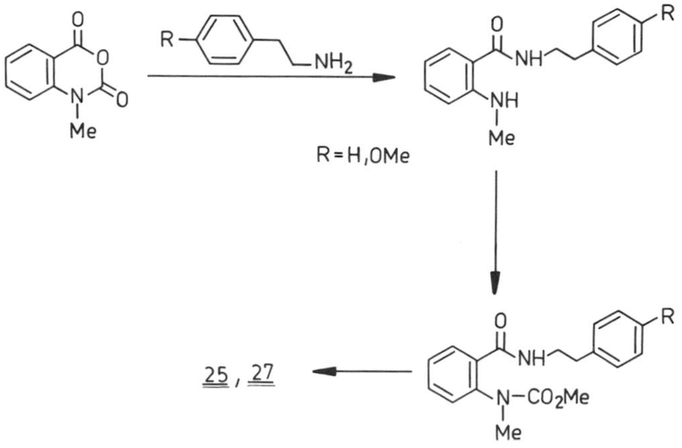

SCHEME 4. Synthetic route to **25** and **27**.

quinazoline, 2,4-dimethylquinazoline, 2-ethyl-4-methylquinazoline, 2-hydroxy-methyl-4-methylquinazoline, 4-methylquinazoline-2-carbonamide, and 2-car-boxy-4-methyl-quinazoline. These quinazolines have been known for a long time, and a number of different methods for their preparation exist (33).

III. The Pyrroloquinazolines

A. VASICINE (PEGANINE)

The chemistry and biochemistry of vasicine (**28**) was reviewed by Openshaw in Volume 3 of this treatise (1) and in other places (3,33,60). This chapter deals with work reported since 1953. The preferred name in English literature is vasicine (derived from *Adhatoda vasica*), but the name peganine (derived from the other source, *Peganum harmala*) is also used. The given numbering of the pyrrolo[2,1-b]quinazoline ring system is currently used by most authors, although other systems (see Refs. 61–64) are still prevalent. The same situation exists for the derivatives: deoxyvasicine is identical with deoxypeganine, etc.

In addition to the already reviewed physical properties, the ^{1}H-NMR spectrum exhibits signals for four aromatic protons at $\delta 6.8$–7.3, a one-proton triplet at $\delta 4.80$ ($J = 7$ Hz), two two-proton multiplets centered at $\delta 2.8$ and 3.5 and assigned to the C-2 and C-1 protons, respectively, and a two-proton singlet at $\delta 4.62$ assigned to the C-9 protons (63,65). In the IR spectrum, bands appear at 1038, 1061, 1079, 1113, 1132, 1158, 1178, 1193, 1233, 1290, 1308, 1338, 1488, 1508, 1577, 1603, and 1637 cm^{-1} (66). The UV spectrum shows a maximum at 293 nm (log ϵ, 3.84). MS fragmentations occur at m/z 188 (M$^+$), 187 (base peak), 169, 159, 131, 104, 89, and 77.

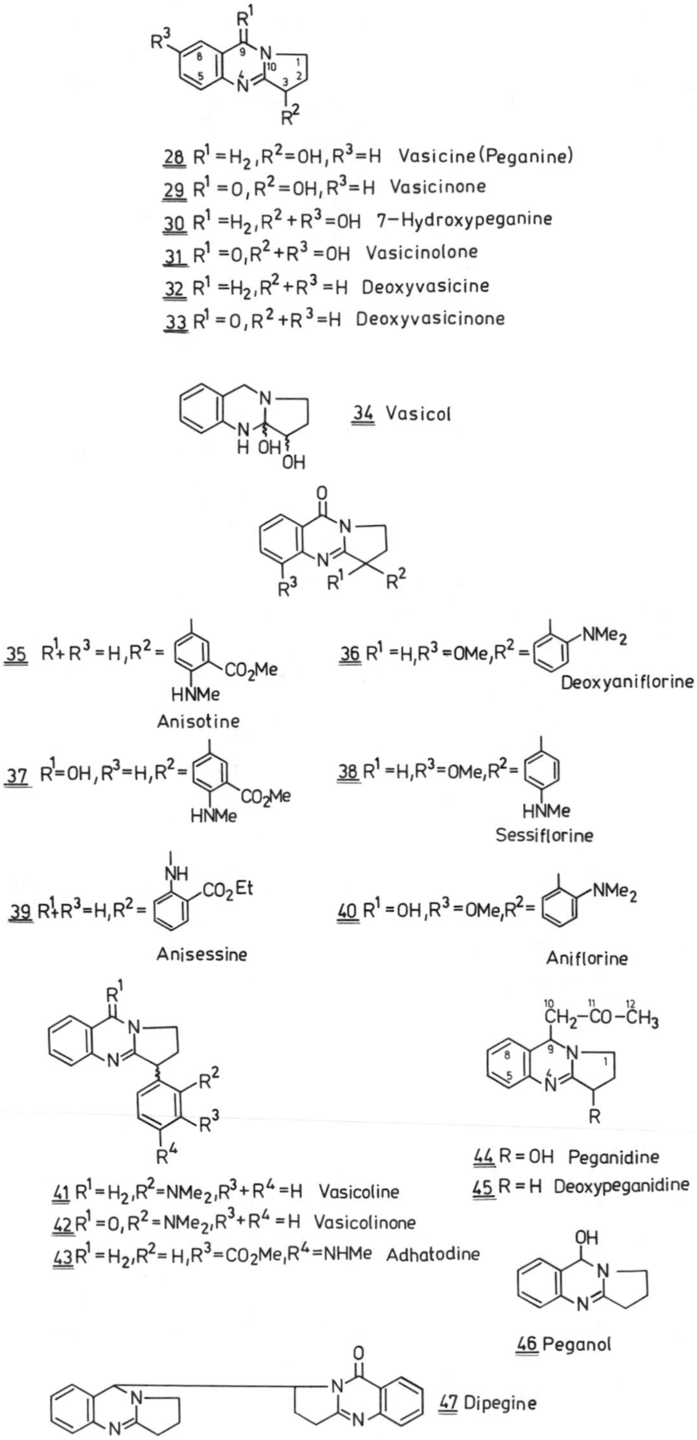

28 $R^1 = H_2, R^2 = OH, R^3 = H$ Vasicine (Peganine)
29 $R^1 = O, R^2 = OH, R^3 = H$ Vasicinone
30 $R^1 = H_2, R^2 + R^3 = OH$ 7-Hydroxypeganine
31 $R^1 = O, R^2 + R^3 = OH$ Vasicinolone
32 $R^1 = H_2, R^2 + R^3 = H$ Deoxyvasicine
33 $R^1 = O, R^2 + R^3 = H$ Deoxyvasicinone

34 Vasicol

35 $R^1 + R^3 = H, R^2 =$

Anisotine

36 $R^1 = H, R^3 = OMe, R^2 =$

Deoxyaniflorine

37 $R = OH, R^3 = H, R^2 =$

38 $R^1 = H, R^3 = OMe, R^2 =$

Sessiflorine

39 $R^1 + R^3 = H, R^2 =$

Anisessine

40 $R^1 = OH, R^3 = OMe, R^2 =$

Aniflorine

41 $R^1 = H_2, R^2 = NMe_2, R^3 + R^4 = H$ Vasicoline
42 $R^1 = O, R^2 = NMe_2, R^3 + R^4 = H$ Vasicolinone
43 $R^1 = H_2, R^2 = H, R^3 = CO_2Me, R^4 = NHMe$ Adhatodine

44 R = OH Peganidine
45 R = H Deoxypeganidine

46 Peganol

47 Dipegine

FIG 2. Structures of pyrroloquinazolines.

The crystal and molecular structures of (−)-**28** hydrochloride have been determined by X-ray crystallography; (−)-**28** possesses the *R* configuration at C-3, and the pyrrolidine ring is in the envelope conformation (*67*). ^{13}C-NMR spectra of **28**, vasicinone (**29**), and related compounds have been reported (*68*), and the mass spectral fragmentations of **28**, vasicinone (**29**), deoxyvasicine (**32**), and deoxyvasicinone (**33**) (Fig. 2) have been discussed (*69,70*). The mass spectral behavior of 2,3-polymethylene-3,4-dihydroquinazolin-4-ones (*71*) and of 2,3-polymethylene-1,2,3,4-tetrahydroquinazolin-4-ones (*72*) have been studied. Integrated intensities of the 1480–1630-cm[−1] IR bands of the skeletal vibrations of the heteroaromatic rings of pyrroloquinazoline alkaloids have been correlated with electron densities and substituent effects (*73*).

The seasonal distribution of **28** in different parts of *Adhatoda vasica* (*74,75*), *Sida cordifolia* (*76*), *Linaria genistifolia* (*77*), *L. vulgaris* (*77*), *Peganum harmala* (*78,79*), and *Galega officinalis* (*80,81*) has been determined. A high-performance liquid chromatographic determination of vasicine and vasicinone has been developed (*82*).

Supplementary to the syntheses of **28** by Späth *et al.* in the 1940s, several other preparations have been recorded. Southwick and Casanova (*83*) prepared **28** in a six-step sequence starting from *o*-nitrobenzylamine (Scheme 5). The intermediate 1-(*o*-aminobenzyl)-3-hydroxy-2-oxopyrrolidone evidently underwent cyclization rapidly and spontaneously to give the 3,4-dihydroquinazoline system. This synthetic scheme was applied readily to the synthesis of 7-hydroxyvasicine. The Schöpf–Oechler scheme (*84*) for the synthesis of **28** from *o*-aminobenzaldehyde and γ-amino-α-hydroxybutyraldehyde under "physiological conditions" has been realized, and **28** was isolated in 39% overall yield (*85*). A facile synthesis of **28** was developed by Möhrle and Gundlach (*86*): condensation of *o*-nitrobenzyl chloride with 3-hydroxypyrrolidine and reduction of the

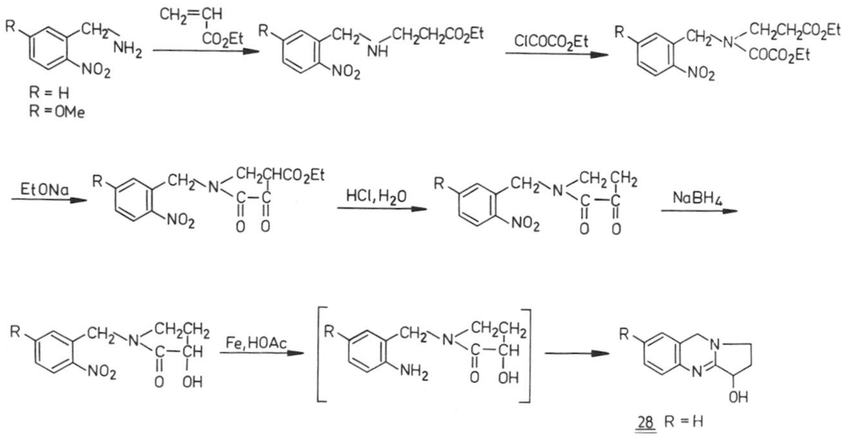

SCHEME 5. Synthesis of vasicine (**28**) according to Southwick and Casanova (*83*).

nitro function followed by treatment with mercuric acetate–EDTA gave **28** and not the other possible cyclization product.

B. VASICINONE

Vasicinone (**29**) was found as a minor constituent in some plant species (see Table III). Physical data are as follows: UV [λ_{max}, nm]: 227, 272, 302, 315; ^{1}H NMR (CF$_3$COOH): δ7.3–7.8 (aromatic protons), 8.05 (C-8 H), 5.56 (C-3 H), 2.64 + 2.21 (C-2 H$_2$), 3.92 + 4.32 (C-1 H$_2$). In each case **29** was accompanied by vasicine (**28**). It remains unclear whether **29** is already present in the plant or formed by autooxidation of vasicine (**28**). Pure **28** could be converted to **29** by autooxidation (*87*) or with 30% hydrogen peroxide (*88*). Vasicinone isolated from the crude total alkaloids was predominantly *l*-**29**. The fact that the crude total alkaloids, when oxidized, yield *l*- and *dl*-forms of **29** indicates that the racemization and oxidation take place simultaneously. A new synthesis of **29** was reported by Onaka (*89*) (Scheme 7). Reaction of anthranilic acid with *O*-methylbutyrolactim formed the pyrrolidinoquinazoline, which, with NBS, afforded the 2-monobromide. Treatment of this compound with AcONa–AcOH gave acetylvasicinone, which was converted to *dl*-**29** in an overall yield of 17%.

C. 7-HYDROXYPEGANINE

7-Hydroxypeganine (**30**) was isolated by Späth and Kesztler-Gandini (*90*) from the leaves of *Adhatoda vasica* and synthesized by Kuffner *et al.* (*91*) using the method Southwick and Casanova (*83*) developed for the synthesis of vasicine and its methoxy derivatives. Rajagopalan *et al.* (*92*) have reported the isolation of "vasicinol" from the roots of the same plant. Bhatnagar *et al.* (*93*) have converted vasicinol to the monomethyl ether [mp 211–212°C, [α]$_D^{20}$ +2° (1.0, 5% acetic acid); λ_{max} (ethanol) 296 nm (log ϵ, 4.03]. This derivative was identical with an authentic sample of 7-methoxypeganine. Vasicinol, therefore, is 7-hydroxypeganine. The appearance of broad but weak bands at 1870 and 1818 cm^{-1} in **30** indicate that N-10 may undergo bonded interaction with the phenolic hydroxyl group (*93*).

D. VASICOL

Vasicol (**34**) occurs in the roots of *Adhatoda vasica* (*64*). The molecular formula and spectral data suggest that this alkaloid differs from vasicine (**28**) by extra H and OH functions. These functions could be best placed at the N-4 and C-3a positions in order to explain the IR, ^{1}H-NMR, and mass spectra. The mass spectrum shows M$^+$ at *m/z* 206, a base peak at *m/z* 106, and other principal fragment ions at *m/z* 162, 161, and 133. The IR spectrum shows prominent bands at 3330–3360, 1603, and 1493 cm^{-1}. The ^{1}H-NMR spectrum displays signals at δ2.1 and 3.16 (multiplets, 2H each) for C-2 H$_2$ and C-1 H$_2$ protons, respectively. D$_2$O exchange simplifies a complex multiplet centered at δ4.2 (6H) into two broad singlets at δ4.33 and 4.36 (1H each) attributed to the C-9 H$_a$ and H$_b$ proton, respectively. The peak at δ4.5 (1H, t, partially obscured by the C-9 H$_2$ signal) is attributed to the C-3 methine proton, and the two multiplets centered at δ6.63 and 7.03 (integrating for four protons) are attributed to aromatic protons. The appearance of the benzylic protons (C-9 H$_2$) as two broad singlets instead of the one observed in vasicine may be attributed to their slightly nonequivalent environment arising from the presence of the C-3a OH.

Passing HCl gas into the dry methanolic solution of vasicol gave vasicine hydrochloride by dehydration of the tertiary OH at C-3a. The reaction of vasicol with POCl$_3$–C$_5$H$_5$N gave chlorodeoxyvasicine possibly by dehydration followed by chlorination. On the basis of these and further results obtained by methylation and acetylation, and in agreement with the spectral data, vasicol may be represented as 2,3,4,9-tetrahydropyrrolo[2,1-b]quinazolin-3,3a(1H)-diol (**34**). This structure was confirmed by synthesis. Heating a mixture of vasicine and water in a sealed tube at 140–150°C for 16 hr gave **34** in a yield of 60%.

E. VASICINOLONE

Jain and Sharma (*94*) have isolated vasicinolone (**31**) from the roots of *Adhatoda vasica*. The compound exhibits absorptions in the UV spectrum λ_{max} at 225, 276, and 325 nm. The IR spectrum shows prominent bands at 1490, 1600, 1624, 1670, and 3400 cm^{-1}. ^{1}H-NMR signals appear at δ2.4 (2H, m, C-2 H$_2$), 4.2 (2H, m, C-1 H$_2$), 5.6 (1H, t, C-3 H), and 7.6 (3H, m, C-5 H, C-6 H, and C-8 H). The mass spectrum shows a molecular ion M$^+$ at m/z 218 (base peak) and important fragment ions [m/z (%)] at 162 (83), 135 (53), 131 (13), 119 (16), 106 (14), and 69 (57). Acetylation gave a diacetyl derivative (mp 204°C) exhibiting ν_{max} at 1220 and 1750 cm^{-1} in the IR spectrum. In the ^{1}H-NMR spectrum, two new signals at δ2.2 (3H, s, C-3 OCOCH$_3$) and 2.42 (3H, s, C-7 OCOCH$_3$) are seen. Vasicinolone has been synthesized from 7-hydroxypeganine (**30**) with H$_2$O$_2$ (30%).

F. DEOXYVASICINE

Deoxyvasicine (**32**) was isolated from the aboveground part of *Peganum harmala* (*95*). ^{1}H-NMR peaks appear at δ3.46 (t, C-1 H$_2$), 2.02 (m, C-2 H$_2$), 2.75 (t, C-3 H$_2$), 4.45 (s, C-9 H$_2$), 6.5–7.0 (m, aromatic protons). Also characterized were the picrate (mp 203–204°C), perchlorate (mp 244–245°C), nitrate (mp 137–138°C, dec.), and hydrochloride (mp 250°C). Siddiqui (*96*) reported the isolation of a crystalline compound, C$_{11}$H$_{12}$N$_2$, from the seeds of *Peganum* *harmala*. This compound was possibly deoxyvasicine, but unequivocal identification has not been accomplished.

The first total syntheses of **32** were performed by Späth and co-workers (*97,98*) and by Hanford and Adams (*99*). Another synthesis was described by Munoz and Madronero (*100*). From a biogenetic point of view, some syntheses under "physiological conditions" are very interesting, for example, (1) from δ¹-pyrroline and *o*-aminobenzaldehyde at pH 5 (*84,101,102*) or (2) from α-keto-δ-aminovaleric acid and *o*-aminobenzaldehyde (*103*). Compound **32** can also be prepared by reduction of deoxyvasicinone (*88,104,137*) or by starting from vasicine (via chlorodeoxyvasicine) (*105*) or from vasicinone (via chlorovasicinone) (*88*). Sattarova *et al.* (*106*) reported a method for the quantitative determination of deoxyvasicine hydrochloride.

G. Deoxyvasicinone

Deoxyvasicinone (**33**) was isolated from *Peganum harmala* (*107–109*), *P. nigellastrum* (*110*), *Mackinlaya macrosciadia* (*111*), and *Linaria* species (*112*). The distribution of **33** in *Peganum harmala* has been determined (*78,79*). Physical data are as follows: UV [λ_{max} (log ϵ)]: 224 (4.2), 267 (3.63), 272 (3.59), 302 (3.4), 314 (3.33) nm; IR: strong bands at 1626 and 1681 cm^{-1} (amide carbonyl and imino moieties); ^{1}H NMR (CF$_3$COOH): δ4.2 (C-1 H$_2$), 2.29 (C-2 H$_2$), 3.43 (C-3 H$_2$), 8.00 (C-8 H), 7.3–7.8 (aromatic protons) (*62,63,113*).

Two early syntheses of **33,** known since 1935, and three further syntheses were performed:

Treatment of γ-phenoxybutyrylchloride with *o*-aminobenzamide (*88*)
From isatoic anhydride and pyrrolidone (*114*)
Starting from anthranilic acid and γ-aminobutyric acid (*113*), with *O*-methylbutyrolactim (*89,115*), or with 2-pyrrolidone (*137*)
From *o*-aminobenzamide and γ-chlorobutyrylchloride (*157*)
Treatment of the sulfinamide anhydride of anthranilic acid with *O*-methylbutyrolactim (*30*) (see Scheme 1)

Synthesis of **33** is also possible by oxidation of deoxyvasicine with hydrogen peroxide (see Refs. *88,* and *104*).

H. Vasicoline, Vasicolinone, and Adhatodine

It is remarkable that the new alkaloids vasicoline, vasicolinone, and adhatodine as well as anisotine (**35**) were found only in young plants of *Adhatoda vasica* in very low yield (*65*). The isolation of pure vasicoline and adhatodine was difficult because both bases are oxidized to vasicolinone and anisotine by atmospheric oxygen. Therefore, structures were assigned mainly on the basis of spectral properties.

1. Vasicoline

The UV absorption maxima of vasicoline (41) are observed at 212 (log ϵ, 4.10), 224 (4.07), 232 (3.94) and 293 (3.81) nm. The IR spectrum shows the presence of a C=N moiety (1621 cm^{-1}). Other prominent absorptions are at 1592 and 1572 cm^{-1}); an amide band is missing. The ^{1}H-NMR spectrum indicates the presence of a dimethylamino function ($\delta 2.67$, s, 6H). The signals assigned to the C-9 methylene protons appear as a singlet at $\delta 4.60$ (2H) overlapping with a triplet at $\delta 4.50$ (1H) arising from the C-3 methine hydrogen. The remaining methylene hydrogens at C-1 and C-2 appear as multiplets at $\delta 3.1$–3.4 (2H), 1.6–2.1 (1H), and 2.25–3.8 (1H). The aromatic hydrogen signals appear as a multiplet at $\delta 6.8$–7.3 (8H). The MS contains a molecular ion at m/z 291 and prominent peaks at m/z 276, 259, 247, 185, 171, 169, 144, 130, and 106. A mechanistic interpretation of the fragmentation pattern of 41 is not yet clear. Reduction with LiAlH$_4$–THF formed the 1,2-dihydro compound.

2. Vasicolinone

IR bands in the spectrum of vasicolinone (42) show the presence of an aromatic ester (1678 cm^{-1}), an aromatic amide (1664 cm^{-1}), and a C=N moiety (1618 cm^{-1}). In the ^{1}H-NMR spectrum the aromatic hydrogens are observed at $\delta 8.31$ (d, $J = 9$ Hz, C-8 H) and at $\delta 7.05$–7.7 (m, 7H). The spectrum also shows the dimethylamino function at $\delta 2.64$ (s, 6H). The C-3 proton appears as a triplet at $\delta 5.04$ (1H, $J = 9$ Hz) which is coupled to the C-2 H$_2$ (m, $\delta 1.9$–2.9, 2H). The two protons attached to C-2 couple to the two hydrogens at C-1 (m, $\delta 4.0$–4.55). Other data are as follows: MS: m/z 305, 290, 273, 261, 185, 144; UV [λ_{max}, nm (log ϵ)]: 227 (4.33), 268 (3.91), 305 (3.52), 317 (3.42); UV [λ_{min}, nm (log ϵ)]: 250 (3.69), 297 (3.33), 313 (3.30). The UV spectrum shows two chromophores isolated from each other. It is very similar to that of an equimolar mixture of N,N-dimethyl-o-toluidine and deoxyvasicinone (33). Consistent with the structure 42 for vasicolinone, the compound gives no acetyl derivative. Kaneko et al. (213) have reported a route to vasicolinone from 2-chloro-4-quinazolinone through an acid-catalyzed cyclization of 2-chloro-3-indolylethylquinazolin-4-one.

3. Adhatodine

The IR spectrum of adhatodine (43) exhibits maxima at 1678 (aromatic methoxy carbonyl) and 3378 cm^{-1} (=NH). UV absorptions [λ_{max}, nm (log ϵ)]

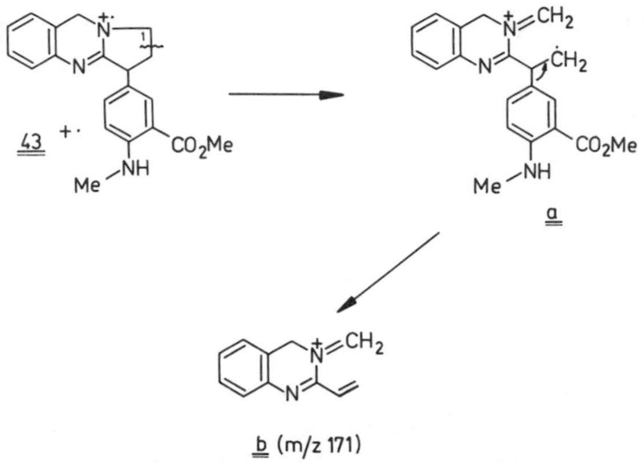

<u>b</u> (m/z 171)

SCHEME 6. Fragmentation pathway of adhatodine (**43**).

appear at 225 (4.51), 262 (4.03), 300 (3.86), and 361 (3.70). Treatment with acetic anhydride gives a monoacetyl derivative. Oxidation of **43** led to anisotine (**35**). The mass spectrum has a base peak at m/z 335 and characteristic ions at m/z 334, 302, 277, 276, 248, 247, 246, 218, 171, 150.5, and 136.5. In the mass spectrum of anisotine (**35**) the most intense fragment peak is m/z 185, compared to m/z 171 for the corresponding deoxo compound, adhatodine (**43**). The structure of this ion is **b** (Scheme 6).

I. ANIFLORINE AND DEOXYANIFLORINE

Aniflorine and deoxyaniflorine along with the alkaloids anisotine (**35**), anisessine (**39**), and sessiflorine (**38**) were obtained as minor, optically inactive alkaloids from *Anisotes sessiliflorus* C.B.Cl. (*63*). The major alkaloid in this plant is vasicine (**28**).

Aniflorine (**40**) shows IR absorptions at 2.9 (hydroxyl group) and at 6.02 μm (amide group) and the absence of an ester carbonyl. The UV absorption pattern is similar to those of anisotine (**35**) and anisessine (**39**), thereby indicating a close relationship to both alkaloids: peaks [λ_{max}, nm (ϵ)] at 207 (31300), 235 (19900), 286 (7000), 312 (6600), and 324 (5800). The ^{1}H-NMR spectrum shows a six-proton singlet at δ2.80 for the dimethylamino group. The aromatic methoxyl gives a singlet at δ3.88. This group is attached to the quinazolinone aromatic ring, as concluded from the mass spectrum. The position of the methoxy group is evident from the C-8 H spin–spin coupling pattern which shows ortho and meta coupling (J = 8 and 1 Hz), as would only be possible if the methoxy is in the 5 position. The signals for the other aromatic protons appear at δ7.05–7.30. The position of the hydroxyl group becomes evident from the absence of the C-3 H signal. The tertiary nature of this group was in accordance with the resistance of

40 to acetylation and CrO_3 oxidation. The para position for the $N(CH_3)_2$ group is ruled out by the absence of a symmetrical A_2B_2 pattern. The mass spectrum shows the transfer of two hydrogens from the N-dimethyl and from the hydroxyl group as shown by deuterium labeling. Such a process would only be possible if the $N(CH_3)_2$ group is situated ortho to the point of attachment of the phenyl ring to C-3. The hydrogenation of aniflorine with Raney nickel in ethanol led to deoxyaniflorine (**36**), which was found to be identical with the naturally occurring alkaloid.

J. ANISESSINE

The mass spectrum of anisessine (**39**) shows the loss of ethanol (m/z 303). Other prominent peaks in the spectrum are m/z 302, 275, 200, and 183. The compound shows typical UV data for a quinazolinone (λ_{max} at 207, 225, 253, 300, 311, and 340 nm). The IR spectrum indicates the presence of an =NH (2.95 μm), and amide and aromatic ester carbonyls (6.05 and 5.95 μm, respectively). The presence of an ethyl ester was evident from the characteristic A_2X_3 ethyl pattern, a triplet centered at δ1.37 ($J = 7$ Hz), and a quartet centered at δ4.38 ($J = 7$ Hz). No N-CH_3 signal was present, but a secondary amine was indicated by the disappearance, on equilibration with deuterium oxide, of a one-proton doublet at δ8.48 ($J = 3$ Hz) and the collapse of a multiplet centered at δ5.83 (C-3 H) to a triplet representing the four lines of the X-proton of an ABX system where $J_{AX} = J_{BX} = 4$ Hz. A total of eight aromatic protons suggested that the phenyl substituent on the deoxyvasicinone skeleton was disubstituted. The analysis of the aromatic proton signals showed the C-8 low-field proton signal at δ8.27 ($J = 4$ Hz). Another proton appeared at low field δ8.00 as a double doublet ($J = 4$ and 1 Hz), typical of an aromatic proton next to an ester grouping coupled to ortho and meta protons. Two aromatic proton signals at higher field δ6.9 (doublet, $J = 4$ Hz) and δ6.73 (triplet, $J = 4$ Hz) indicated the presence of protons ortho and para to the amino group (data cited from Ref. *63*).

Oxidation of anisessine with potassium permanganate gave ethyl anthranilate. Onaka (*89*) prepared **39** from anthranilic acid, ethyl anthranilate, and O-methylbutyrolactim according to Scheme 7.

K. ANISOTINE

Anisotine (**35**) was obtained as a minor, optically inactive substance from *Anisotes sessiliflorus* C.B.Cl. (*63*) and also from *Adhatoda vasica* (*65*). The UV spectrum (λ_{max} at 212, 226, 259, 301, 312, and 356 nm) is very similar to that of 4(3*H*)-quinazolinone. It remained practically unaltered in acid or alkaline solution. The IR spectrum indicates the presence of an =NH, two carbonyl functions, and an aromatic carbomethoxy group.

In the mass spectrum of **35,** loss of methanol from the molecular ion showed that the ester group is situated ortho with respect to the amino substituent on the aromatic ring. This interpretation was supported by the loss of methanol-*d* (M − 33) from the *N*-methyl-d compound obtained by treatment with D_2O. In the NMR spectrum seven aromatic protons indicate the presence of a second aromatic ring. A signal at δ8.27 is characteristic for the C-8 proton situated peri to the amide carbonyl in a 4(3*H*)-quinazolinone moiety. The OCH_3 signal appears at δ3.80. A doublet at δ2.9 (*J* = 5 Hz) ($NHCH_3$) and an amino proton signal between δ7.20–7.90 demonstrate the presence of the indicated groups.

Potassium permanganate oxidation of anisotine afforded **37.** In the NMR spectrum this derivative shows no signal for the C-3 proton in the δ3.9–4.6 region.

L. SESSIFLORINE

The IR spectrum of sessiflorine (**38**) shows an amide carbonyl (6.05 μm). The ^{1}H-NMR spectrum has signals at δ3.94 (aromatic CH_3) and δ2.95 (tertiary methylamino). The double doublet for the C-8 H appears at δ7.89 (*J* = 8 and 1 Hz). The UV maxima [λ_{max} (ε)] occur at 212 (34700), 239 (30300), 286 (9200), 315 (8800), and 327 (6600) nm. The first structure assignment (**49**) was proposed by Arndt *et al.* (*63*). Onaka (*89*) developed a general three-step route for the synthesis of pyrrolidino[2,1-b]quinazolinones (see Scheme 7). The treatment of **48** with diazomethane and NBS followed by reaction with methylaniline

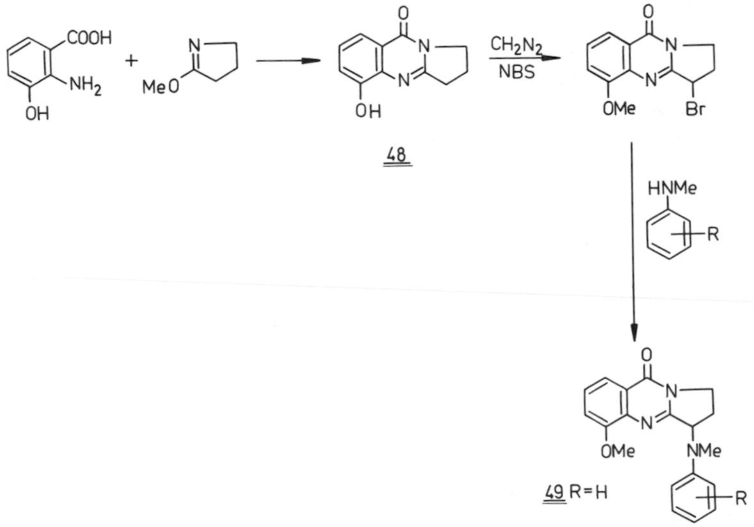

SCHEME 7. Synthesis of pyrrolidino[2,1-*b*]quinazolines according to Onaka (*89*).

yielded **49**. The comparison of this compound with the natural sessiflorine showed significant differences. Synthetic **49** showed no =NH IR absorption and a one-proton triplet at δ5.47 (J = 8.2 Hz) for the C-3 H. The IR spectrum of natural sessiflorine showed the =NH band at 3290 cm^{-1}, and its ^{1}H-NMR spectrum an absorption at δ4.66 (t, 1H, J = 6 Hz). The presence of an A_2 part of A_2B_2 pattern at δ6.7–7.0 (2H) and the deshielded nature of the N-methyl signal (δ2.95) favor structure **38** for sessiflorine.

M. PEGANIDINE, ISOPEGANIDINE, DEOXYPEGANIDINE, DIPEGINE, AND PEGANOL

Close examination of *Peganum harmala*, especially by Soviet scientists, resulted in the discovery and spectral detection of some new quinazoline alkaloids in low yield from different plant parts. Peganidine (**44**) (*116*) shows UV absorptions [λ$_{max}$ (log ε)] of 226 (4.04) and 297 (3.96) nm and IR bands at 1350, 1700, and 2870 cm^{-1}. The ions in the mass spectrum are m/z 244, 187, 169, 159, 131, 104, and 77, similar to vasicine. Compound **44** yielded an oxime (mp 85–87°C) and a semicarbazone (mp 204–206°C). Deoxypeganidine (**45**) (*117*), upon potassium permanganate oxidation, yielded vasicinone. Dipegine (**47**) (*118*) is the first dimeric quinazoline alkaloid. Physical data are as follows: UV (λ$_{max}$, nm (log ε)]: 226 (4.48), 277 (4.14), 305 (4.05), 317 (3.89); IR: 1590, 1620, 1660 cm^{-1}; MS: m/z 356 (M$^+$), 185 (3%), 171 (100%). Isopeganidine was isolated and described as a "racemic diastereomer of peganidine" (*118*). It is possible that both peganidine and dipegine are artifacts formed during the isolation procedure. Peganol (**46**) (*119*) shows only one absorption maximum in UV at 275 nm (log ε, 3.96). Table I lists the ^{1}H-NMR values of these alkaloids. Further details can be found in the original papers and in Telezhenetskaya and Yunusov (*7*).

IV. The Pyrido[2,1-*b*]quinazolines

Leaves and stems of the two *Mackinlaya* species, *M. macrosciadia* (F. Muell.) F. Muell. from Queensland and a New Guinea species *M. subulata* Philipson [now revised to *M.* cf. *klossii* Philipson (*111*)], contain 6,7,8,9-tetrahydro-11*H*-pyrido[2,1-*b*]quinazoline (**50**) and 6,7,8,9-tetrahydropyrido[2,1-*b*]quinazolin-11-one (**51**) (*120,121*). Both alkaloids were identified by a combination of physical and chemical methods, and their structures were confirmed by direct comparison with synthetic samples.

The mass spectrum of **50** shows peaks at m/z 186 (M$^+$, 85%), 185 (base peak), 144 (13%), 89 (13%), 77 (25%), and 51 (29%). In the ^{1}H-NMR spectrum, the C-1 proton is similar in chemical shift to the C-2, C-3, and C-4

TABLE I

NMR Spectra of Some Quinazoline Alkaloids[a]

Alkaloid	Aromatic protons	C-1	C-2	C-3	C-9	C-10	C-12
Peganidine (44)	6.6–7.1 (m)	3.73 (m) 3.35 (m)	2.02 (m) 2.38 (m)	5.02 (t)	5.02 (t)	3.07 (d)	1.86 (s)
Isopeganidine	6.6–7.0 (m)	3.47 (m)	2.05 (m) 2.45 (m)	5.05 (m)	5.05 (m)	2.95 (d)	1.82 (s)
Deoxypeganidine (45)	6.9–7.2 (m)	3.42 (m)	2.00 (m) 2.60 (m)	3.42 (m)	5.14 (t)	2.76 (m)	1.95 (s)
Peganol (46)	7.5–8.0	4.71 (t)	2.40 (m)	3.40 (t)			

[a] Figures are δ values in ppm, in CF_3COOH, except in the case of deoxypeganidine ($CDCl_3$).

protons. The signals appear at δ6.67–7.13 (multiplet). The C-11 methylene protons give a sharp singlet at δ4.33. The multiplet at δ2.80–3.17 can be attributed to the C-9 protons. For **50** some syntheses have been recorded (see Ref. *100*). The catalytic hydrogenation of **50** or reduction with sodium borohydride in ethanol gave a quantitative yield of the 5,5a-dihydro compound.

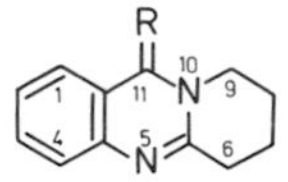

50 R = H$_2$

51 R = O

The mass spectrum of **51** shows peaks at m/z 200 (M$^+$, base peak), 199 (85%), 185 (54%), 130 (13%), 116 (21%), 90 (27%), 89 (21%), 77 (25%), and 76 (28%). In the ^{1}H-NMR spectrum, the low-field multiplet at δ8.12–8.33 has been assigned to the C-1 proton which shows the expected ortho (J = 8 Hz) and meta (J = 2 Hz) couplings. The paramagnetic shift of 0.7 ppm relative to the C-2, C-3, and C-4 protons is due to the anisotropy of the lactam carbonyl group at C-11. The broad two-proton triplet at δ3.90–4.20 has been assigned to the C-9 methylene protons. The second broad triplet at δ2.80–3.15 (2H) can be assigned to the C-6 methylene protons. Compound **51** can be prepared by the method of Stephen and Stephen (*122*, see Ref. *123*). Kametani *et al.* (*28*) have developed a convenient synthesis of **51** from a sulfinamide anhydride and O-methylpiperidone in 82.4% yield according to Scheme 1. Reduction of **51** with sodium borohydride afforded the 5,5a-dihydro compound, which on acetylation gave the N-acetyl compound.

V. The Indoloquinazolines

A. CANDIDINE

Candidine was isolated from culture solutions of *Candida lipolytica* (*124*) and identified as 6,12-dihydro-6-(3-oxoindolid-2-ene)-12-oxoindolo[2,1-b]quinazoline (**52**) (*125*). Physical data are as follows: MS: m/z 364 (M + 1, 23%), 363 M$^+$, 100%), 335 (37%), 259 (10%), 258 (11%), 242 (9%); UV–Vis [λ$_{max}$, nm (log ε)]: 244 (4.34), 250 (4.36), 282 (4.21), 538 (3.95), 573 (4.11); IR (KBr): 3210, 2730, 1690, 1650, 1625, 1600, 1460, 1355, 1315, 740 cm^{-1}.

Compound **52** was synthesized by condensation of tryptanthrin (**53**) with N-acetylindoxyl followed by mild hydrolysis. Candidine was briefly described in 1922 by Martinet and Grosjean as a condensation product of indoxyl and tryptanthrin.

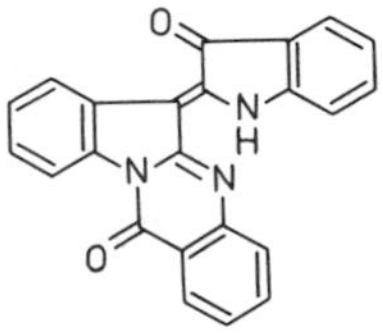

52

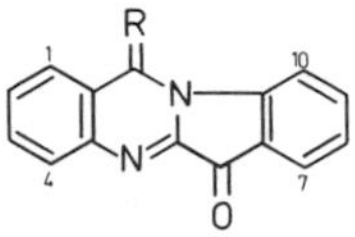

53 R = O
55 R = OH,H

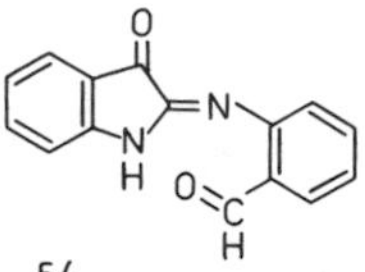

54

B. TRYPTANTHRIN

Tryptanthrin (**53**) is an antibiotic produced by the yeast *Candida lipolytica* only in the presence of large amounts of L-tryptophan (*131*). It was also isolated from the fruit of the cannonball tree *Couroupita guianensis* Aubl. (*126*). Honda *et al.* identified tryptanthrin as the active principle in the leaves of *Strobilanthes cusia* O. Kuntze (*127*) and as a component of *Polygonum tinctorium* and *Isatis tinctoria* (*129*). It is a lipophilic, neutral, yellow compound. The mass spectrum shows intense peaks at *m/z* 248, 220, 192, 102, and 46. The IR bands at 1688 and 1725 cm^{-1} are characteristic for carbonyl groups. The aromatic character of tryptanthrin is also indicated by the UV spectrum. For details readers are referred to the original literature (*130,131*). The 90- and 100-MHz ^{1}H-NMR spectra were not very informative. Jarrah and Thaller (*132*) have compared the 300-MHz ^{1}H-NMR spectrum of **53** with those of synthetic tryptanthrin derivatives. At this frequency, spin–spin decoupling experiments made it possible to assign the signals to the individual protons (Table II). X-ray analysis (*130,133*) allows assignment of indolo-[2,1-*b*]quinazolin-6,12-dione as the structure for tryptanthrin with an approximately planar arrangement of nearly all atoms.

This compound has been known since 1892 and was previously synthesized in 1915 by oxidation of isatin with neutral potassium permanganate (*134*). Bird (*135*) has synthesized **53** from O-methylisatin and o-aminobenzaldehyde. The isatin-α-(2-formylanil) **54** (appears to exist predominantly as **55**) was converted to **53** by oxidation with CrO$_3$. The reaction of *N*-sodioisatin with o-nitrobenzoyl chloride gave 1-(2′-nitrobenzoyl)isatin, and its reduction afforded **53** (*136*). Bergman *et al.* (*126,153*) have developed a simple method: heating of isatin, isatoic anhydride, and diisopropyl carbodiimide in pyridine with *N*-methylpiperidine as catalyst afforded **53** in high yield. For information on other

TABLE II

Chemical Shifts (τ) and Coupling Constants (J Hz) for Protons
on C-1–C-4 and C-7–C-10 of Tryptanthrin

C-1	C-2	C-3	C-4	C-7	C-8	C-9	C-10
1.52	2.28	2.10	1.92	2.04	2.53	2.17	1.33
ddd	ddd	ddd	ddd	ddd	ddd	ddd	dm
8, 1.5,	8, 7.5,	8, 7.5,	8, 1,	7.5, 1.5,	7.5, 7.5,	8, 7.5,	8
0.5	1	1.5	0.5	0.5	1	1.5	

syntheses as well as cultivation methods for **53** and various halogen- and NO$_2$-substituted analogs, see Refs. *124, 132, 147, 154,* and *155.*

VI. Biosynthesis

In practically all theories of the biosynthesis of quinazolines, anthranilic acid or an equivalent is regarded as one of the building blocks. Robinson (*158*) postulated that anthranilic acid (N-methylated where necessary), ammonia or amine, and formic acid or phenylacetic acid derived from phenylalanine units or their equivalents build up the simple substituted quinazoline derivatives. Thus vasicine (**28**) may be derived from anthranilic acid and proline, or closely related metabolites. Febrifugine (**9**) possibly could be formed from anthranilic acid, formic acid, ammonia, a C$_3$-unit, and lysine or equivalent compounds. For 6,7,8,9-tetrahydropyrido[2,1-*b*]quinazolin-11-one (**51**) and related alkaloids, it has been postulated that their biosynthesis starts from anthranilic acid and lysine (*121*). Pakrashi and Bhattacharyya (*19*) suggested that, at least in the case of 4(3*H*)-quinazolinones, N-methylation of anthranilic acid precedes its condensation with other compounds.

Schöpf and Oechler (*84*) postulated that vasicine (**28**) is formed in the plant from *o*-aminobenzaldehyde and α-hydroxy-γ-aminobutyraldehyde. The feasibility of this scheme was confirmed by the successful synthesis of deoxyvasicine (**32**) from *o*-aminobenzaldehyde and γ-aminobutyraldehyde in 18% yield (*84*) and by the synthesis of vasicine (**28**) under very mild physiological conditions in a yield of 39% using α-hydroxy-γ-aminobutyraldehyde (*85*). According to Skurský (*102*), the conversion of the quaternary base formed as an intermediate in this reaction to deoxyvasicine and deoxyvasicinone is also catalyzed by pea homogenates.

Modifications of Schöpf's synthesis of deoxyvasicine have been described by Macholán (*103*) and Skurský (*101*). Instead of γ-aminobutyraldehyde Macholán

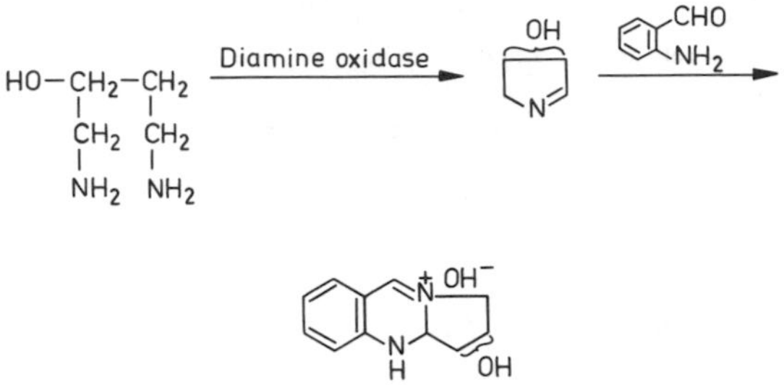

SCHEME 8. Conversion of 2-hydroxyputrescine to a pyrrolidino[2,1-b]quinazoline derivative.

used α-keto-δ-aminovaleric acid and obtained via several intermediates pegane-3a-carboxylic acid which, upon oxidation, is easily converted to deoxyvasicine (**32**). In the plant the initially formed deoxyvasicine could be hydroxylated to vasicine (**28**), or, alternatively, α-keto-β-oxy-δ-aminovaleric acid could react with *o*-aminobenzaldehyde to give vasicine (**103**). Simple quinazolines and quinazolinones have been obtained by Macholán (*103*) by reaction of *o*-aminobenzaldehyde with methylamine and glyoxylic acid. The significance of this reaction for the elaboration of febrifugine (**9**) has been discussed. Skurský (*101*) has reacted Δ¹-pyrroline with *o*-aminobenzaldehyde to give deoxyvasicine (**32**).

Macholán (*159*) has incubated 2-hydroxyputrescine in the presence of *o*-aminobenzaldehyde in phosphate buffer at pH 6.5–7 with purified diamine oxidase and obtained a 1,2-dihydroquinazolinium compound (Scheme 8). Only one product was formed, although, since there are two amino groups in the substrate molecule, two reaction products would have been expected. Apparently, however, there is a preferential formation of the 4-hydroxy-Δ¹-pyrroline (*160*), which is converted spontaneously into pyrrole if no *o*-aminobenzaldehyde is present.

The biosynthesis of quinazoline alkaloids using labeled compounds was intensively studied in the 1960s, and it is interesting to note that no experimental work has been reported, to our knowledge, since the last investigations in 1973 (*161*). This topic has been thoroughly reviewed (*162–164*), and for a more detailed description readers are referred to the cited reviews.

In 1960 it was demonstrated that aseptically grown plants of *Peganum harmala* are capable of incorporating [*carboxy*-^{14}C]anthranilic acid into vasicine (**28**). With *Adhatoda vasica* plants it was shown that the incorporation of anthranilic acid involves retention of the carboxyl group (*165*). The origin of the "non-anthranilic acid moiety" is still controversial. It was found that in *Adhatoda vasica* C-1, C-2, and N-10 of (**28**) may be derived from aspartic acid and C-3 and C-3a from a C_2 unit (*161,166*) (Scheme 9). This was supported experimentally by the incorporation of [3-^{14}C]aspartic acid, [3-^{14}C]malic acid, and [2,3-^{14}C] succinic acid into **28** with localization of the label at C-1 and C-2. The

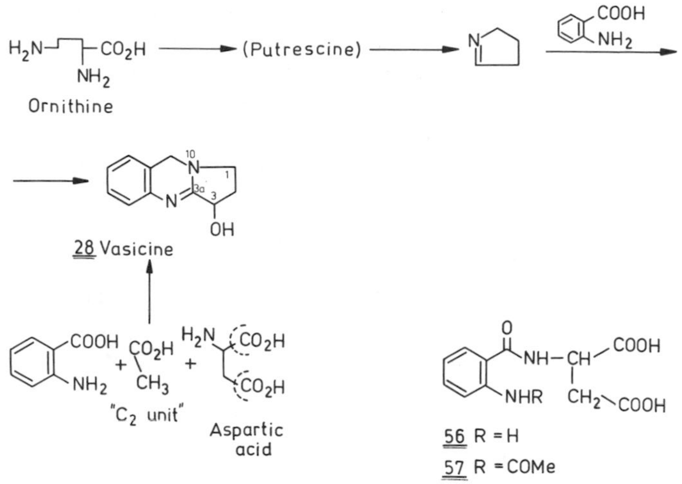

SCHEME 9. Possible biosynthetic pathways of vasicine (**28**).

incorporation of succinic acid could proceed via aspartic acid since succinic acid can easily be converted to oxalacetic acid in the citric acid cycle. The specific incorporation of [2'-^{14}C,^{15}N]N-acetylanthranilic acid and especially of anthranoyl[3-^{14}C]aspartic acid (**56**) is in accordance with the other results. The most important precursor of **28** may possibly be acetylanthranoyl aspartic acid (**57**). [^{15}N]Anthranilic acid and [carbonyl-^{15}N]glutamine are converted into **28** with a high specific rate of incorporation. These compounds apparently provide the N-4 of **28**. Members of the α-ketoglutaric acid family were incorporated only nonspecifically. N-Formylanthranilic acid and N-methylanthranilic acid were not incorporated. Likewise it has been shown that no direct hydroxylation of deoxyvasicine (**32**) had taken place and that vasicinone (**29**) was poorly incorporated in **28**.

It has been found that in *Peganum harmala* [2-^{14}C]ornithine and labeled proline, putrescine, and related compounds are more or less specifically incorporated into the pyrrolidino ring system of **28** (*167*). This result suggested that the ornithine was being decarboxylated to putrescine, a symmetrical molecule which would be incorporated with equal labeling at C-1 and C-3a. The results can be rationalized by postulating that a symmetrical intermediate such as putrescine is involved in the pathway from ornithine to **28** or that [2-^{14}C]ornithine serves as a source of [1-^{14}C]acetate (via α-ketoglutaric, succinic, fumaric, or malic acid) which is then incorporated by way of N-acetylanthranilic acid. The investigations do not indicate which of the alkaloids is formed first as a result of the condensation to the pyrroloquinazoline system. Any interconversions between the alkaloids in *Peganum harmala* have been masked by their rapid metabolism.

Pegamine (**8**) may be a potential intermediate for deoxyvasicinone (**33**). Adhatodine (**43**), vasicoline (**41**), vasicolinone (**42**), anisotine (**35**), and sessiflorine (**38**) may arise from pyrroloquinazolines and derivatives of anthranilic acid.

Arborine (**1**) is produced from phenylalanine and anthranilic acid (*168*). After administration of [U-^{14}C]phenylalanine, 92% of the radioactivity is located in the phenylacetic acid part of **1**. After feeding [*Me*-^{14}C]methionine, the *N*-methyl group of **1** was specifically labeled. The same result was obtained after the application of [*Me*-^{14}]*N*-methylanthranilic acid. These results seem to indicate that anthranilic acid is methylated prior to the condensation with phenylalanine which leads to **1**. The N-1 atom of **1** is provided by the nitrogen of anthranilic acid. The experiment with [3-^{14}C,^{15}N]phenylalanine shows clearly that with the exception of the carboxyl group the aliphatic side chain of phenylalanine is incorporated into arborine. *N*-Methyl-*N*-phenyl[^{14}COCH$_2$] acetylanthranilamide was transformed very efficiently in *Glycosmis* into arborine. However, it is doubtful whether the *N*-phenylacetyl derivative of *N*-methylanthranilamide is a true precursor in arborine biosynthesis. Surprisingly, it was shown that the transformation of *N*-methyl-*N*-phenylacetylanthranilamide to **1** also takes place in pea plants. Apparently the ring closure of *N*-methyl-*N*-phenylacetylanthranilamide is catalyzed by an enzyme or enzymatic system that is unspecific and widely distributed (*169*). In contrast to these results, another laboratory has postulated that the incorporation of phenylalanine proceeds via phenylacetic acid and *N*-methyl-*N*-phenylacetylanthranilic acid (*170*).

Sakar and Chakraborty (*42*) postulated that the isolation of glycophymoline (**5**), glycosminine (**6**), arborine (**1**), and glycomide (phenylacetanilide) provides circumstantial evidence for the formation of arborine (**1**) and glycophymoline (**5**) from the precursor of glycomide via glycosminine (**6**). Arborine may be considered to be the N-methylated product of glycosminine and glycophymoline, the enol ether of glycosminine.

It has been demonstrated that [carboxy-^{14}C]anthranilic acid is also a precursor of the animal alkaloids glomerin (**2**) and homoglomerin (*171*).

The simple quinazoline derivatives produced by *Pseudomonas* are formed in the course of tryptophan degradation ("quinazoline pathway"). Tryptophan is first converted to formylkynurenine and then to *N*-formylaminoacetophenone, which forms 4-methylquinazoline with ammonia or loses the formyl group to give 2-aminoacetophenone. After reacylation and cyclization with ammonia this latter product yields the other derivatives of 4-methylquinazoline (*58,59*).

The antibiotic tryptanthrin (**53**) has been biosynthesized from 1 mol of tryptophan and 1 mol of anthranilic acid. Upon feeding tryptophan and substituted anthranilic acids, or substituted tryptophans and anthranilic acid, the expected derivatives of **53** were isolated. The enzymes involved in the biosynthesis of **53** had no specificity for these substrates, with the exception of bromotryptophan. The anthranilic acid moiety used during biosynthesis results from tryptophan degradation (*131,172*).

From the preceding remarks it should be obvious that remarkable progress has been achieved in our understanding of the main pathways of quinazoline alkaloid formation although a number of problems concerning the detailed mechanisms at the enzymatic level remain, and several of the reaction mechanisms are speculative. Little progress has been made on the isolation of enzymes responsible for the formation of this type of alkaloid; many problems remain to be solved.

VII. Biological Activity of Quinazoline Alkaloids and Their Analogs

The fact that several quinazoline alkaloids and the plants containing such types of alkaloids show a variety of biological activities has spurred the preparation and pharmacological and biological evaluation of a great number of quinazoline derivatives. Intensive research in the quinazoline field is still in progress. This topic has been reviewed by Amin *et al.* (*173*), Gupta *et al.* (*174*), Johne (*175,176*), Süsse and Johne (*177*), and by Yakhontov *et al.* (*178*).

Some plant species containing quinazoline alkaloids have long been used in the indigenous medicine of some countries. *Adhatoda vasica* Nees (Sanskrit: *Vasaka;* family Acanthaceae) has been used in India as an expectorant and a mild bronchial antispasmodic. This plant also shows benefical effects in diarrhea and dysentery and possesses antiseptic, antiperiodic, and anthelmintic properties (*179*). Vasicine (**28**) has been reported to be a bronchodilator, a respiratory stimulant, and hypotensive in action (*180*). Its activity against histamine-induced bronchospasm is specific but brief. This alkaloid has come into prominence as an uterine stimulant and uterotonic abortifacient (*181*). Some discrepancies in the results of the pharmacological tests of vasicine and related alkaloids and in their interpretations need further clarification. The text by Atal (*182*) is a leading reference on the pharmacology of vasicine; for further details see also Ref. *183*. Zutschi *et al.* (*184*) investigated the absorption and distribution of ^{14}C-labeled vasicine in mice.

Vasicinone (**29**), the autooxidation product of vasicine, produces definite bronchodilatation against histamine-induced bronchospasm, slight hypotension, and a positive inotropic action with increased coronary flow (*152,156,183,185*). Prakash *et al.* (*140*) postulated that the combination of sympathomimetic amines and vasicinone in some *Sida* species would account for their major therapeutic uses in the Indian system of medicine, for example, in the treatment of asthma and other chest ailments. Vasicinone has been used as a model for the development of the expectorant 2-amino-3,5-dibromo-*N*-cyclohexyl-*N*-methylbenzene-methanamine.

7-Hydroxypeganine (**30**) produces transient hypotension in cats, contraction of isolated guinea pig intestine, and depression of isolated guinea pig heart, all of which can be blocked by atropine. It also produces slight contraction of isolated guinea pig tracheal chain, but antagonizes the histamine-induced bronchospasm in guinea pigs *in vivo*. It shows mild anticholinesterase activity (*183*).

Deoxyvasicine (**32**) and deoxyvasicinone (**33**) show cholinergic activity (*186*). Plugar *et al*. have investigated the metabolism of deoxyvasicine (**32**), deoxyvasicinone (**33**), and vasicinone (**29**) in animals (*186,187*). The authors found in rats a transformation of deoxyvasicine to deoxyvasicinone and vasicinone. Deoxyvasicinone was metabolized to an isovasicinone with an α-OH group at C-1 and 2-(2′-ethoxycarbonyl)-ethylquinazolin-4-one.

Although the claims made for the biological activity of vasicine and vasicinone are controversial, some analogs of both alkaloids have been synthesized. 6,7-Methylenedioxyvasicinone possesses uterine stimulant and bronchodilator activities comparable to those of vasicine, but with more marked hypotensive and cardiac depressant activities (*188*). A number of analogs of 7-hydroxyvasicinone, in which ring C ranges from five- to nine-membered, were synthesized (*189*). Except for a decrease in spontaneous locomotor activity in mice, none of these compounds possess any significant pharmacological activity.

Several authors have reported that deoxyvasicinone (**33**) and related fused quinazolinones undergo reaction at the C-3 position in the pyrrolidine ring (see Ref. *190*). Dunn *et al*. (*190,191*) have investigated reactions of **33** with chloroformate esters, benzoyl chloride, benzoic anhydride, and other electrophiles.

Gupta *et al*. (*192*) have synthesized vasicine analogs by condensation of deoxyvasicine (**32**) with aromatic aldehydes. Doria *et al*. (*193*) have found that some 3-benzylidene-1,2,3,9-tetrahydro-9-oxopyrrolo[2,1-b]quinazolinecarboxylic acids and 6-benzylidene-6,7,8,9-tetrahydro-11-oxo-11*H*-pyrido[2,1-*b*]quinazolinecarboxylic acids showed antiulcer and other pharmacologically interesting activities.

Soviet workers have condensed deoxyvasicinone (**33**) with a number of aromatic aldehydes (*194*). The same group synthesized 7-nitro-, amino-, and 3,3-dibromodeoxyvasicinone (*195*) and methylenebis(7,7′-deoxyvasicinone) and its homologs by cyclocondensation of 5,5′-methylenedianthranilic acid with the corresponding lactams (*196*). Sharma and Jain (*197*) reported on the condensation of deoxyvasicinone (**33**) with aromatic aldehydes. 9-Thio analogs of deoxyvasicinone, its derivatives and homologs were obtained by boiling of the corresponding oxygen compounds with P_2S_5 in *m*-xylene. The desulfurated derivatives were obtained by heating with Zn in 10% HCl (*198*).

Oripov *et al*. (*199*) reported on some reactions of 3-hydroxy- and 3-dimethylaminoformylidenepyrrolo[2,1-*b*]quinazoline derivatives. Karimov *et al*. (*200,201*) reported on the synthesis of methoxy- and oxy-substituted deoxyvasicinones and deoxyvasicines. Some compounds derived from deoxyvasicinone showed fungicidal activity (*202*). The preparation of a specific antibody to deoxyvasicinone (**33**) for the immunoassay of Δ^1-pyrroline has been described by Sakamoto and Samejima (*203*).

Febrifugine (9) was the first alkaloid outside the *Cinchona* group known to possess marked antimalarial activity. It is about 100 times as active as quinine against *Plasmodium lophurae* in ducks and against the trophozoites of *Pl. cynomolgi* in monkeys, and it is also very active against *Pl. gallinaceum* in chicks. The high degree of toxicity renders it unsuitable for clinical use in malaria (*18*). Febrifugine is also known for its antipyretic and emetic activities and is a very effective coccidiostatic. It also shows anti-moth activity (*204*).

The side-chain modification of febrifugine (see references in Ref. *205*) and the synthesis of pyridine analogs of this alkaloid (*206*) were found to be unfruitful. Structural modification of febrifugine was achieved through the synthesis of methylenedioxy analogs by Chien and Cheng (*205*). The 5,6-, 6,7-, and 7,8-methylenedioxy analogs were found to be active against *Pl. berghei*. Their toxicities in mice are much lower than that of febrifugine. The therapeutic indices of these compounds are comparable with those of the parent compound. 6-Chloro-7-bromofebrifugine is prophylactically active against *Eimeria tenella* when added to fodder in an amount of 0.0003% (*207,208*).

Budĕšinsky *et al.* (*209*) have prepared a number of 3-(3-amino-2-hydroxypropyl)- and 3-(3-aminoacetonyl)-quinazolin-4-ones in which the 3-piperidinol part of febrifugine has been replaced by readily available bases (aniline, pyrrolidine, etc.). These compounds show a decrease in coccidiostatic activity and marked activity against *Eimeria tenella, Nippostrongylus brasiliensis,* and *Fasciola hepatica.*

Glycosmis arborea (Sanskrit: *Ash-shoura;* family Rutaceae), in the Ayurvedic system of medicine, is useful in flatulence, cold, rheumatism, intestinal worms, anemia, fever, and jaundice. Arborine (1), the major alkaloid, inhibits the action of acetylcholine peripherally on rat uterus, on guinea pig ileum, and also on skeletal muscle. It shows marked inhibitory action against pituitrin on the rat uterus. The fall of blood pressure in cats *in vivo* is central in origin, being absent in cat spinal preparations but present in vagotomized animals (*16,210*). Mukherjee and Dey (*211*) have investigated the changes in amino acid metabolism in rat brain following administration of arborine (1).

Glycosminine (6) is a moderate inhibitor of the serine protease human leukocyte elastase. It shows considerable specitivity, since no inhibition of porcine pancreatic elastase, cathepsin G, or chymotrypsin was observed (*212*). Glomerin (2) and homoglomerin are constituents of the defensive secretion of *Glomeris marginata*. This millipede discharges its secretion in response to pinching, tapping, or, on occasion, even mere prodding. The secretion contains also other, possibly macromolecular, components.

Tryptanthrin (53) is a highly specific antimicrobial agent active against dermatophytes, with an effect similar to that of griseofulvin against *Trichophyton mentagrophytes*. Its antibiotic activity depends on the composition of the growth medium. This alkaloid is also the active principle in the leaves of *Strobilanthes*

TABLE III
Quinazoline Alkaloids

Alkaloid (formula)	Composition	Melting point (°C)	$[\alpha]_D$ (°)	Source
4-Methylquinazoline[a]	$C_9H_8N_2$			Pseudomonas sp. (58,59)
Glycorine (3)	$C_9H_8N_2O$	145–147		Glycosmis arborea (Roxb.) DC. (43,44) G. bilocularis Thw. (11)
Glycosmicine (4)	$C_9H_8N_2O_2$	270–271		Glycosmis arborea (Roxb.) DC. (43,44)
2-Carboxy-4-methyl-quinazoline[a]	$C_{10}H_8N_2O_2$			Pseudomonas sp. (58,59)
4-Methylquinazoline-2-carbonamide[a]	$C_{10}H_9N_3O$			Pseudomonas sp. (58,59)
2,4-Dimethylquinazoline[a]	$C_{10}H_{10}N_2$			Pseudomonas sp. (58,59)
Glomerin (2)	$C_{10}H_{10}N_2O$	204		Glomeris marginata Vill. (39,41)
2-Hydroxymethyl-4-methyl-quinazoline	$C_{10}H_{10}N_2O$			Pseudomonas sp. (58,59)
Chrysogine (7)	$C_{10}H_{10}N_2O_2$	189–190	-26 ± 4	Penicillium chrysogenum (38)
Deoxyvasicinone (33)	$C_{11}H_{10}N_2O$	110–111		Linaria sp. (112) Mackinlaya macrosciadia (F. Muell.) F. Muell. (111) Peganum harmala L. (107–109) P. nigellastrum (110) Adhatoda vasica Nees (152)
Vasicinone (29)	$C_{11}H_{10}N_2O_2$	l: 201–202 dl: 211–212	-129	Linaria sp. (112) Nitraria sibirica (138) Peganum harmala L. (78,87,108,109) P. nigellastrum (110) Sida sp. (139,140)
Vasicinolone (31)	$C_{11}H_{10}N_2O_3$	279		Adhatoda vasica Nees (94)
2-Ethyl-4-methyl-quinazoline[a]	$C_{11}H_{12}N_2$			Pseudomonas sp. (58,59)
Deoxyvasicine (32)	$C_{11}H_{12}N_2$	86–87		Peganum harmala L. (95)
Vasicine (peganine) (28)	$C_{11}H_{12}N_2O$	dl: 209–210		Anisotes sessiliflorus C.B.Cl. (63) Galega orientalis Lam. (143) Linaria sp. (see Ref. 144)

TABLE III (*Continued*)

Alkaloid (formula)	Composition	Melting point (°C)	$[\alpha]_D$ (°)	Source
				Peganum nigellastrum (*110*)
				Sida sp. (*139,140*)
		l: 211–212	−254	*Adhatoda vasica* Nees (*141*)
				Peganum harmala L. (*145*)
		d: 170–173	+162.5	*Galega officinalis* L. (*142*)
Homoglomerin	$C_{11}H_{12}N_2O$	149		*Glomeris marginata* Vill. (*39,41*)
Peganol (**46**)	$C_{11}H_{12}N_2O$	178–180		*Peganum harmala* L. (*119*)
7-Hydroxypeganine (**30**)	$C_{11}H_{12}N_2O_2$	272–273	+45.8	*Adhatoda vasica* Nees (*90*)
				Peganum harmala L. (*108,109*)
				Sida sp. (*139,140*)
Pegamine (**8**)	$C_{11}H_{12}N_2O_2$	160–161		*Peganum harmala* L. (*55*)
Vasicol (**34**)	$C_{11}H_{14}N_2O_2$	204–206 (HCl)	−17.34	*Adhatoda vasica* Neees (*64*)
50	$C_{12}H_{14}N_2$	85–87		*Mackinlaya macrosciadia* (F. Muell.) F. Muell.
				M. subulata Philipson (*120,121*)
51	$C_{12}H_{12}N_2O$	99		*Mackinlaya macrosciadia* (F. Muell.) F. Muell.
				M. subulata Philipson (*120,121*)
Deoxypeganidine (**45**)	$C_{14}H_{16}N_2O$	76–79		*Peganum harmala* L. (*117*)
Peganidine (**44**)	$C_{14}H_{16}N_2O_2$	189–190		*Peganum harmala* L. (*116*)
Isopeganidine	$C_{14}H_{16}N_2O_2$	169–170		*Peganum harmala* L. (*118*)
Tryptanthrin (**53**)	$C_{15}H_8N_2O_2$	268 (261)		*Candida lipolytica* (*131*)
				Couroupita guaianensis Aubl. (*126,146*)
				Isatis tinctoria (*129*)
				Leucopaxillus cerealis var. *piceina* (Peck) (*132*)
				Polygonum tinctorum (*129,147*)

(*continued*)

TABLE III (*Continued*)

Alkaloid (formula)	Composition	Melting point (°C)	$[\alpha]_D$ (°)	Source
Glycosminine (6)	$C_{15}H_{12}N_2O$	249		*Strobilanthes cusia* (*127,147*) *Glycosmis arborea* (Roxb.) DC. (*12,43*) *Ruta* sp. (*48*)
Arborine (1)	$C_{16}H_{14}N_2O$	155–156		*Glycosmis arborea* (Roxb.) DC. (*9,10,13*) *G. bilocularis* Thw. (*11*)
Glycophymoline (5)	$C_{16}H_{14}N_2O$	165		*Glycosmis pentaphylla* (*42*)
Febrifugine (9)	$C_{16}H_{19}N_3O_3$	139–140 154–156	+28	*Dichroa febrifuga* Lour. (*148–150*) *Hydrangea umbellata* Rheder (*151*)
Isofebrifugine	$C_{16}H_{19}N_3O_3$	129–131	+120	*Dichroa febrifuga* Lour. (*148–150*) *Hydrangea umbellata* Rheder (*151*)
25	$C_{17}H_{16}N_2O_2$	100–102		*Zanthoxylum arborescens* (Rose) (*56*)
27	$C_{18}H_{18}N_2O_3$	133–134		*Zanthoxylum arborescens* (Rose) (*56*)
Vasicolinone (42)	$C_{19}H_{19}N_3O$	152		*Adhatoda vasica* Nees (*65*)
Sessiflorine (38)	$C_{19}H_{19}N_3O_2$	195–197		*Anisotes sessiliflorus* C.B.Cl. (*63*)
Vasicoline (41)	$C_{19}H_{21}N_3$	135		*Adhatoda vasica* Nees (*65*)
Anisotine (35)	$C_{20}H_{19}N_3O_3$	189–190		*Adhatoda vasica* Nees (*65*) *Anisotes sessiliflorus* C.B.Cl. (*63*)
Anisessine (39)	$C_{20}H_{19}N_3O_3$	170		*Anisotes sessiliflorus* C.B.Cl. (*63*)
Adhatodine (43)	$C_{20}H_{21}N_3O_2$	183		*Adhatoda vasica* Nees (*65*)
Deoxyaniflorine (36)	$C_{20}H_{21}N_3O_2$	168–172		*Anisotes sessiliflorus* C.B.Cl. (*63*)
Aniflorine (40)	$C_{20}H_{21}N_3O_3$	197		*Anisotes sessiliflorus* C.B.Cl. (*63*)
Dipegine (47)	$C_{22}H_{20}N_4O$	221–223		*Peganum harmala* L. (*118*)
Candidine (52)	$C_{23}H_{13}N_3O_2$	267		*Candida lipolytica* (*124*)

[a] The alkaloids were identified by TLC.

cusia O. Kuntze, which has a long tradition in Japan as a remedy against dermatophytic infections, notably athlete's foot (*128,153*).

Addendum

Chowdhury and Bhattacharyya (*214*) have reported the isolation of 1,2,3,9-tetrahydro-5-methoxypyrrolo[2,1-*b*]quinazolin-3-ol from the leaves of *Adhatoda vasica*. Physical data are as follows: $C_{12}H_{14}N_2O_2$ (M 218), mp 224–225°C; UV [λ_{max}] 307 nm (log ϵ, 3.85); IR: 3470 (OH), 1630 ($=C=N-$), 1605, 1500 (aromatic residue), 1250 (aromatic ether), 845 cm^{-1} (substituted benzene derivative); ^{1}H NMR: δ6.60–7.0 (m, 3*H*, aromatic protons), 4.71 (t, 1*H*, C-3 H), 4.50 (s, 2*H*, C-9 H_2), 3.80 (s, 3*H*, aromatic OCH_3), 3.21 (m, 2*H*, C-1 H_2), 2.18 (m, 2*H*, C-2 H_2).

Acknowledgment

I thank Prof. Manfred G. Reinecke, Texas Christian University, Fort Worth, Texas, for his help in the preparation of the manuscript.

REFERENCES

1. H. T. Openshaw, *Alkaloids (N.Y.)* **3**, 101 (1953).
2. H. T. Openshaw, *Alkaloids (N.Y.)* **7**, 247 (1960).
3. S. Johne and D. Gröger, *Pharmazie* **25**, 22 (1970).
4. S. Johne, *Prog. Chem. Org. Nat. Prod.* **46**, 159 (1984).
5. J. R. Price, *Prog. Chem. Org. Nat. Prod.* **13**, 330 (1956).
6. A. Prakash and S. Ghosal, *J. Sci. Ind. Res.* **42**, 309 (1983).
7. M. V. Telezhenetskaya and S. Yu. Yunusov, *Khim. Prir. Soedin.*, 731 (1977).
8. M. P. Jain, V. N. Gupta, and C. K. Atal, *Indian Drugs* **21**, 313 (1984).
9. D. Chakravarti, R. N. Chakravarti, and S. C. Chakravarti, *J. Chem. Soc.*, 3337 (1953).
10. R. N. Chakravarti and S. C. Chakravarti, *J. Proc. Inst. Chemists (India)* **24**, 96 (1952).
11. I. H. Bowen, K. P. W. C. Perera, and J. R. Lewis, *Phytochemistry* **17**, 2125 (1978).
12. A. Chatterjee and S. Ghosh Majumdar, *J. Am. Chem. Soc.* **76**, 2459 (1954).
13. A. Chatterjee and S. Ghosh Majumdar, *J. Am. Chem. Soc.* **75**, 4365 (1953).
14. D. Chakravarti, R. N. Chakravarti, L. A. Cohen, B. Das Gupta, S. Dutta, and H. K. Miller, *Tetrahedron* **16**, 224 (1961).
15. S. C. Pakrashi and J. Bhattacharyya, *Tetrahedron* **24**, 1 (1968).
16. R. N. Chakravarti, *Bull. Calcutta School Trop. Med.* **11**, 37 (1963).
17. S. C. Pakrashi, J. Bhattacharyya, L. F. Johnson, and H. Budzikiewicz, *Tetrahedron* **19**, 1011 (1963).
18. D. Chakravarti and R. N. Chakravarti, *J. Proc. Inst. Chemists (India)* **39** (Part III), 131 (1967).
19. S. C. Pakrashi and J. Bhattacharyya, *J. Sci. Ind. Res.* **24**, 293 (1963).
20. D. Chakravarti, R. N. Chakravarti, and S. C. Chakravarti, *Sci. Cult.* **18**, 553 (1953).

21. D. Chakravarti, R. N. Chakravarti, and S. C. Chakravarti, *Experientia* **9**, 333 (1953).
22. S. C. Pakrashi, A. De, and S. Chattopadhyay, *Indian J. Chem.* **6**, 472 (1968).
23. E. Ziegler, W. Steiger, and Th. Kappe, *Monatsh. Chem.* **100**, 948 (1969).
24. P. Bhattacharyya, M. Sakar, T. Roychowdhury, and D. P. Chakraborty, *Chem. Ind.*, 532 (1978).
25. H. Möhrle and C.-M. Seidel, *Arch. Pharm. (Weinheim, Ger.)* **309**, 503; 572 (1976).
26. D. P. Chakraborty, A. K. Mandal, and S. K. Roy, *Synthesis*, 977 (1981).
27. T. Kametani, C. V. Loc, T. Higa, M. Ihara, and K. Fukumoto, *J. Chem. Soc., Perkin Trans. 1*, 2347 (1977).
28. T. Kametani, C. V. Loc, T. Higa, M. Koizumi, M. Ihara, and K. Fugumoto, *J. Am. Chem. Soc.* **99**, 2306 (1977).
29. T. Kametani, T. Higa, K. Fukumoto, and M. Koizumi, *Heterocycles* **4**, 23 (1976).
30. T. Kametani, T. Higa, C. V. Loc, M. Ihara, M. Koizumi, and K. Fukumoto, *J. Am. Chem. Soc.* **98**, 6186 (1976).
31. T. Kametani, C. V. Loc, M. Ihara, and K. Fugumoto, *Heterocycles* **9**, 1585 (1978).
32. T. Kametani, C. V. Loc, T. Higa, M. Koizumi, M. Ihara, and K. Fukumoto, *Heterocycles* **4**, 1487 (1976).
33. W. L. F. Armarego, *in* "The Chemistry of Heterocyclic Compounds, Vol. 24/I: Quinazolines." (A. Weissberger, ed.) Wiley (Interscience), New York.
34. R. K. Hill and A. G. Edwards, *Chem. Ind.*, 858 (1962).
35. T. Reichstein, H. R. Rosenberg, and R. Eberhardt, *Helv. Chim. Acta* **18**, 721 (1935).
36. D. F. Barringer, Jr., G. Berkelhammer, S. D. Carter, L. Goldman, and A. E. Lanzilotti, *J. Org. Chem.* **38**, 1933 (1973).
37. D. F. Barringer, Jr., G. Berkelhammer, and R. S. Wayne, *J. Org. Chem.* **38**, 1937 (1973).
38. H. Hikino, S. Nabetani, and T. Takemoto, *Yakagaku Zasshi* **93**, 619 (1973); *Chem. Abstr.* **79**, 40922q (1973).
39. Y. C. Meinwald, J. Meinwald, and Th. Eisner, *Science* **154**, 390 (1966).
40. H. Schildknecht and W. F. Wenneis, *Z. Naturforsch.* **21b**, 552 (1966).
41. H. Schildknecht, W. F. Wenneis, K. H. Weis, and U. Maschwitz, *Z. Naturforsch.* **21b**, 121 (1966).
42. M. Sarkar and D. P. Chakraborty, *Phytochemistry* **18**, 694 (1979).
43. S. C. Pakrashi and J. Bhattacharyya, *J. Sci. Ind. Res.* **21b**, 49 (1962).
44. S. C. Pakrashi and J. Bhattacharyya, *Ann. Biochem. Exp. Med.* **23**, 123 (1963).
45. H. Culbertson, J. C. Decius, and B. E. Christensen, *J. Am. Chem. Soc.* **74**, 4834 (1952).
46. J. Bhattacharyya and S. C. Pakrashi, *Heterocycles* **14**, 1469 (1980).
47. M. Vincent, J. Maillard, and M. Benard, *Bull. Soc. Chim. Fr.*, 119 (1963).
48. T. N. Vasudevan and M. Luckner, *Pharmazie* **23**, 520 (1968).
49. Y. Hagiwara, M. Kurihara, and N. Yoda, *Tetrahedron* **25**, 783 (1969).
50. S. Palazzo, L. I. Giannola, and S. Caronna, *Atti Accad. Sci. Lett. Arti Palermo, Parte 1*, **34**, 339 (1976); *Chem. Abstr.* **89**, 43309n (1978).
51. R. P. Rhee and J. D. White, *J. Org. Chem.* **42**, 3650 (1977).
52. I. Ganjian and I. Lalezari, *Synth. Commun.* **14**, 33 (1984).
53. M. Sakar and D. P. Chakraborty, *Phytochemistry* **16**, 2007 (1977).
54. J. Bhattacharyya and S. C. Pakrashi, *Heterocycles* **12**, 929 (1979).
55. Kh.N. Khashimov, M. V. Telezhenetskaya, Ya. V. Rashkes, and S. Yu. Yunusov, *Khim. Prir. Soedin.*, 453 (1970).
56. D. L. Dreyer and R. C. Brenner, *Phytochemistry* **19**, 935 (1980).
57. J. A. Grina, M. R. Ratcliff, and F. R. Stermitz, *J. Org. Chem.* **47**, 2648 (1982).
58. S. Mann, *Arch. Mikrobiol.* **56**, 324 (1967).
59. S. Mann, *Arch. Hygiene Bakteriol.* **151**, 474 (1967).

60. H. G. Boit, in "Fortschritte der Alkaloid-Chemie seit 1933," p. 331. Akademie-Verlag, Berlin, 1950; "Ergebnisse der Alkaloid-Chemie bis 1960," p. 741. Akademie-Verlag, Berlin, 1961.
61. E. Späth, Monatsh. Chem. 72, 115 (1939).
62. M. V. Telezhenetskaya and S. Yu. Yunusov, Khim. Prir. Soedin., 731 (1977).
63. R. R. Arndt, S. H. Eggers, and A. Jordaan, Tetrahedron 23, 3521 (1967).
64. K. L. Dhar, M. P. Jain, S. K. Koul, and C. K. Atal, Phytochemistry 20, 319 (1981).
65. S. Johne, D. Gröger, and M. Hesse, Helv. Chim. Acta 54, 826 (1971).
66. J. Holubek and O. Strouf (eds.), "Spectral Data and Physical Constants of Alkaloids," Vol. I. Publishing House of the Czechoslovakian Academy of Sciences, Prague 1965.
67. K. Szulzewsky, E. Höhne, S. Johne, and D. Gröger, J. Prakt. Chem. 318, 463 (1976).
68. S. Johne, B. Jung, D. Gröger, and R. Radeglia, J. Prakt. Chem. 319, 919 (1977).
69. A. K. Bhatnagar and S. P. Popli, Indian J. Chem. 4, 291 (1966).
70. Ya. V. Rashkes, M. V. Telezhenetskaya, V. N. Plugar, and S. Yu. Yunusov, Khim, Prir. Soedin., 378 (1977).
71. V. N. Plugar, Ya. V. Rashkes, and Kh. M. Schakhidoyatov, Khim. Prir. Soedin., 180 (1979).
72. V. N. Plugar, Ya. V. Rashkes, and Kh. M. Schakhidoyatov, Khim. Prir. Soedin., 414 (1978).
73. E. L. Kristallovich, M. R. Yagudaev, M. V. Telezhenetskaya, and S. Yu. Yunusov, Khim, Prir. Soedin., 122 (1984).
74. D. Gröger and S. Johne, in "Festschrift K. Mothes," p. 205. G.-Fischer-Verlag, Jena, 1965.
75. E. M. Huq, M. Ikram, and S. A. Warsi, Pakistan J. Sci. Ind. Res. 10, 224 (1967); Chem. Abstr. 69, 16807d (1968).
76. A. A. L. Gunatilaka, S. Sotheeswaran, S. Balasubramaniam, A. I. Chandrasekara, and H. T. B. Sriyani, Planta Med. 39, 66 (1980).
77. K. G. Lupu, Rast. Resur. 9, 206 (1973); Chem. Abstr. 79, 15906e (1973).
78. Kh.N. Khashimov, M. V. Telezhenetskaya, N. N. Sharakhimov, and S. Yu. Yunusov, Khim. Prir. Soedin., 382 (1971).
79. N. V. Plekhanova and S. T. Aktanova, Issled. Flory Kirgizii na Alkaloidonosnost, Akad. Nauk. Kirg. SSR, Inst. Organ. Khim., 57 (1965); Chem. Abstr. 64, 11550g (1966).
80. J. Schäfer and M. Stein, Biol. Zentralbl. 88, 755 (1969).
81. J. Schäfer and M. Stein, Naturwissenschaften 54, 205 (1967).
82. K. R. Brain and B. B. Thapa, J. Chromatogr. 258, 183 (1983).
83. P. L. Southwick and L. Casanova, J. Am. Chem. Soc. 80, 1168 (1958).
84. C. Schöpf and E. Oechler, Justus Liebigs Ann. Chem. 523, 1 (1936).
85. N. J. Leonhard and M. J. Martell, Tetrahedron Lett., 44 (1960).
86. H. Möhrle and P. Gundlach, Tetrahedron Lett., 3249 (1970).
87. D. R. Mehta, J. S. Naravane, and R. M. Desai, J. Org. Chem. 28, 445 (1963).
88. R. C. Morris, W. E. Hanford, and R. Adams, J. Am. Chem. Soc. 57, 951 (1935).
89. T. Onaka, Tetrahedron Lett., 4387 (1971).
90. E. Späth and F. Kesztler-Gandini, Monatsh. Chem. 91, 1150 (1960).
91. F. Kuffner, G. Lenneis, and H. Bauer, Monatsh. Chem. 91, 1152 (1960).
92. T. R. Rajagopalan, S. Bhattacharji, and M. L. Dhar, Proc., Symp. Drugs Antibiot. (Defense Research Laboratory, Kanpur), 121 (1961).
93. A. K. Bhatnagar, S. Bhattacharji, and S. Popli, Indian J. Chem. 3, 525 (1965).
94. M. P. Jain and V. K. Sharma, Planta Med. 46, 250 (1982).
95. Kh. N. Khashimov, M. V. Telezhenetskaya, and S. Yu. Yunusov, Khim. Prir. Soedin. 456 (1969).
96. S. Siddiqui, Pakistan J. Sci. Ind. Res. 5, 207 (1962); Chem. Abstr. 59, 5213 (1963).
97. E. Späth, F. Kuffner, and N. Platzer, Ber. Dtsch. Chem. Ges. 68, 497 (1935).
98. E. Späth and N. Platzer, Ber. Dtsch. Chem. Ges. 69, 255 (1936).

99. W. E. Hanford and R. Adams, *J. Am. Chem. Soc.* **57**, 921 (1935).
100. G. G. Munoz and R. Madronero, *Chem. Ber.* **95**, 2182 (1962).
101. L. Skurský, *Z. Naturforsch.* **14b**, 474 (1959).
102. L. Skurský, *Collect. Czech. Chem. Commun.* **30**, 2080 (1965).
103. L. Macholán, *Collect. Czech. Chem. Commun.* **24**, 550 (1959).
104. T. P. Ghose, A. Krishna, K. S. Narang, and J. N. Ray, *J. Chem. Soc.*, 2740 (1932).
105. T. P. Ghose, *Quart. J. Indian Chem. Soc.* **4**, 1 (1927).
106. A. H. Sattarova, E. K. Dobronravova, and T. T. Shakirov, *Khim. Prir. Soedin.* 84 (1985).
107. A. Al-Shamma, S. Drake, D. L. Flynn, L. A. Mitscher, Y. H. Park, G. S. R. Rao, A. Simpson, J. K. Swayze, T. Veysoglu, and S. T. S. Wu, *J. Nat. Proc.* **44**, 745 (1981).
108. N. I. Koretskaya, *Zhur. Obshchei. Khim.* **27**, 3361 (1957).
109. N. I. Koretskaya and L. M. Utkin, *Zhur. Obshchei. Khim.* **28**, 1087 (1958).
110. D. Batsuren, M. V. Telezhenetskaya, S. Yu. Yunusov, and T. Baldan, *Khim. Prir. Soedin.*, 418 (1978).
111. N. K. Hart, S. R. Johns, and J. A. Lamberton, *Aust. J. Chem.* **24**, 223 (1971).
112. N. V. Plekhanova and G. P. Sheveleva, *Issled. Flory Kirgizii na Alkaloidononost, Akad. Nauk Kirg. SSR, Inst. Organ. Khim.*, 54 (1965). *Chem. Abstr.* **64**, 11550e (1966).
113. A. Chatterjee and M. Ganguly, *Phytochemistry* **7**, 307 (1968).
114. E. Späth and N. Platzer, *Ber. Dtsch. Chem. Ges.* **68**, 2221 (1935).
115. S. Petersen and E. Tietze, *Justus Liebigs Ann. Chem.* **623**, 166 (1959).
116. Kh. N. Khashimov, M. V. Telezhenetskaya, and S. Yu. Yunusov, *Khim. Prir. Soedin.*, 599 (1969).
117. B. Kh. Zharekeev, M. V. Telezhenetskaya, and S. Yu. Yunusov, *Khim. Prir. Soedin.*, 279 (1973).
118. B. Kh. Zharekeev, Kh. N. Khashimov, M. V. Telezhenetskaya, and S. Yu. Yunusov, *Khim. Prir. Soedin.* 264 (1974).
119. M. V. Telezhenetskaya, Kh. N. Khashimov, and S. Yu. Yunusov, *Khim. Prir. Soedin.*, 849 (1971).
120. S. R. Johns and J. A. Lamberton, *J. Chem. Soc., Chem. Commun.*, 267 (1965).
121. J. S. Fitzgerald, S. R. Johns, J. A. Lamberton, and A. H. Redcliffe, *Aust. J. Chem.* **19**, 151 (1966).
122. T. Stephen and H. Stephen, *J. Chem. Soc.*, 4694 (1956).
123. W. L. Mosby, *in* "Heterocyclic Systems with Bridgehead Nitrogen Atoms" (A. Weissberger, ed.), Part 2, p. 1153. Wiley (Interscience), New York, 1961.
124. H.-P. Fiedler, *Thesis Univ. Tübingen (FRG) 1974*.
125. J. Bergman and U. Tilstam, *Tetrahedron* **41**, 2883 (1985).
126. J. Bergmann, B. Egestad, and J.-O. Lindström, *Tetrahedron Lett.*, 2625 (1977).
127. G. Honda and M. Tabata, *Planta Med.* **36**, 85 (1979).
128. G. Honda, M. Tabata, and M. Tsuda, *Planta Med.* **37**, 172 (1979).
129. G. Honda, V. Tosirisuk, and M. Tabata, *Planta Med.* **38**, 275 (1980).
130. M. Brufani, W. Fedeli, F. Mazza, A. Gerhard, and W. Keller-Schierlein, *Experientia* **27**, 1249 (1971).
131. F. Schindler and H. Zähner, *Arch. Mikrobiol.* **79**, 187 (1971).
132. M. Y. Jarrah and V. Thaller, *J. Chem. Res. (S)*, 186 (1980); *J. Chem. Res. (M)*, 2601 (1980).
133. W. Fedeli and F. Mazza, *J. Chem. Soc., Perkin Trans. 2*, 1621 (1974).
134. P. Friedländer and W. Roschdestwensky, *Ber. Dtsch. Chem. Ges.* **48**, 1841 (1915).
135. C. W. Bird, *Tetrahedron* **19**, 901 (1963).
136. R. Kikumoto and T. Kobayashi, *Tetrahedron* **22**, 3337 (1966).
137. Kh. Shakhidoyatov, A. Irishbaeva, and Ch. Sh. Kadyrov, *Khim. Prir. Soedin.*, 681 (1974).
138. Z. Osmanov, A. A. Ibragimov, and S. Yu. Yunusov, *Khim. Prir. Soedin.*, 126 (1982).
139. S. Ghosal, R. B. P. S. Chauhan, and R. Mehta, *Phytochemistry* **14**, 830 (1975).

140. A. Prakash, R. K. Varma, and S. Ghosal, *Planta Med.* **43**, 384 (1981).
141. D. Hooper, *Pharm. J.* **18**, 841 (1888).
142. K. Schreiber, O. Aurich, and K. Pufahl, *Arch. Pharm. (Weinheim, Ger.)* **295**, 271 (1962).
143. H. Köhler, *Biol. Zentralbl.* **88**, 165 (1969).
144. S. Johne and D. Gröger, *Pharmazie* **23**, 35 (1968).
145. A. D. Rosenfeld and D. G. Kolesnikov, *Ber. Dtsch. Chem. Ges.* **69**, 2022 (1936), and references cited therein.
146. A. K. Sen, S. B. Mahato, and N. L. Dutta, *Tetrahedron Lett.*, 609 (1974).
147. Takeda Chemical Industries, Ltd. *Jpn. Kokai Tokkyo Koho* 8047,684 (1980); *Chem. Abstr.* **93**, 186 400d (1980).
148. T. Q. Chou, F. Y. Fu, and Y. S. Kao, *J. Am. Chem. Soc.* **70**, 1765 (1948).
149. C. S. Jang, F. Y. Fu, C. Y. Wang, K. C. Huang, G. Lu, and T. C. Chou, *Science* **103**, 59 (1946).
150. J. B. Koepfli, J. F. Mead, and J. A. Brockman, *J. Am. Chem. Soc.* **71**, 1048 (1949).
151. F. Ablondi, S. Gordon, J. Morton II, and J. H. Williams, *J. Org. Chem.* **17**, 14 (1952).
152. A. H. Amin and D. R. Metha, *Nature (London)* **184**, 1317 (1959).
153. J. Bergman, J.-O. Lindström, and U. Tilstam, *Tetrahedron* **41**, 2879 (1985).
154. E. Fiedler, H.-P. Fiedler, A. Gerhard, W. Keller-Schierlein, W. A. König, and H. Zähner, *Arch. Microbiol.* **107**, 249 (1976).
155. L. A. Mitscher, W. C. Wong, T. De Meulenaere, J. Sulko, and S. Drake, *Heterocycles* **15**, 1017 (1981).
156. G. W. Cambridge, A. B. A. Jansen, and D. A. Jarman, *Nature (London)* **196**, 1217 (1962).
157. R. Landi-Vittory and F. Gatta, *Gazz. Chim. Ital.* **99**, 59 (1969).
158. R. Robinson, "The Structural Relations of Natural Products." Oxford Univ. Press (Clarendon), London and New York, 1955.
159. L. Macholán, *Naturwissenschaften* **52**, 186 (1965).
160. L. Macholán, L. Rozprimova, and S. Sedláčková, *Biochim. Biophys. Acta* **136**, 258 (1967).
161. S. Johne, K. Waiblinger, and D. Gröger, *Pharmazie* **28**, 403 (1973).
162. D. Gröger, *in* "Biosynthese der Alkaloide" (K. Mothes and H. R. Schütte, eds.), p. 551. Deutscher Verlag der Wissenschaften, Berlin, 1969.
163. D. Gröger, *Lloydia* **32**, 221 (1969).
164. M. Luckner and S. Johne, *in* "Biochemistry of Alkaloids" (K. Mothes, H. R. Schütte, and M. Luckner, eds.), p. 328. Deutscher Verlag der Wissenschaften, Berlin, and Verlag Chemie, Weinheim, 1985.
165. D. Gröger and K. Mothes, *Arch. Pharm. (Weinheim, Ger.)* **293**, 1049 (1960).
166. K. Waiblinger, S. Johne, and D. Gröger, *Phytochemistry* **11**, 2263 (1972).
167. D. J. Liljegren, *Phytochemistry* **7**, 1299 (1968): **10**, 2661 (1971).
168. D. Gröger and S. Johne, *Z. Naturforsch.* **23b**, 1072 (1968).
169. S. Johne, K. Waiblinger, and D. Gröger, *Eur. J. Biochem.* **15**, 415 (1970).
170. D. G. O. Donovan and H. Horan, *J. Chem. Soc. C*, 2466 (1970).
171. H. Schildknecht and W. F. Wenneis, *Tetrahedron Lett.*, 1815 (1967).
172. E. Fiedler, H. P. Fiedler, A. Gerhard, W. Keller-Schierlein, W. A. König, and H. Zähner, *Arch. Mikrobiol.* **107**, 249 (1976).
173. A. H. Amin, D. R. Metha, and S. S. Samarth, *Prog. Drug. Res.* **14**, 218 (1970).
174. C. M. Gupta, A. P. Bhaduri, and N. M. Khanna, *J. Sci. Ind. Res.* **30**, 101 (1971).
175. S. Johne, *Pharmazie* **36**, 583 (1981).
176. S. Johne, *Prog. Drug Res.* **26**, 259 (1982).
177. M. Süsse and S. Johne, *Z. Chem.* **21**, 431 (1981).
178. L. N. Yakhontov, S. S. Libermann, G. P. Zhikhareva, and K. K. Kuzmina, *Khim.-Farm. Zh.* **11** (5), 14 (1977).
179. See references cited in Ref. *183*.

180. O. P. Gupta, M. L. Sharma, B. J. Ray Ghatak, and C. K. Atal, *Indian J. Med. Res.* **66,** 680 (1977).
181. O. P. Gupta, K. K. Anand, B. J. Ray Ghatak, and C. K. Atal, *Indian J. Exp. Biol.* **16,** 1075 (1978).
182. C. K. Atal, ''Chemistry and Pharmacology of Vasicine—A New Oxytocic and Abortifacient.'' Ray Bandu Industrial Co., New Delhi, 1980.
183. P. K. Lahiri and S. N. Pradhan, *Indian J. Exp. Biol.* **2,** 219 (1964).
184. U. Zutshi, P. G. Rao, A. Soni, O. P. Gupta, and C. K. Atal, *Planta Med.* **40,** 373 (1980).
185. M. B. Bhide, P. Y. Naik, S. S. Mahajani, R. B. Ghooi, and R. S. Joshi, *Bull. Haffkine Inst.* **2** (1), 6 (1974); *Chem. Abstr.* **82,** 38618p (1975).
186. V. N. Plugar, T. T. Gorovits, N. Tulyaganov, and Ya. V. Rashkes, *Khim. Prir. Soedin.*, 250 (1977).
187. V. N. Plugar, Ya. V. Rashkes, and N. Tulyaganov, *Khim. Prir. Soedin.*, 201 (1981).
188. R. L. Sharma, R. K. Gupta, B. K. Chowdhury, K. L. Dhar, and C. K. Atal, *Indian J. Chem.* **18B,** 449 (1979).
189. G. Devi, R. S. Kapil, and S. P. Popli, *Indian J. Chem.* **14B,** 354 (1976).
190. A. D. Dunn, E. L. M. Guy, and K. I. Kinnear, *J. Heterocycl. Chem.* **20,** 779 (1983).
191. A. D. Dunn and K. I. Kinnear, *J. Heterocycl. Chem.* **21,** 603 (1984); *ibid.,* **22,** 311 (1985).
192. V. N. Gupta, M. P. Jain, C. K. Atal, and M. Bhardawaj, *Indian J. Chem.* **21B,** 74 (1982).
193. G. Doria, C. Passarotti, R. Magrini, R. Sala, P. Sberze, M. Tibolla, G. Arcari, R. Ceserani, and R. Castello, *Il Farmaco* **39,** 968 (1984).
194. Kh. M. Shakhidoyatov, Ya. Yamankulov, and Ch. Sh. Kadyrov, *Khim. Prir. Soedin.*, 552 (1977).
195. Kh. M. Shakhidoyatov, A. Irisbaeva, E. Oripov, and Ch. Sh. Kadyrov, *Khim. Prir. Soedin.*, 557 (1976).
196. Kh. M. Shakhidoyatov and Ch. Sh. Kadyrov, *Khim. Prir. Soedin.*, 544 (1977).
197. V. K. Sharma and M. P. Jain, *Indian J. Chem.* **21B,** 75 (1982).
198. Kh. M. Shakhidoyatov and Ch. Sh. Kadyrov, *Khim. Prir. Soedin.*, 668 (1977).
199. E. Oripov, L. M. Yun, Kh. M. Shakhidoyatov, and Ch. Sh. Kadyrov., *Khim. Prir. Soedin.*, 603 (1978).
200. A. Karimov, M. V. Telezhentskaya, and S. Yu. Yunusov, *Khim. Prir. Soedin.*, 498 (1982).
201. A. Karimov, V. N. Telezhenetskaya, and S. Yu. Yunusov, *Khim. Prir. Soedin.*, 396 (1983).
202. Kh. M. Shakhidoyatov, E. O. Oripov, L. M. Yun, M. Ya. Yamankulov, and Ch. Sh. Kadyrov, *Fungitsidy,* 66 (1980); *Chem. Abstr.* **94,** 192253v (1981).
203. S. Sakamoto and K. Samejima, *Chem. Pharm. Bull.* **27,** 2220 (1979); *ibid.,* **28,** 916 (1980).
204. E. S. Zabolotnaya and L. N. Safronich, *Lekarstv. Rasteniya* (**15**) 356, (1969); *Chem. Abstr.* **76,** 32240p (1972).
205. P.-L. Chien and C. C. Cheng, *J. Med. Chem.* **13,** 867 (1970).
206. M. Fishman and Ph. A. Cruickshank, *J. Med. Chem.* **13,** 155 (1970).
207. E. Waletzky, G. Berkelhammer, and S. Kantor, U.S. Patent 3,320,124 (1967); *Chem. Abstr.* **68,** 39647v (1968).
208. P. Yvore, N. Foure, J. Aycardi, and G. Bennejean, *Recl. Med. Vet.* **150,** 495 (1974).
209. Z. Buděšinsky, P. Lederer, and J. Danek, *Collect. Czech. Chem. Commun.* **42,** 3473 (1977).
210. M. L. Chatterjee and M. S. De, *Bull. Calcutta School Trop. Med.* **8,** 102 (1960).
211. A. Mukherjee and P. K. Dey, *Indian J. Exp. Biol.* **8,** 263 (1970).
212. T. Teshima, J. C. Griffin, and J. C. Powers, *J. Biol. Chem.* **257,** 5085 (1982).
213. Ch. Kaneko, T. Chiba, K. Kasai, and Ch. Miwa, *Heterocycles* **23,** 1385 (1985).
214. B. K. Chowdhury and P. Bhattacharyya, *Phytochemistry* **24,** 3080 (1985).

CHAPTER 3

THE NAPHTHYL ISOQUINOLINE ALKALOIDS*

GERHARD BRINGMANN

Organisch-Chemisches Institut der Universität Münster
Orléansring 23
D-4400 Münster, West Germany

I. Introduction

Despite their great structural variety, isoquinoline alkaloids were, for a long time, considered as a vast, but biosynthetically uniform family of natural products, the common key step of their biogenesis always being the Mannich-type condensation of phenylethylamines with aldehydes or α-ketoacids (*1*). The preparative imitation of this reaction principle, the Pictet–Spengler-type isoquinoline synthesis (*2*), has been the basis for countless biomimetic alkaloid syntheses (*3*).

The unique substitution pattern of ancistrocladine (**1**), however, which was first isolated by Govindachari from the Indian liana *Ancistrocladus heyneanus* Wall. (Ancistrocladaceae) (*4*), can barely be brought into line with such a "conventional" isoquinoline biosynthesis. Its unprecedented structure, for which it was termed "the most unusual of all the isoquinoline alkaloids" (*5*), makes obvious that its biosynthesis must also differ from that of all other tetrahydroisoquinoline alkaloids by starting not from aromatic amino acids, but from polyketide precursors, as first proposed by Govindachari (*6*).

Ancistrocladine (**1**) is the most prominent representative of a young and growing group of now more than 20 related natural products, the naphthyl isoquinoline alkaloids, one of the very few really novel structural types discovered in the past one or two decades. The structural features common to all of these fascinating alkaloids—especially the hindered naphthalene isoquinoline linkage, which gives rise to highly stable atropisomers—constitute a synthetic challenge, successfully accepted only recently on the basis of biomimetic-type syntheses (*7*).

---

* Dedicated to Prof. T. R. Govindachari, formerly at the CIBA Research Centre, Bombay, now at Madras, who did excellent, pioneering work in this field.

THE ALKALOIDS, VOL. 29

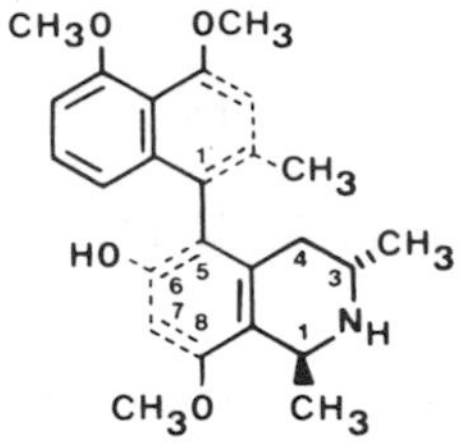

FIG. 1

II. Alkaloids from Ancistrocladaceae

Ancistrocladus is the only genus of the Ancistrocladaceae, comprising about 20 species of lianas and shrubs found in the tropical rain forests of Africa and Asia (*8*). Though the presence of toxic alkaloids [e.g., in *A. vahlii* (= *A. hamatus*)] had already been mentioned in the last century (*9*), and had later been confirmed (e.g., for *A. tectorius*) (*10*), systematic chemical investigations began only in the early 1970s, carried out initially by an Indian and a French group.

A. *Ancistrocladus heyneanus*

Due to the brilliant, pioneering work by Govindachari and his colleagues at the CIBA Research Centre in Bombay, *Ancistrocladus heyneanus* Wall. is the most thoroughly investigated of all the Ancistrocladaceae. From the roots of this liana, a new alkaloid was isolated in nearly 1% yield, named ancistrocladine, and analyzed as $C_{25}H_{29}O_4N$, as also confirmed by the mass spectrum ($M^+ = 407$) (*4,6*). It possesses three methoxyls, one aromatic methyl, and two secondary methyl groups, as well as one secondary amino and one phenolic hydroxy function. The UV spectrum strongly hints of a condensed aromatic ring system.

The ingenious elucidation of naphthyl isoquinoline alkaloids has been reviewed very proficiently by Govindachari and Parthasarathy (*11*). Nonetheless, some of the numerous degradation reactions on which the structural proposal **1** for ancistrocladine was founded are briefly sketched in Schemes 1 and 2.

Oxidative degradation of ancistrocladine (**1**) with $KMnO_4$ to the naphthalene carboxylic ester **2**, which was independently synthesized, accounted for more than half of the carbon atoms of the molecule and simultaneously demonstrated the site of the linkage to the other molecular moiety. The structure and substitution pattern of the tetrahydro isoquinoline part of **1** became evident mainly from degradation products like the γ-lactone **7**, the fully aromatic isoquinoline **4**, as well as the benzofurane **9** (see Scheme 2), obtained by a Claisen rearrangement

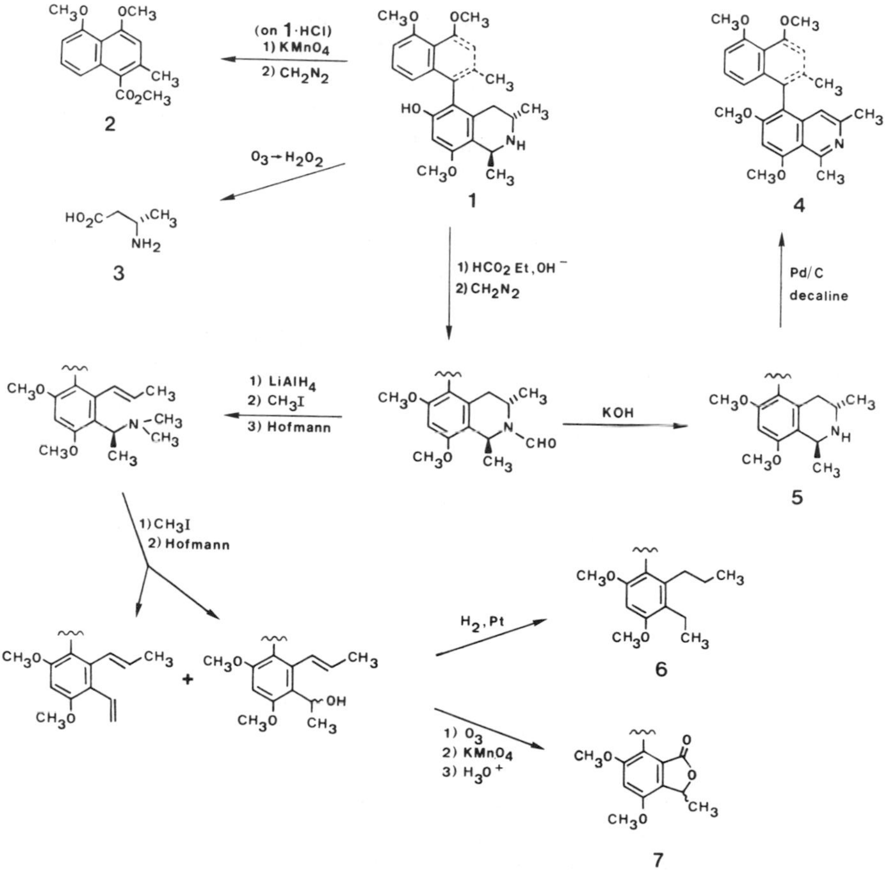

SCHEME 1. Some degradation reactions of (−)-ancistrocladine (1).

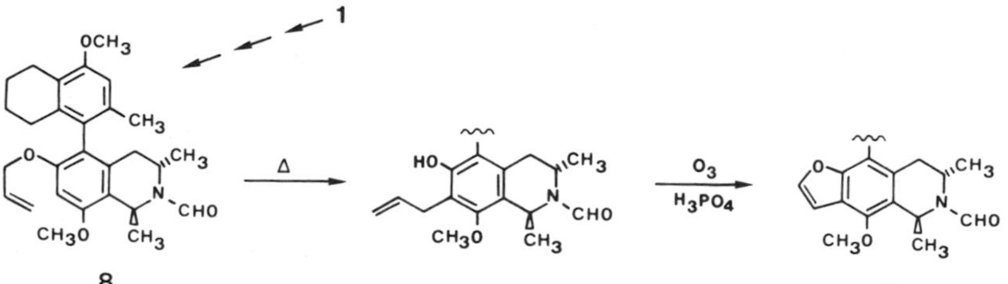

SCHEME 2. Evidence of a free aromatic position at C-7 by Claisen rearrangement.

of the allyl ether **8.** The position of the naphthyl substituent was deduced from NMR high-field shifts, for example, the methoxy group at C-6 of **5** and the benzylic CH_2 of the *n*-propyl side chain in **6.**

The configuration at C-3 was demonstrated to be *S* by exhaustive ozonolysis of **1** to the known β-amino acid **3** (*12*). The relative configuration at C-1 compared with C-3 was shown to be trans by NMR methods, and the absolute configuration at the biphenyl linkage was elucidated by application of the exciton chirality method (*13*) on the fully aromatized chiral naphthyl isoquinoline **4** (*12*). The structure **1** thus deduced for ancistrocladine was later confirmed by X-ray analysis (*12*).

From the same plant, 1,2-dehydro ancistrocladine (**14**), named ancistrocladinine (for formula and data, see Table I, Section II,F) (*14*), as well as *O*-methylancistrocladine (**5**) (*15*) were also isolated by Govindachari and coworkers and structurally determined. Besides this 5–1′ type, the 7–1′-coupled alkaloid ancistrocladisine (**18**) (*16*), and the 7-2′ product ancistrocladidine (**21**) (*17*) were also found in *A. heyneanus* and were completely elucidated (*18*) with all their chiral centers. The quality and reliability of this work can be judged from the fact that the two latter structures **18** and **21** were, much later (*19*), confirmed by X-ray analysis.

B. *Ancistrocladus hamatus*

Ancistrocladus hamatus (Vahl) Gilg is a related climber, found in Sri Lanka. Chemical investigation by Govindachari's group showed that this plant also contains ancistrocladine (**1**) as the major alkaloid, accompanied by its atropisomer, called hamatine (**10**). Compound **10,** on dehydrogenation of its methyl ether, gives an aromatic isoquinoline, enantiomeric to **4,** whereas ozonolysis yields the same β-amino acid **3,** as ancistrocladine (**1**) (*20,21*).

C. *Ancistrocladus tectorius*

From the roots of *A. tectorius* (Lour.) Merr., another Asian liana species, ancistrocladine (**1**) was again isolated, along with its fully aromatized dehydrogenation product (±)-**15,** named ancistrocladeine (*22*). In view of the fact that the main alkaloid **1** has a defined stereochemistry at the biphenyl axis and keeps its configuration even during the dehydrogenation reaction in refluxing decaline to optically active **15** ($[\alpha]_D^{25} = +21.3°$) (*14*), it is astonishing that **15,** as isolated by Cavé and his group (*22*), was reported to be racemic ($[\alpha]_{578}^{20} = 0°$).

Later, Chinese chemists found ancistrocladine (**1**) and hamatine (**10**) in the stems and twigs of the same species, together with a new alkaloid, named ancistrocline. This compound, for which, by spectroscopic evidence alone, the rough structure **11** was proposed, remarkably displays a relative cis configuration of the two methyl groups at C-1 and C-3 (*23*).

In 1985 Cordell and co-workers (*24*) found another new alkaloid in the leaves of the same plant, named ancistrotectorine (**22**). The structure, which shows a rare 7–2′ linkage of the two molecular parts, was clearly established by spectroscopic and X-ray evidence.

D. *Ancistrocladus ealaensis*

From the roots of the African liana *A. ealaensis,* Cavé and co-workers (*25–27*) isolated, apart from racemic ancistrocladeine [(±)-**15**], four alkaloids, called ancistroealaensine, ancistrocladonine, ancistrine, and ancistine. For these, the gross formulas **12, 13, 19,** and **20** have been proposed, which now remain to be confirmed and refined by convincing spectroscopic methods, degradations, or total syntheses. No doubtless assignments of any configuration have been established so far. The biphenyl linkage especially was not taken into account as a source of additional stereoisomerism.

E. *Ancistrocladus congolensis*

Cavé's group also investigated the root and stem bark of yet another African liana, *A. congolensis* (Leonard) (*28*). They reported the noteworthy isolation of (−)-ancistrocladine [(−)-**1**] and its optical antipode, named (+)-ancistrocladine [''(+)-**1**''], which separated from each other on an alumina column! Moreover, *O*-methylancistrocladine (**5**) and two new alkaloids were found, called ancistrocongolensine and ancistrocongine. Their proposed gross structures **16** and **17,** which represent unique 1,2-dihydroisoquinolines, remain to be confirmed.

F. TABLE OF *Ancistrocladus* ALKALOIDS

Table I summarizes the occurrence, the structures, and some physical data of *Ancistrocladus* alkaloids.

III. Alkaloids from Dionchophyllaceae

A. *Triphyophyllum peltatum*

Beside the Ancistrocladaceae, only one other plant family so far has been found to produce naphthyl isoquinoline alkaloids (see also Section IV,B), the Dionchophyllaceae (*31–33*). *Triphyophyllum peltatum* (Hutch. et Dalz.) Airy Shaw is a large liana (*34*), endemic to the rain forests of West Africa. From the twigs of this plant, Bruneton and Cavé isolated a new alkaloid, named triphyophylline (*31*), which, in contrast to all known *Ancistrocladus* alkaloids (see Section II), has only *one* oxygen function on the isoquinoline part of the molecule. No degradation reactions have been described for any of the *Triphyophyllum* alkaloids. Based exclusively on spectral evidence, triphyophylline was assigned the structure **23** (see Table II). The stereochemistry of the two

TABLE I
Naphthyl Isoquinoline Alkaloids from *Ancistrocladus* Plants

Formula (mol wt), structure	Occurrence	mp (°C)	$[\alpha]_D$ (solvent)	Methods for structural elucidation
5-1' Coupled Alkaloids				
Ancistrocladine (**1**) $C_{25}H_{29}O_4N$ (407)	*A. heyneanus*; roots (*4,6,12,30*)	265–267 (*6*)	−25.5° (methanol) (*6*)	Spectroscopy (*4,6,12*)
	A. hamatus; roots (*15,20*)	262–263 (*22*)	−25° (pyridine) (*22,29*)	Exciton chirality method (*12*)
	A. tectorius; roots (*22*), stems, and twigs (*23*)	260–263 (*23*)	−20.5° (CHCl$_3$) (*23*)	Extended degradation (*4,6,30*)
	A. congolensis; roots and stems (*28*)			X-Ray (*12*)
				Partial synthesis from **14** (*11*)
				Total synthesis (*73*)

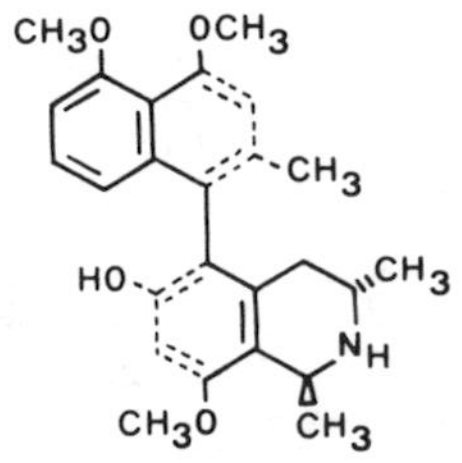

1

| (+)-Ancistrocladine [''(+)-**I**''] $C_{25}H_{29}O_4N$ (407) | *A. congolensis*; roots and stems (*28,29*) | 258–259 (*28,29*) | +27° (methanol) (*28*) | Spectroscopy (*28,29*) |

146

Hamatine (10)
$C_{25}H_{29}O_4N$ (407)

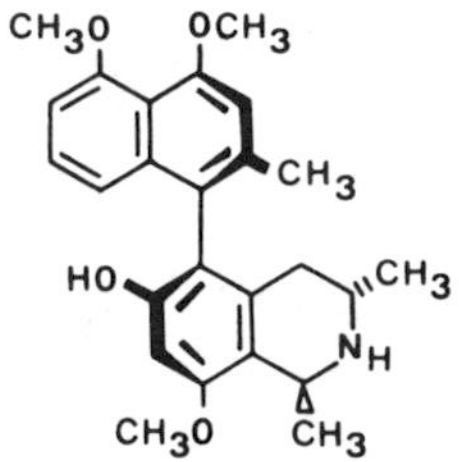

A. *hamatus*; roots (20)
A. *tectorius*; stems and twigs (23)

250–252 (20,23)

+77.4° (CHCl₃) (20)
+66.1° (CHCl₃) (23)

Spectroscopy (20,21)
Exciton chirality method (20)
Degradation (20,21)
Total synthesis (73)

10

O-Methylancistrocladine (5)
$C_{26}H_{31}O_4N$ (421)

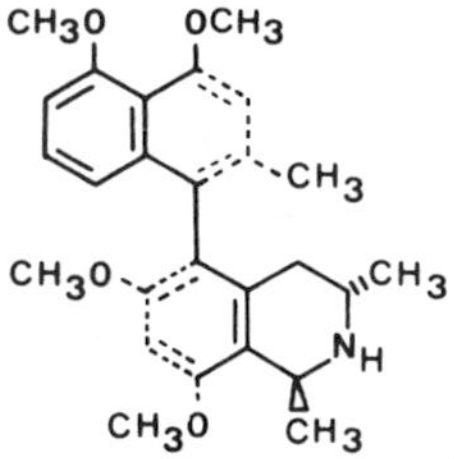

A. *heyneanus*; roots (15)
A. *congolensis*; roots and stems (28)

200–202 (6)
196–197 (28)

−56.1° (hydrochloride) (6)
−35° (pyridine) (28)

Spectroscopy (28)
Partial synthesis from **1** (6,29)

5

(continued)

TABLE I (Continued)

Formula (mol wt), structure	Occurrence	mp (°C)	[α]$_D$ (solvent)	Methods for structural elucidation
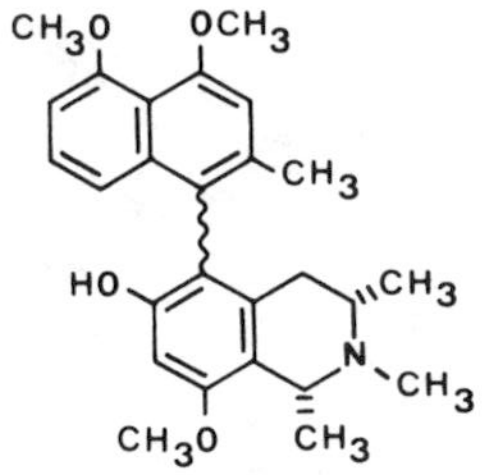 Ancistrocline (11)[a] C$_{26}$H$_{31}$O$_4$N (421)	A. tectorius; stems and twigs (23)	227–228 (23)	+61.7° (CHCl$_3$) (23)	Spectroscopy (23)
Ancistroealaensine (12)[b,c,d] C$_{26}$H$_{29}$O$_4$N (419) (25) C$_{26}$H$_{31}$O$_4$N (421) (26,29) 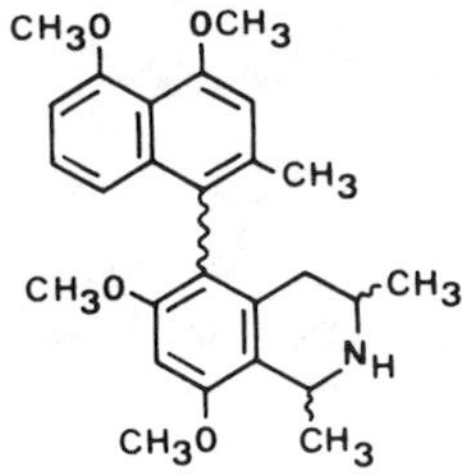	A. ealaensis; roots and stems (25,26,29)	84 (25,26) 176 (29)	−26° (methanol) (26) −32° (methanol) (29)	Spectroscopy (25,26,29) Degradation (26,29)

11

12

Ancistrocladonine (13)[b,c,d]
$C_{26}H_{29}O_4N$ (419) (25)
$C_{27}H_{33}O_4N$ (435) (26,29)

A. ealaensis; roots and stems (25,26,29)

82 (25,26)
135 (29)

+20° (methanol) (25,26)
+45° (methanol) (29)

Spectroscopy (25,26,29)
Degradation (26,29)

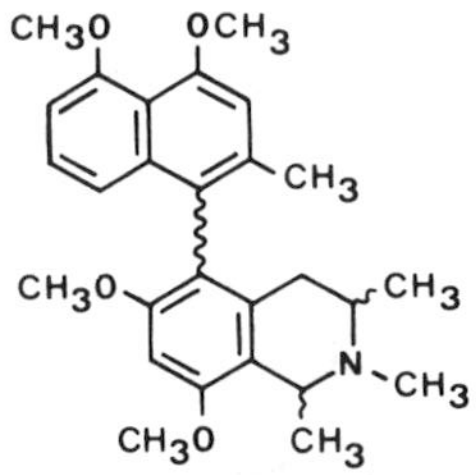

13

Ancistrocladinine (14)
$C_{25}H_{27}O_4N$ (405)

A. heyneanus; roots (14)

235–238 (14)

−321.8° (pyridine) (14)

Spectroscopy (14)
Transformation to **1** (11) and
to (+)−**15** (14)

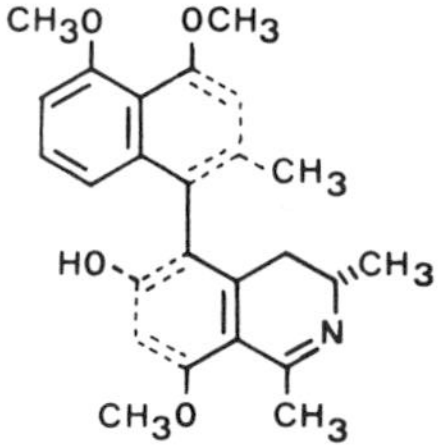

14

(continued)

149

TABLE I (*Continued*)

Formula (mol wt), structure	Occurrence	mp (°C)	$[\alpha]_D$ (solvent)	Methods for structural elucidation
Ancistrocladeine [(±)−**15**] (*22,27,29*)[c] $C_{25}H_{25}O_4N$ (403) 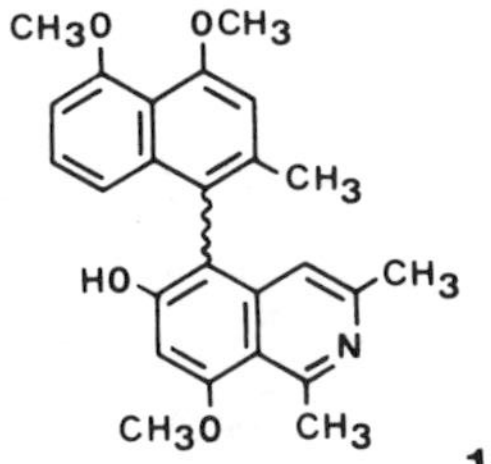	*A. tectorius*; roots (*22,29*) *A. ealaensis*; roots (*27,29*)	275–277 (*22,27,29*) 285–288 (synthesized from **1**, Ref. *14*)	0° (methanol) (*22,27, 29*) +21.3° (pyridine) (synthesized from **1**, Ref. *14*)	Spectroscopy (*22*) Partial synthesis from **1** (*14,22,29*) and from **14** (*14*)
Ancistrocongolensine (**16**) (*28,29*)[c] $C_{24}H_{25}O_4N$ (391) 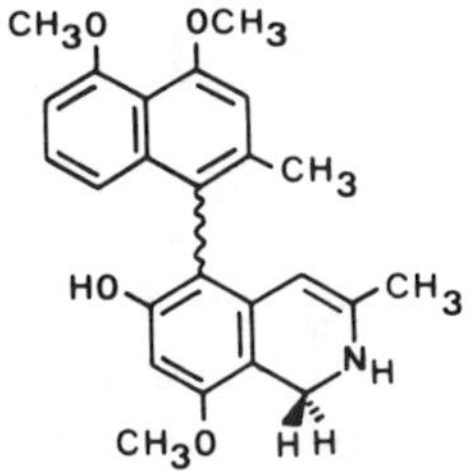	*A. congolensis*; roots and twigs (*28,29*)	258 (*28,29*)	0° (methanol) (*28,29*)	Spectroscopy (*28,29*)

150

Ancistrocongine (**17**)[c]
(28,29)
$C_{22}H_{21}O_3N$ (347)

A. congolensis, roots and twigs (28,29)

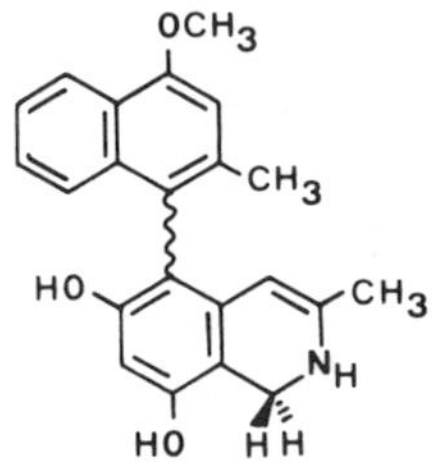

298 (28,29)

0° (methanol) (28,29)

Spectroscopy (28,29)

17

7-1′ Coupled Alkaloids

Ancistrocladisine (**18**)
$C_{26}H_{29}O_4N$ (419)

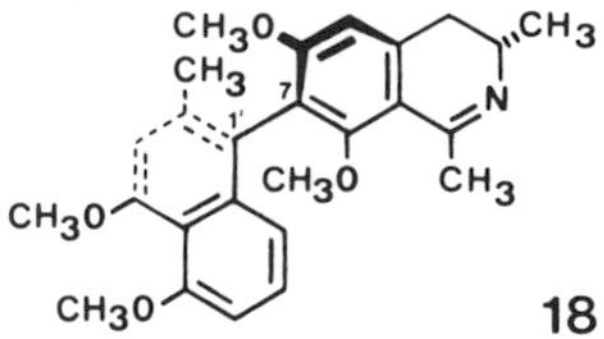

A. heyneanus; roots (16)

178–180 (16)

−16.1° (hydrochloride, in CHCl₃) (16)

Spectroscopy (16)
Exciton chirality method (18)
Degradation (16,18)
X-Ray (19)

18

(*continued*)

151

TABLE I (*Continued*)

Formula (mol wt), structure	Occurrence	mp (°C)	[α]$_D$ (solvent)	Methods for structural elucidation
Ancistrine (**19**) (27)[b,c] C$_{25}$H$_{29}$O$_4$N (407) 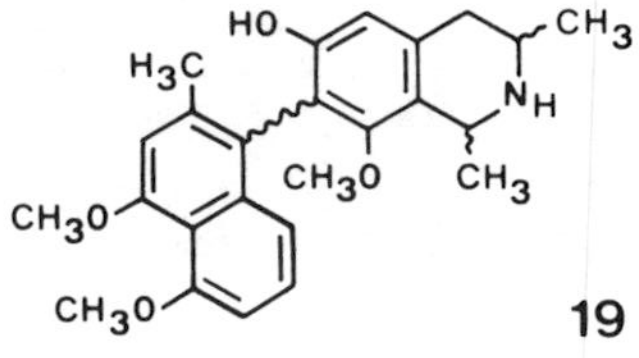 **19**	*A. ealaensis*; roots and twigs (27)	230–231 (27)	−35° (methanol) (27)	Spectroscopy (27) Same[e] O-methylation product as **20** (27)
Ancistine (**20**) (27)[b,c] C$_{25}$H$_{29}$O$_4$N (407) 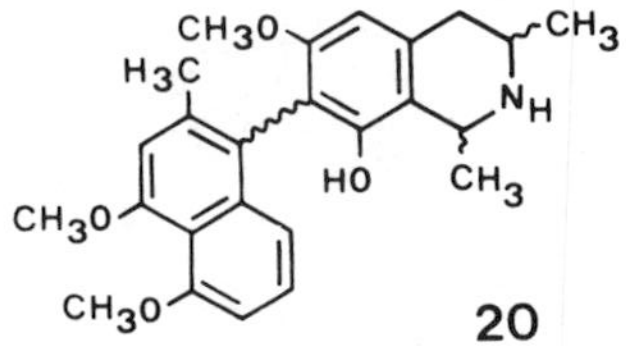 **20**	*A. ealaensis*; roots and twigs (27)	275–276 (27)	−34° (CHCl$_3$–methanol, 1:1) (27)	Spectroscopy (27) Same[e] O-methylation product as **19** (27)

7-2' Coupled Alkaloids

Ancistrocladidine (21)
$C_{25}H_{27}O_4N$ (405)

A. heyneanus; roots (17)

245–247 (17)

−149.7° (CHCl$_3$) (17)

Spectroscopy (17)
Exciton chirality method (18)
Degradation (18)
X-Ray (19)

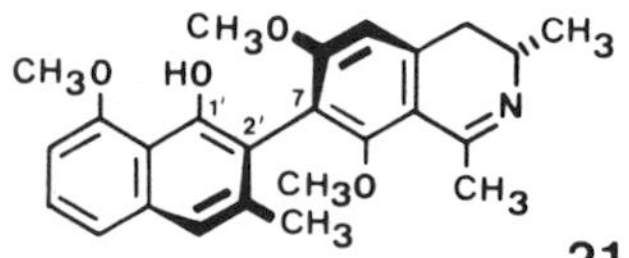

21

Ancistrotectorine (22)
$C_{26}H_{31}O_4N$ (421)

A. tectorius; leaves (24)

134–140 (24)

0° (CHCl$_3$) (24)

Spectroscopy (24)
X-Ray (24)

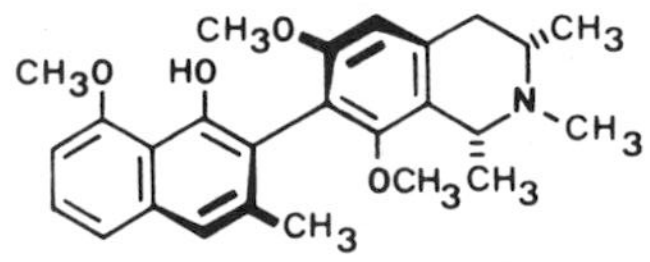

22

[a] Postulated *relative* stereochemistry.

[b] No relative or absolute configuration established.

[c] Complete neglect of atropisomerism.

[d] Compounds 12 and 13 have been proposed to be stereoisomeric mixtures (11,21).

[e] Identity of O-methylation products was shown on the basis of one single peak (δ = 3.58) in the 60-MHz NMR of 10 mg of a crude reaction mixture (27,29).

153

methyl groups was interpreted to be trans, based on the magnetic inequivalence of the benzylic protons in N-benzyltriphyophylline (33) (!).

By similar arguments, the structure of the isomeric alkaloid isotriphyophylline was interpreted as **24,** with a relative cis configuration of the two secondary methyl groups, because its N-benzyl derivative showed a "singlet" in its 60-MHz NMR (33). The possibility that isotriphyophylline might be, for example, the atropisomer of **23** was not regarded.

From the same plant, O-methyltriphyophylline (**25**), N-methyltriphyophylline (**26**), O-methyl-1,2-didehydrotriphyophylline (**28**), and O-methyltetradehydrotriphyophylline (**29**) were also isolated and elucidated by similar methods (32,33). Astonishingly, reduction of **28** with NaBH$_4$, which is expected (11,14) to give the corresponding cis product, was reported (32) to yield **25,** described as the methylether (32) of the trans-configurated (33) alkaloid **23.** The structure of **29,** however, was recently (7) confirmed by total synthesis (see Section V,D). The first naphthyl isoquinoline alkaloid to possess a functionalized methyl group on the naphthalene moiety, triphyopeltine, was proposed to have the gross structure **27** (31), which will have to be confirmed by future thorough investigation.

B. *Dionchophyllum tholonii*

Bruneton's last communication (33) on naphthyl isoquinoline alkaloids includes work on the related plant D. *tholonii* Baill., also belonging to the Dionchophyllaceae. It is not clear, however, which of the alkaloids **24, 25, 26,** or **29,** mentioned above, occur in *this* species.

C. Table of *Triphyophyllum* (and *Dionchophyllum*) Alkaloids

Table II summarizes the occurrence, structures, and physical data of alkaloids isolated from Dionchophyllaceae and includes the methods applied for their structural elucidation.

IV. Biosynthesis of Naphthyl Isoquinoline Alkaloids

A. The Conception of Acetogenin Isoquinoline Alkaloids

1. General Considerations

The published structures of *Ancistrocladus* alkaloids (see Section II) show a broad variation concerning the hydrogenation degree of the isoquinoline part, the

TABLE II

Naphthyl Isoquinoline Alkaloids from *Triphyophyllum* and *Dionchophyllum* Plants

Formula (mol wt), structure	Occurrence	mp (°C)	$[\alpha]_D$ (solvent)	Methods for structural elucidation
Triphyophylline (23) (31,33)[a,c] C$_{24}$H$_{27}$O$_3$N (377) 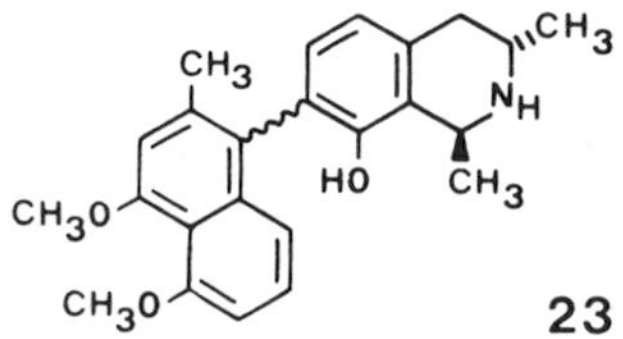 23	T. peltatum; twigs (31)	215 (31)	−14° (CHCl$_3$) (31)	Spectroscopy (31,33) Same dehydrogenation product as 24 (33) Total synthesis of rac. 23 confirming the constitution as well as the relative configuration at C-1 and C-3
Isotriphyophylline (24) (33)[a,c] C$_{24}$H$_{27}$O$_3$N (377) 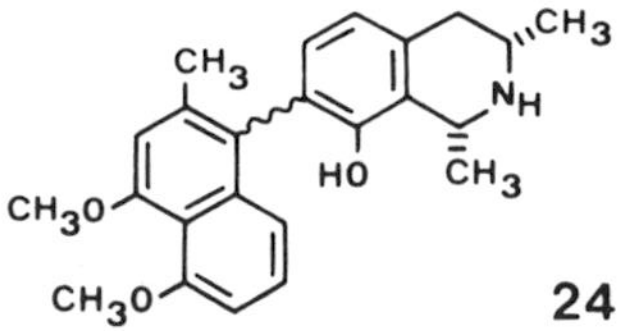 24	T. peltatum; twigs and leaves (33) D. tholonii; twigs and leaves (33)	256 (33)	−22° (CHCl$_3$) (33)	Spectroscopy (33) Same dehydrogenation product as 23 (33)
O-Methyltriphyophylline (25) (32)[a,c] C$_{25}$H$_{29}$O$_3$N (391)	T. peltatum; twigs (32)	—	−30° (CHCl$_3$) (32)	Spectroscopy (32) Partial synthesis from 28 (32)[d]

(continued)

TABLE II (*Continued*)

Formula (mol wt), structure	Occurrence	mp (°C)	[α]$_D$ (solvent)	Methods for structural elucidation
N-Methyltriphyophylline (**26**) (*33*)[a,c,e] C$_{25}$H$_{29}$O$_3$N (391)	*T. peltatum*; twigs and leaves (*33*) *D. tholonii*; twigs and leaves (*33*)	185 (*33*)	+70° (CHCl$_3$) (*33*)	Spectroscopy (*33*) Partial synthesis from **23** (*33*)
Triphyopeltine (**27**) (*31*)[b,c,f] C$_{23}$H$_{25}$O$_4$N (379)	*T. peltatum*; twigs (*31*)	241 (*31*)	−95° (pyridine) (*31*)	Spectroscopy (*31*)

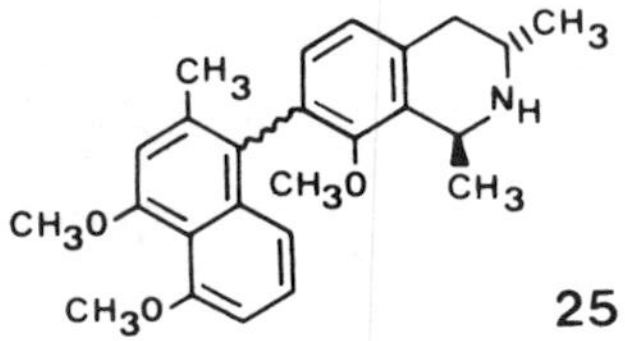

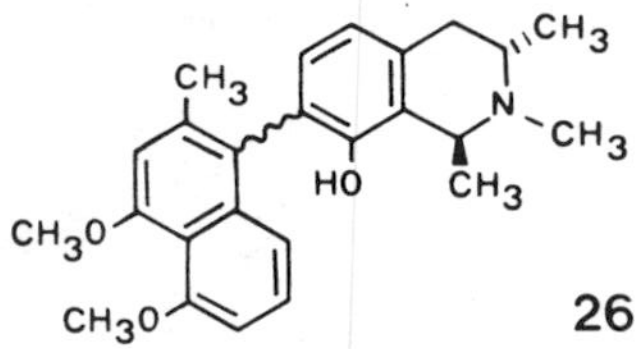

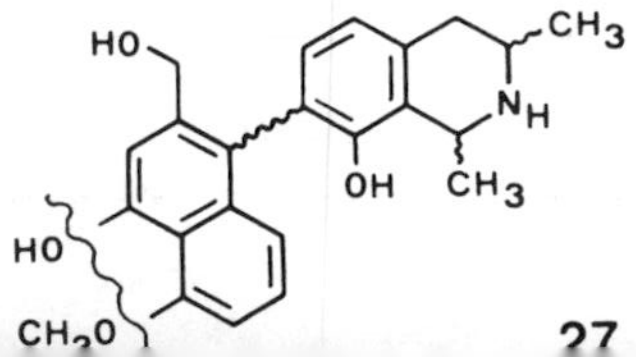

25

26

27

O-Methyl-1,2-didehy-drotriphyophylline (**28**) (32)[b,c] $C_{25}H_{27}O_3N$ (389)	*T. peltatum*; twigs (32)	amorphous (32)	0° (solvent not specified) (32)	Spectroscopy (32) Conversion to **25** (32)[d]

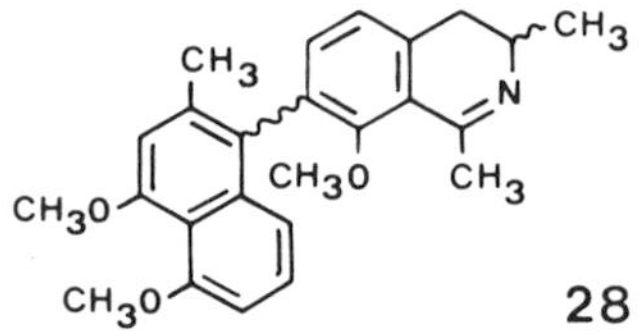

28

O-Methyltetradehydrotri-phyophylline (**29**) (33)[c,e] $C_{25}H_{25}O_3N$ (387)	*T. peltatum*; twigs and leaves (33) *D. tholonii*; twigs and leaves (33)	166 (33,7)	0° (CHCl₃) (33)	Spectroscopy (33) Partial synthesis from **23** (33) Total synthesis (7)

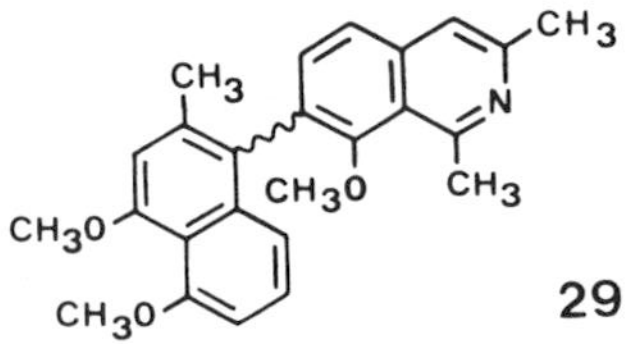

29

[a] Postulated *relative* stereochemistry.
[b] No relative or absolute configuration established.
[c] Complete neglect of atropisomerism.
[d] Astonishing that reduction with NaBH₄ should lead only to the trans isomer **25**.
[e] Formula miscalculated in Ref. 33.
[f] Regiochemistry not established.

stereochemistry at the aliphatic chiral centers and at the biphenyl linkage, as well as the coupling position of the two molecular moieties. Nonetheless, three structural features are common to all of these alkaloids: the most unusual methyl group at C-3 of the isoquinoline, the meta oxygenation pattern at C-6 and C-8, and the characteristic naphthalene substituent at C-5 or C-7. These structural peculiarities, however, do not fit into a conventional scheme of isoquinoline biosynthesis from aromatic amino acids. In contrast, they strongly point to a unprecedented biogenetic pathway to isoquinoline alkaloids from acetate units, via β-polycarbonyl precursors.

2. *Ancistrocladus* Alkaloids

A possible biosynthetic route to acetogenin isoquinoline alkaloids is sketched in Scheme 3. According to this hypothesis (*35,36*), the common precursor to *both* molecular moieties of all the *Ancistrocladus* alkaloids could be the β-pentaketone **30** (or the corresponding β-polycarbonyl acid), which, for its part, would arise from six acetate units. Aldol-type condensation of **30** and aromatiza-

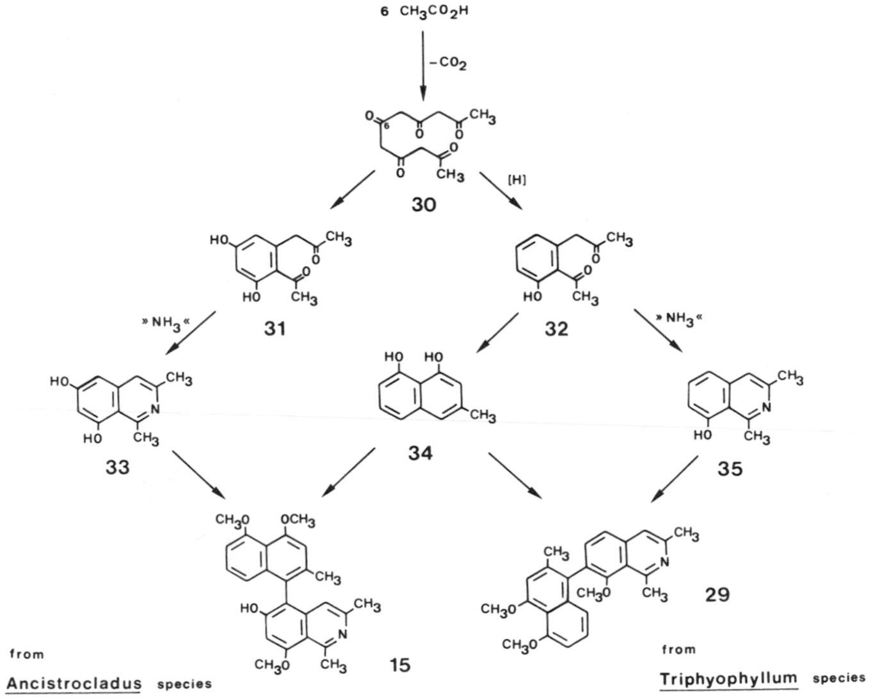

SCHEME 3. Proposed biogenetic scheme for naphthyl isoquinoline alkaloids (*35,36*).

tion to the diketo resorcinol **31,** followed by an incorporation of free or bound ammonia, would lead to the isoquinoline **33** with the characteristic oxygenation pattern. The formation of naphthalene **34** could best be rationalized by a reduction of the central carbonyl function (C-6) of **30,** followed by *two* subsequent aldol condensation steps, this time via the less oxygenated diketo phenol **32.** Final coupling of the two molecular moieties **33** and **34** to the complete alkaloid ancistrocladeine (**15**), by phenol oxidation, would simultaneously explain the other coupling types (7–1′ and 7–2′) of similar alkaloids.

3. *Triphyophyllum* Alkaloids

A possible biosynthesis of the structurally related *Triphyophyllum* alkaloids fits impressingly well into this concept. Their lack of an oxygen function at C-6 perfectly corresponds to a joint biogenesis of both molecular halves of **29** from the same, less oxygenated diketone **32**—a remarkably economical, converging biosynthesis which, despite the structural diversity of the two molecular moieties **34** and **35,** manages with only the one monocyclic precursor **32.**

4. The Nitrogen Source

The especially widespread occurrence of the less oxidized corresponding di- and tetrahydroisoquinolines in both families (see Sections II and III) requires more detailed considerations on the biological source of the nitrogen, also in view of the uncommon use of the toxin ammonia outside the primary metabolism (*37*). Moreover, formation of fully conjugated isoquinolines according to Scheme 3 would necessitate a subsequent hydrogenation of the resulting stable aromates **33** and **35,** a biochemically unusual reaction type (*3*), which also, chemically, requires relatively hard reaction conditions (e.g., Zn–HCl, see Section V,C,3).

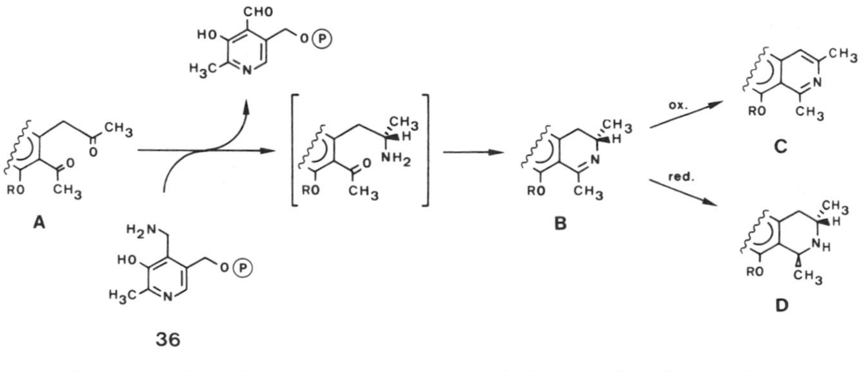

SCHEME 4. A possible nitrogen source for naphthyl isoquinoline alkaloids (*38*).

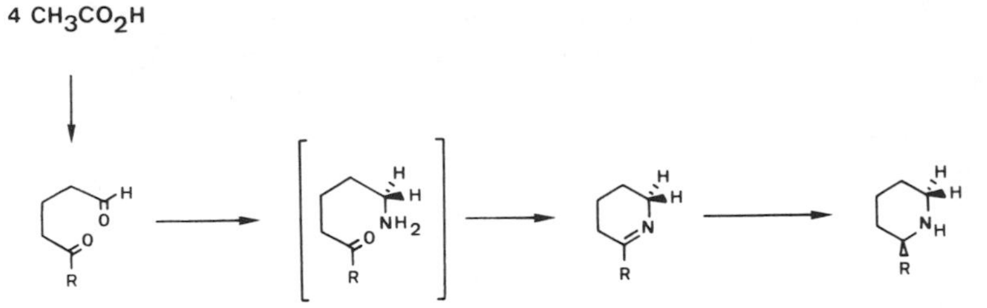

GERHARD BRINGMANN

4 CH₃CO₂H

SCHEME 5. Nitrogen incorporation into coniine (**39**) (*39, 39a*).

A possible way to avoid intermediate stable isoquinolines like **33** and **35** is *reductive* amination of the more reactive carbonyl function of **A** (representing **31** or **32**), [which is biochemically plausible, e.g., with pyridoxylamine (**36**)], leading directly to the dihydroisoquinolines **B,** which then might very easily be further oxidized to isoquinolines **C** or reduced to tetrahydroisoquinolines **D** (see Scheme 4). This possible pathway closely resembles the known (*39, 39a*) bio-synthesis of the hemlock alkaloid coniine (**39**), where reductive amination of the more reactive carbonyl function of the diketo precursor **37** leads to the cyclic imine γ-coniceine (**38**), which is further reduced to **39** (see Scheme 5).

B. TAXONOMIC ASPECTS

The genus *Ancistrocladus* includes more than 20 species of tropical lianas, endemic to the rain forests of West and Central Africa, India and Indonesia (*8*). Due to its peculiar anatomical properties, which do not show a clear relationship to those of other genera, *Ancistrocladus* has been taxonomically categorized, over time, into several different families in the plant kingdom (*40*). Although, in the end, these climbers were classified as a family of their own (*8*), the An-cistrocladaceae, with only the one genus *Ancistrocladus,* their botanical position remains uncertain.

The outstanding taxonomic position of the Ancistrocladaceae clearly corre-sponds to their apparent ingenuity of economically synthesizing their iso-quinolines from acetate units rather than from aromatic amino acids. In this context, the brief tabular mention (*41*) that ancistrocladine (**1**) has been found in *Celosia argentia* (Amaranthaceae), moreover together with the "conventional" biologial amine hordenine, is of great interest, and deserves to be pursued in more detail.

Apart from this preliminary report, *Triphyophyllum peltatum* and *Dioncho-phyllum tholonii* are the only species outside the Ancistrocladaceae that contain naphthyl isoquinoline alkaloids. The characteristic biochemical difference, that

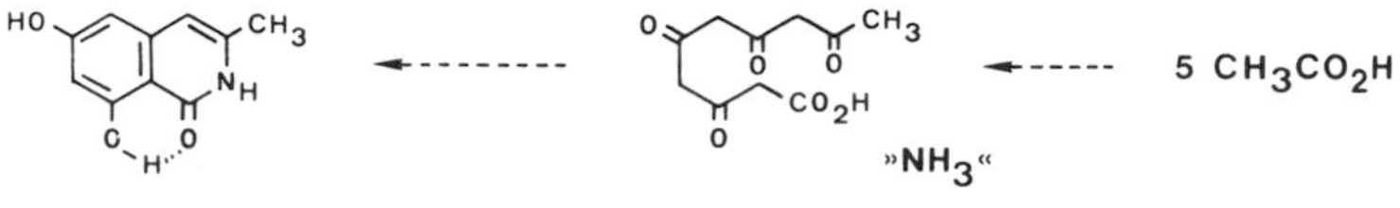

Fig. 2

the oxygen function in the 6-position of the isoquinoline, present in all the *Ancistrocladus* alkaloids, gets lost during biosynthesis in *Dionchophyllum* and *Triphyophyllum,* may help redefine the taxonomical position of the Ancistrocladaceae by pointing to a probable phylogenetic relationship to the Dionchophyllaceae.

Outside these two families, however, no other plants have been found to produce "real" tetrahydro-, dihydro-, and fully aromatic isoquinoline alkaloids in a polyacetate way. Closely related, but chemically a lactam, is siamine (**40**) from *Cassia siamea* Lam. (Leguminosae) (*42*). The structure suggests a biogenesis from a pentaketide precursor **41** (Fig. 2), thus requiring one acetate unit less than each of the naphthyl isoquinoline alkaloid moieties.

C. BIOSYNTHETIC EVIDENCE

1. Feeding Experiments

No systematic investigations on the biosynthesis of naphthyl isoquinoline alkaloids have been published so far. Obviously, one of the principal reasons for this are the highly demanding growth conditions required for at least some of the sensitive Ancistrocladaceae, especially the Indian liana *A. heyneanus,* which, until relatively recently, resisted all cultivation efforts (*43*). Thanks to the collection of seeds, harvested under the guidance of Govindachari and his colleagues in Bombay, plants of this important species could, for the first time, be grown in a green house (*44*). Similarly, some West African lianas were cultivated, as well as *A. hamatus* from Sri Lanka (*45*).

With *A. hamatus,* preliminary feeding experiments were performed, showing

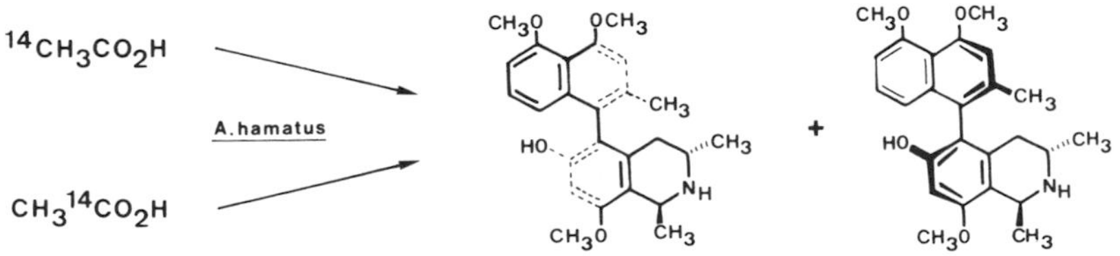

Fig. 3

low, but significant incorporation rates (~0.011%) for both [1-¹⁴C]- and
[2-¹⁴C]acetate into ancistrocladine (**1**) and hamatine (**10**) (Fig. 3). Problems
originated mainly from the presence of the alkaloids nearly exclusively in the
bark of the wooden roots (*45*). Further work to overcome these problems, also on
the basis of tissue cultures, is in progress.

2. Biosynthetically Related Substances in the Same Plants

No direct linear or monocyclic precursors, like **30, 31,** or **32,** which might
further lend credence to the proposed biogenetic scheme (see Scheme 3), have
been detected in Ancistrocladaceae or Dionchophyllaceae. Not even the molecular moieties—isoquinolines or naphthols in their genuine form—have been
found.

The identification of plumbagine (**42**), however, in *Triphyophyllum peltatum*
(*31*), as well as in *Ancistrocladus abbreviatus* Airy Shaw (*46*), co-occurring with
the alkaloids, can be interpreted as a strong indirect indication of a polyacetate
pathway in the plants. By 1972, Zenk and co-workers had already found that the
naphthoquinone **42** is formed from acetate units, albeit in *Drosera* and *Plumbago*
species (*47*). A joint biosynthesis of plumbagine (**42**) and both molecular
moieties, for example, of the alkaloid **23** from the same monocyclic precursor **32**
seems evident (Scheme 6).

In view of the plausibility of acetogenin, non-tyrosine-derived isoquinolines,
it is of interest that, *vice versa,* tyrosine has been found to be incorporated into
plumbagine (**42**) itself, albeit after catabolic degradation (*48*)! Still more oxi-

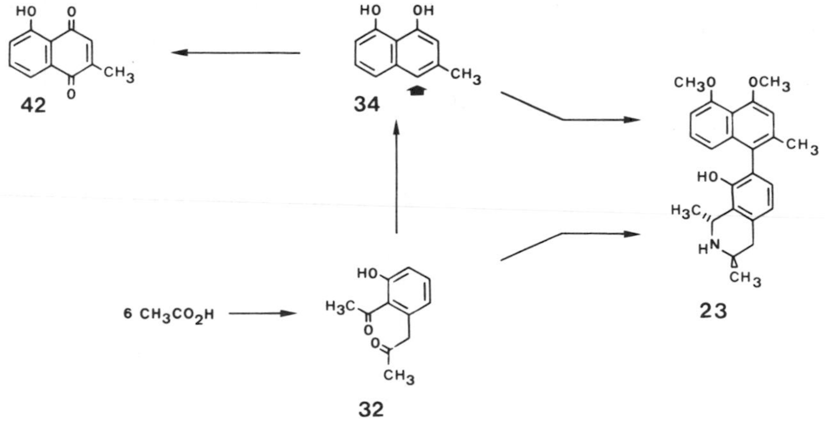

SCHEME 6. Possible biosynthetic relationship between naphthyl isoquinoline alkaloids and co-
occurring naphthoquinones.

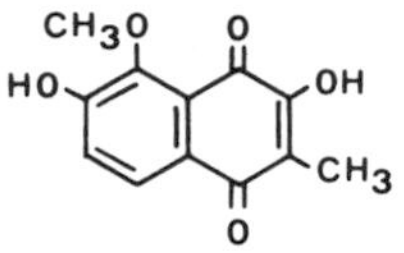

43

FIG. 4

dized and thus biosynthetically further away from diol **34** than plumbagine (**42**) is ancistroquinone (**43**), isolated by Govindachari from the roots of *Ancistrocladus heyneanus* (*49*).

3. Chemical Model Reactions

Additional strong evidence of the chemical plausibility of joint *in vivo* formation of both naphthalenes and isoquinolines from common β-polycarbonyl precursors was obtained from the *in vitro* imitation of the proposed biosynthetic pathway in the laboratory. These model reactions, which simultaneously allowed first total syntheses of naphthyl isoquinoline alkaloids, are depicted in Section V.

V. The Biomimetic Synthesis of Acetogenin Isoquinoline Alkaloids

Following the postulated biogenetic pathway (see Scheme 3), a chemical *in vitro* imitation, aiming at a biomimetic first synthesis of naphthyl isoquinoline alkaloids, had to consist of essentially four crucial steps: (1) synthesis of suitable β-polycarbonyl precursors, (2) their primary cyclization to monocyclic diketones like **31** or **32,** (3) the further differentiation to naphthalenes and isoquinolines, and (4) their mixed coupling to the complete natural alkaloids.

A. Synthesis and Cyclization of β-Pentaketones

1. Terminally Protected β-Polyketones

Synthetic efforts first concentrated on the preparation of the linear precursor **30.** In order to avoid unfavorable different cyclization types, which are also realized in other natural products, the polyketone **30** was specifically built up in its bis-terminally ketalized form **46** (*50*), as shown in Scheme 7.

Condensation of acetone with the protected ester **44** yielded the monoketalized triketone **45,** which, after transformation into its dianion, was further condensed to **46**. This, however, failed to give the resorcinol **48**. The two sterically and electronically inactivating ethylene ketals completely prevented cyclization reac-

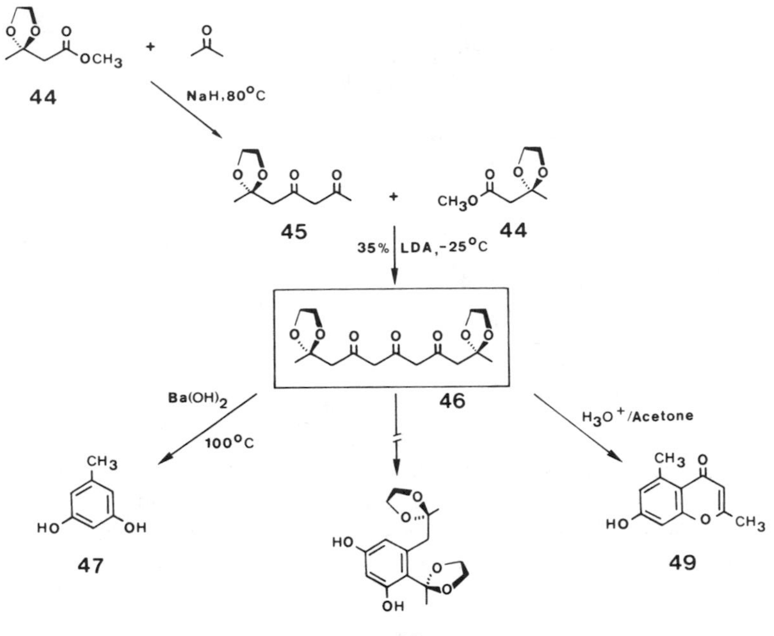

Scheme 7. Synthesis and cyclization behavior of **46** (*50*).

tions, except by base-catalyzed carbon chain cleavage to orcinol (**47**) or by acid-catalyzed deketalization, finally leading to the chromone **49** (*50*). On the other hand, leaving one outer keto function of **46** unprotected, by condensing **44** this time with the free triketone **50** (*50,51*), gave the far more reactive polyketone **51** (Scheme 8). This compound condensed under the reaction and cyclization conditions already, albeit not to **54**, but to its undesired isomer **53**.

The same cyclization behavior of the β-polycarbonyl chain—not to fold in the middle, as postulated for the alkaloid biosynthesis (see Scheme 3), but to "roll up" from an unprotected end—was also observed for the parent compound, the free β-pentaketone **30**, itself, which led *in vitro* directly to doubly cyclized products like **49** (*52*). This type of "chemical" cyclization is sometimes used by nature, as well, and thus could be exploited biomimetically (*50,51,53*) for a first total synthesis of the aloenin aglycon (**52**), a natural product from *Aloe arborescens* var. *natalensis*, in which the carbon chain is similarly folded. The same type of cyclization, using the β-polycarbonyl chain in its complete, non-decarboxylated form **55**, could achieve a still shorter synthesis of **52** (Scheme 9), being even more closely oriented to the biosynthesis (*50,51*).

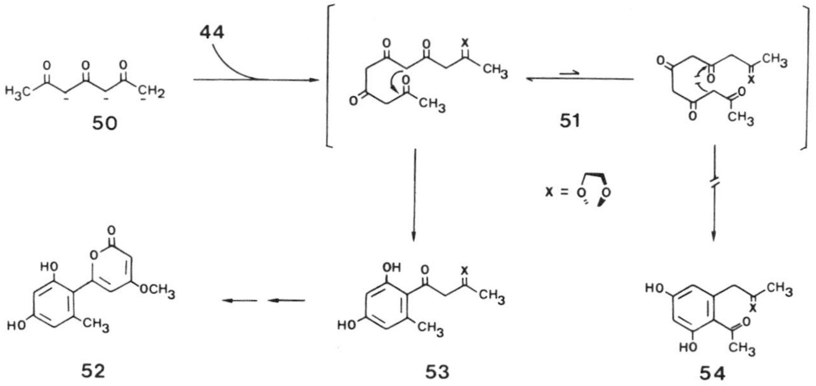

SCHEME 8. Synthesis and cyclization behavior of monoketal **51** (*50,51*).

2. Centrally Modified β-Pentaketones

Neither **30** nor any of the terminally protected polyketones **46, 51,** or **55** could be used as a basis for a total synthesis of naphthyl isoquinoline alkaloids. Synthetic efforts thus focused on a derivation of the *central* carbon atom of the polycarbonyl chain (*54–56*).

a. By Ozonolysis. The idea to set free exactly the required, centrally protected β-pentaketone **57** by ozonolysis of the dihydroindane **56** originates from Birch (*57*), who, however, could not identify any cyclization products thereof. More recent attempts with even longer carbon chains did not lead to definite cyclization products either (*58*).

Nonetheless, this strategy was taken up again (*54,56*). After an improved synthesis of the dihydroindane **56,** cleavage to **57** was carried out using O₂-free (*59*) ozone (Scheme 10). Ring closure could now be brought about very smoothly with various nonnucleophilic bases, more simply, but silica gel-catalyzed by a short column filtration, and indeed gave the desired monocyclic diketone **58.**

The central ketal function, which electronically and sterically governed the

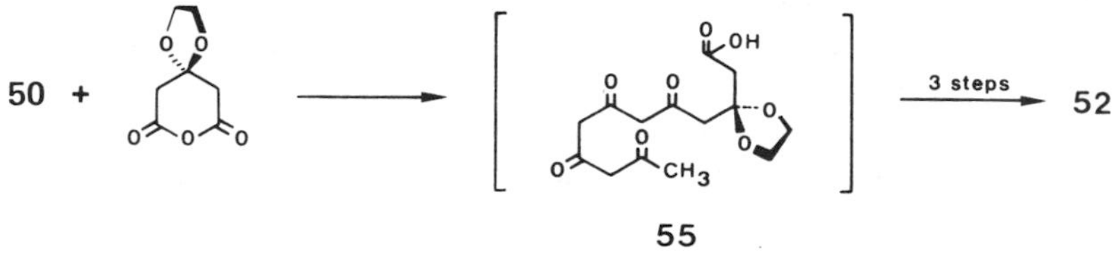

SCHEME 9. Synthesis and cyclization behavior of the polycarbonyl acid **55** (*50,51*).

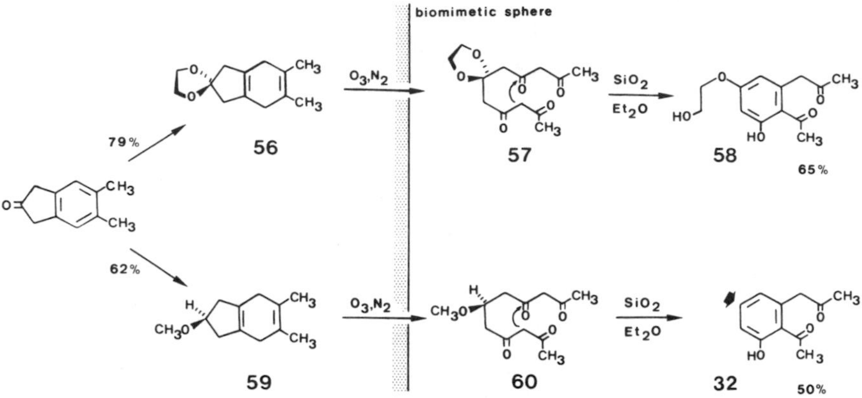

SCHEME 10. Centrally modified β-pentaketones **57** and **60**: synthesis by ozonolysis (*57,54,56*) and cyclization behavior (*54,56*).

cyclization direction, moreover allowed control of the oxidation level of the polyketone, and thus the degree of oxygenation of the resulting aromatic. By ozonolysis of the more reduced indane **59**, also **32**, the oxygen poorer diketo precursor to naphthyl isoquinoline alkaloids (cp. Scheme 3), could be gained very efficiently. A similar influence of a centrally attached protective group on the cyclization behavior of β-polyketones had also been exploited by Harris (*60*) in the synthesis of the mold metabolite emodine. The reactions in Scheme 10 were part of the first natural product synthesis involving an ozonolytic preparation of a β-polycarbonyl precursor.

b. One Pot Syntheses of 32 and 58 Employing Lithium–Potassium Acetone. In view of the favorable cyclization behavior of the centrally modified β-polycarbonyl chains **32** and **58,** on the one hand, and because of the relatively long syntheses of the indane precursors **56** and **59** on the other, efforts were made to develop shorter synthetic routes to these symmetrical polyketones. The synthesis of the diketone **58** in only one step, without the intermediate linear precursor **57** being isolated this time, succeeded (*55,56*) by condensation of the diester **61** with two molecules of the *di*anion **62** of acetone, which is prepared (*61*) by further deprotonation of potassium acetone (**63**) with *n*-butyllithium (Scheme 11). This one-pot preparation of **58,** which failed with conventional acetone enolates, was the first application of this potent nucleophile in organic synthesis. In the same way, the less oxygenated alkaloid precursor **32** was also synthesized in only one step, starting from more reduced diesters like **64** (*55,56*) (see Scheme 12).

c. By Reductive Cleavage of Isoxazoles. Centrally modified β-pentaketones have also been prepared by hydrogenation of isoxazoles like **65** (*62*)

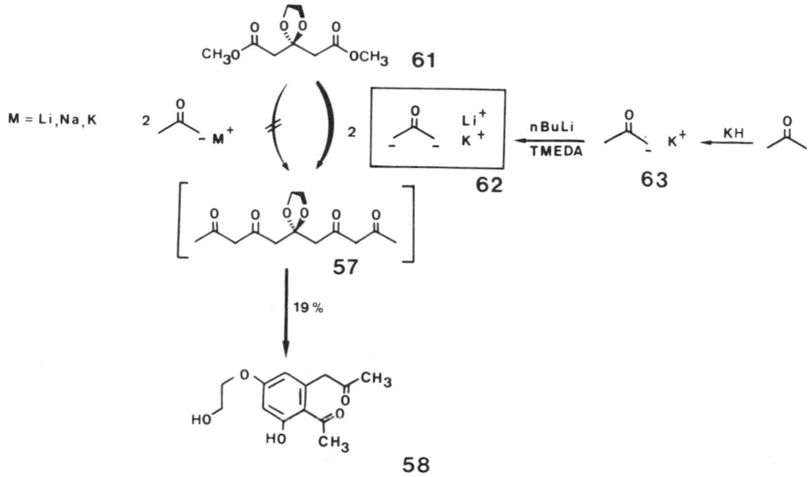

SCHEME 11. One-pot preparation of **58** by ester condensation (*55,56*).

(Scheme 13). Although the bisenamine **66** thus obtained condensed directly to isoquinoline **35**, the reaction did not lead to monocyclic intermediates (like **31, 32,** or **58**), as required for biosynthetic experiments and also for chemical syntheses of naphthalenes (see Section V,C,1) and tetrahydroisoquinolines (see Section V,C,3).

B. NONBIOMIMETIC SYNTHESIS
OF MONOCYCLIC DIKETONES

For future labeling syntheses of potential biosynthetic precursors, as well as for preparation of monocyclic diketones **70,** on a 100- to 1000-g scale, additional, nonbiomimetic syntheses to these important intermediates were elaborated (*63*) (see Scheme 14). The key step of the first synthesis was oxidative cleavage of suitably substituted dimethyl indenes **68,** either directly with permanganate–periodate or via remarkably stable ozonides **69.** The required indenes **68** could be prepared from esters **67,** and thus from very cheap educts like methacrylic acid and

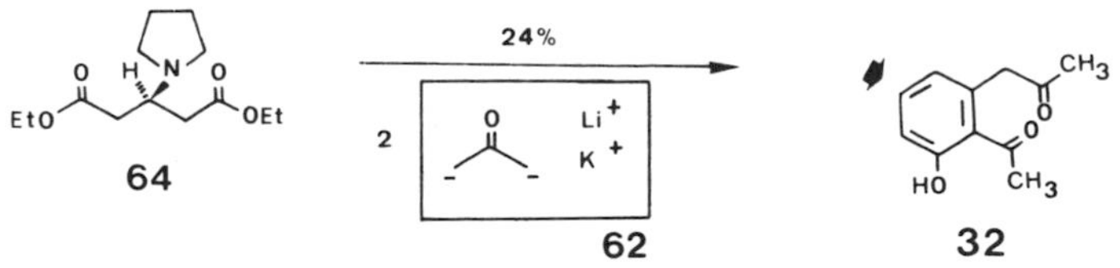

SCHEME 12. One-pot preparation of **32** (*55,56*).

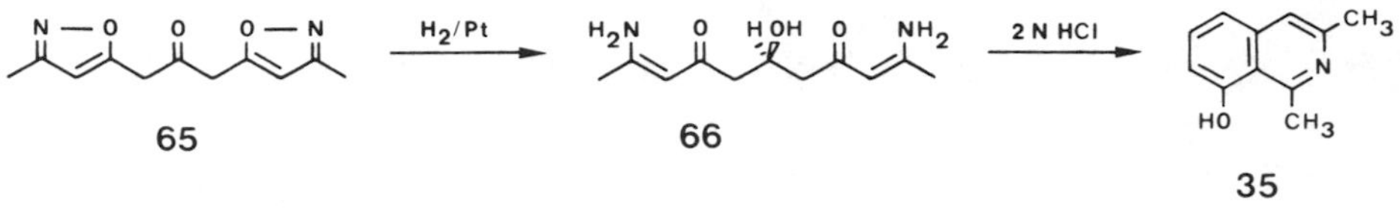

SCHEME 13. The Tanaka cyclization of β-polycarbonyl enamines (62).

phenols. This "indene route" was especially appropriate for the synthesis of *mono*phenolic diketones like **32** (**70,** X = R = H).

The key step of the other synthesis was C-acetylation of aryl propanones like **72,** which are readily accessible from the corresponding benzaldehydes **71.** This "aryl propanone route" was very effective for the synthesis of *bis*phenolic diketones like **70** (X = OR), thus ideally complementing the other route. Both methods also allowed the synthesis of the free diketone **31** (**70,** R = H, X = OH) with two unprotected phenol functions in the aromatic ring, which could not be prepared by O-dealkylation of the hydroxyethyl ether **58,** as obtained in the biomimetic syntheses.

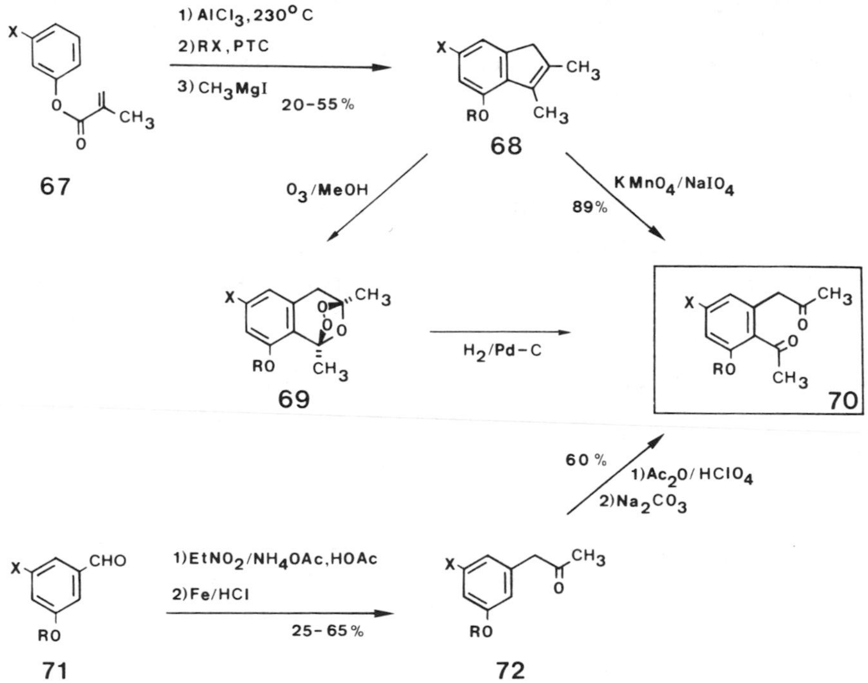

SCHEME 14. Nonbiomimetic pathways to monocyclic alkaloid precursors (63).

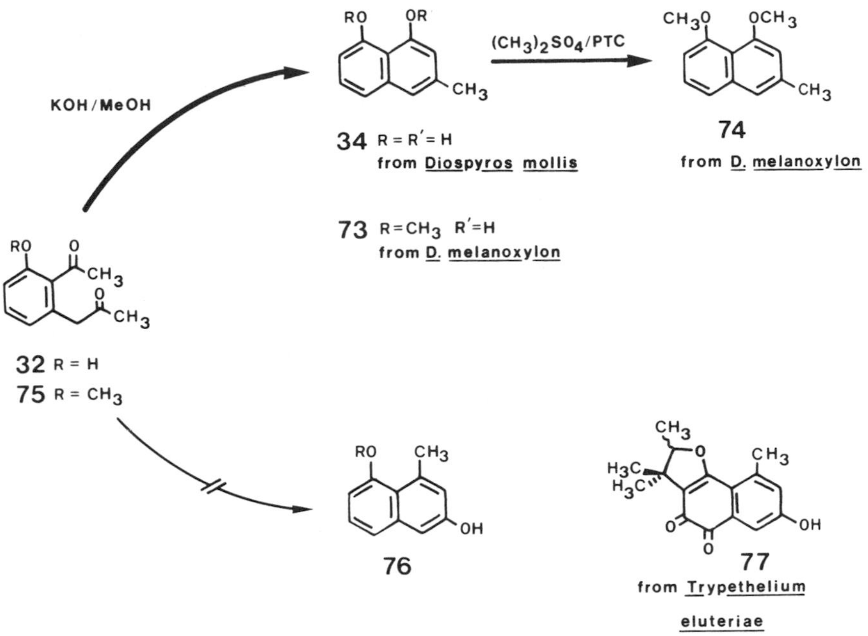

SCHEME 15. Biomimetic synthesis of naturally occurring naphthalenes (36,54,55).

C. NAPHTHALENES AND ISOQUINOLINES FROM JOINT BIOSYNTHETIC PRECURSORS

Thus, the postulated diketo precursors **70** had become accessible by complementary biomimetic and nonbiomimetic synthetic pathways. Starting from these joint monocyclic key intermediates, the biomimetic synthesis of both bicyclic molecular moieties of naphthyl isoquinoline alkaloids was readily achieved (35,36,54,55).

1. Biomimetic Naphthalene Syntheses

Further aldol condensation of the diketone **32** with methanolic KOH quantitatively gave the naphthalene **34,** which was O-methylated to the dimethyl ether **74** (Scheme 15), the isocyclic molecular moiety of most of the *Ancistrocladus* and *Triphyophyllum* alkaloids (36,54,55).

The analogous cyclization of **75** allowed a specific synthesis of the monomethyl ether **73,** the naphthalene part of 7–2' coupled alkaloids like **21** and **22** (36,55). All three naphthalenes **34, 73,** and **74** also have been found free, not coupled to isoquinolines, for example, in tropical Ebenaceae (64,65). The ring closure reaction to **34,** or, respectively, to **73,** was found to be regiospecific; the

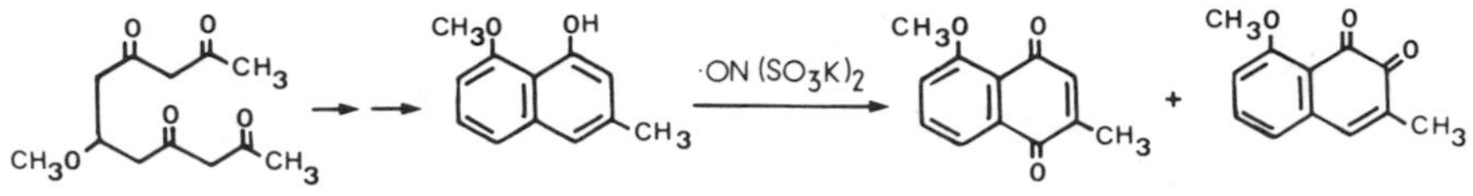

GERHARD BRINGMANN

SCHEME 16. The biomimetic synthesis of acetogenin naphthoquinones (46).

isomeric naphthalene **76**—a potential precursor to the lichen antibiotic (65a) trypethelone (**77**)—could not be detected.

The facile preparation of the correct monomethyl ether **73** from **75** also allowed the biologically imitative oxidation to the corresponding naphthoquinones **78** and **79** (46) (see Scheme 16), a reaction that failed with the diol **34** (64). Such acetogenin naphthoquinones occur not only in alkaloid-containing lianas (cp. Scheme 6) but also in ebony woods (65).

2. Biomimetic Isoquinoline Syntheses

From the monocyclic precursor **75**, the isoquinoline moiety **80** of the *Triphyophyllum* alkaloid **29**, with no oxygen function at C-6, could, be prepared by reaction with aqueous ammonia (36,55) (see Scheme 17). The undesired formation of the aminonaphthalene **81** by simultaneous carbocyclic ring closure could be suppressed by use of ammonium salts as less basic nitrogen sources. In the same way, starting from the higher oxygenated diketone **58**, the isoquinoline moiety **83** of ancistrocladeine (**15**) could also be gained, that is, *with* the oxygen function at C-6 (54,36) (see Scheme 18). A selective cleavage of the hydroxyethyl ether at C-6 of **82** could be achieved with strong bases like potassium hydride.

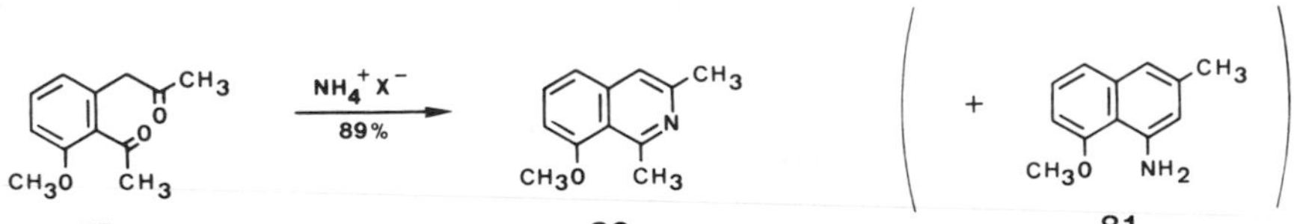

SCHEME 17. Synthesis of the isoquinoline part of the *Triphyophyllum* alkaloid **29**.

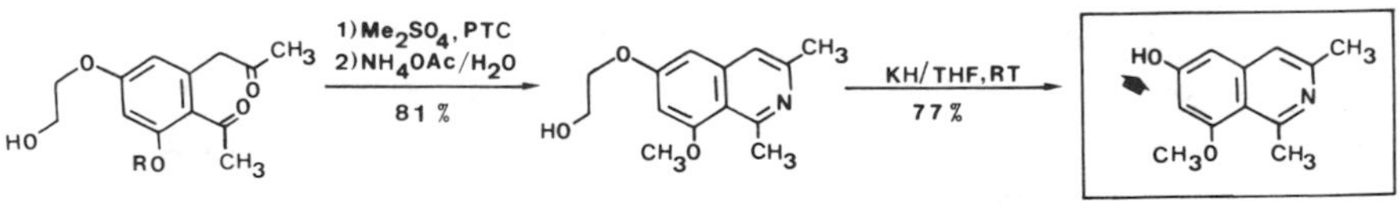

SCHEME 18. Synthesis of the isoquinoline part of the *Ancistrocladus* alkaloid **15** (54, 36).

3. Di- and Tetrahydroisoquinoline Alkaloid Moieties

Starting from aromatic isoquinolines, the corresponding tetrahydroisoquinolines, required for the alkaloid synthesis, could also be prepared (as racemates) by standard procedures (7,35). Thus, reduction of **80** with Zn–HCl gave a cis/trans mixture of **84** and **85** (7:3), whereas catalytic hydrogenation over PtO$_2$ diastereoselectively led to the cis product **84** (>95:5) (see Scheme 19). Obviously because of the steric hindrance exerted by the two methyl groups adjacent to the nitrogen, no preparatively reliable N-benzylation of **80** to a more readily reducible isoquinolinium salt **86** could be achieved (38). By oxidation of **84** or **85** with MnO$_2$, also the corresponding dihydroisoquinoline **87**, molecular moiety of the *Triphyophyllum* alkaloid **28** (see Section III,C) could be gained.

In view of the tiresome zinc reduction of fully aromatized isoquinolines like **80,** attempts were made to avoid these stable intermediates by "biomimetically" incorporating the nitrogen in the sense of a reductive amination with pyridoxamine (**88**) (cp. Scheme 4, Section IV,A,4). *In vitro* reaction of the diketone **75** with the amine form **88** of vitamin B$_6$, however, did not even give traces of the desired product pair **87/89**, but an N-pyridoxylisoquinolinium salt **90**, instead, which as seen before (see Scheme 19) for the N-benzyl compound **86**, would not have been accessible by N-alkylation on the preformed isoquinoline **80**. Reduction of **90** to **91**—now possible with NaBH$_4$ owing to the positive charge—

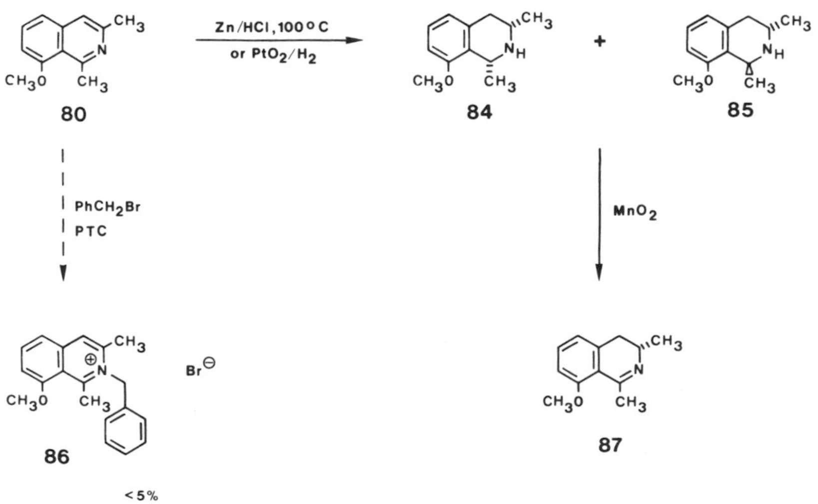

SCHEME 19. Preparation of (racemic) di- and tetrahydro-1,3-dimethylisoquinolines (7,35).

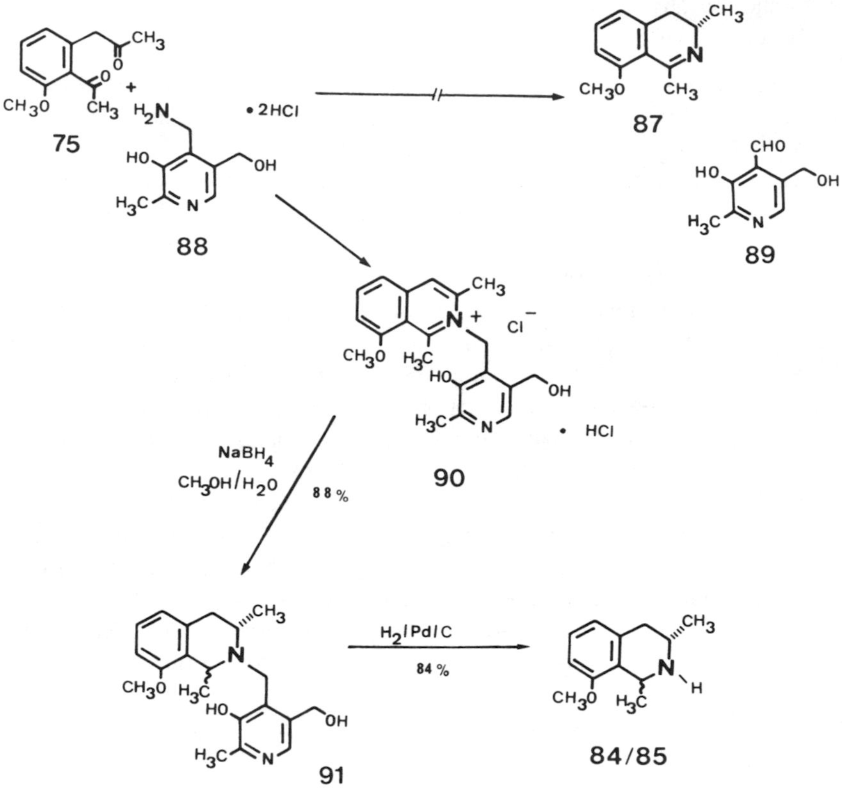

SCHEME 20. Synthesis of tetrahydroisoquinolines with pyridoxamine (88) as nitrogen source (35,38).

followed by hydrogenolytic cleavage of the pyridoxyl group to **84/85** (see Scheme 20), then allowed the mild overall incorporation of the B$_6$ nitrogen, though in a way strongly differing from the proposed (see Scheme 4) biogenetic scheme (38).

The chromatographically and chemically demanding pyridoxyl group which anyhow did not display the required tautomerizing activity could consequently be replaced by the cheaper benzyl residue (see Scheme 21) (35,38). Thus, reaction of diketones like **75** with benzylamine and subsequent reduction of the intermediate isoquinolinium salt **86** provided a very efficient one-pot synthesis of the required tetrahydroisoquinolines **92** and **93,** which were moreover already benzylated for further syntheses.

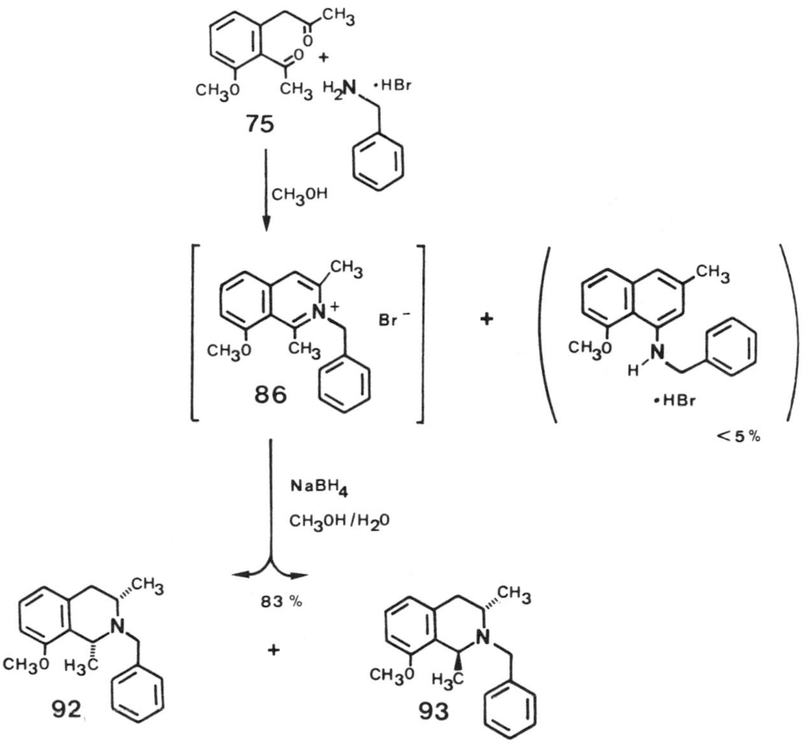

SCHEME 21. One-pot synthesis of N-benzylated 1,3-dimethyltetrahydroisoquinolines from the diketo precursor 75 (*35,38*).

D. THE MIXED ARYL COUPLING TO NAPHTHYL ISOQUINOLINE ALKALOIDS

Following the proposed biogenetic scheme (see Section IV,A), completion of a first total synthesis of entire alkaloids still required a regioselective coupling of sterically highly crowded naphthalenes with isoquinolines. Thus, as expected, direct methods like the classical Ullmann reaction of 94 and 95 failed, giving the desired mixed coupling product 96 only in traces (7) (see Scheme 22).

For this reason, an efficient arylation concept was developed, which tried to imitate a supposed enzymatic prefixation of the two molecular moieties during biosynthesis, by a benzylether-type chemical bridge in the laboratory, sterically forcing the two aromates to the appropriate mutual distance (7) (Scheme 23). Such an auxiliary bridge was easily built up from the already existing "bridge

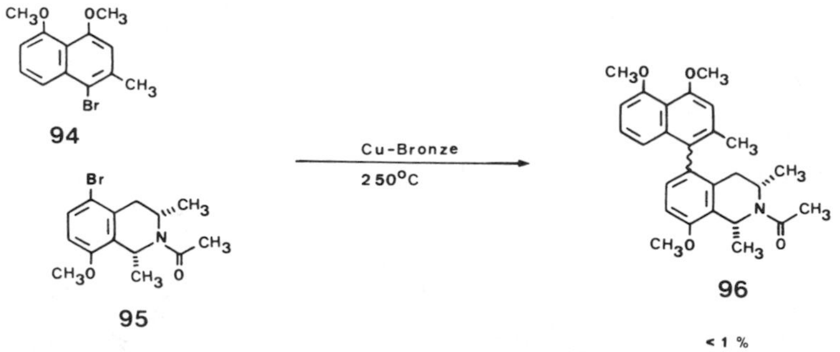

SCHEME 22. Attempted Ullmann coupling to naphthyl isoquinolines.

heads'' methyl (in **74**) and hydroxyl (in **97**). For this purpose, the naphthalene **74** was functionalized in the side chain (step 1, Scheme 23), with simultaneous introduction of an additional halogen atom at C-4, the position in the aromatic ring required for later coupling. Phase-transfer-catalyzed subsequent reaction with **97** (step 2, Scheme 23) gave the desired benzyl analog ether **98.**

Alternatively, starting from the benzaldehyde **101** (*66*), the naphthalene could also be built up nonbiomimetically (*67*) (see Scheme 24) directly in a side-chain functionalized form **103,** with or without halogen in the aromatic ring. The required conversion of the alcohol **102** into the acid-sensitive halide **103** could

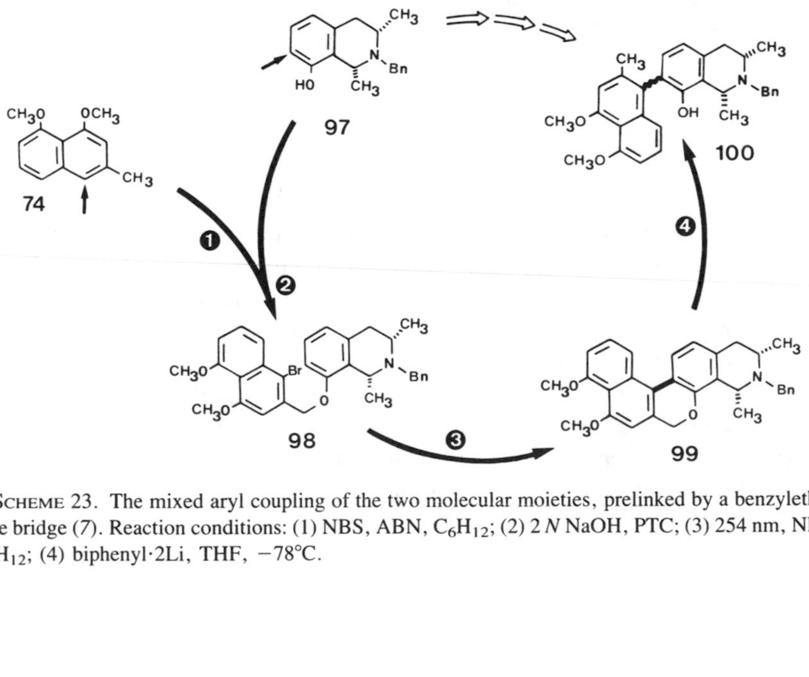

SCHEME 23. The mixed aryl coupling of the two molecular moieties, prelinked by a benzylether-type bridge (*7*). Reaction conditions: (1) NBS, ABN, C_6H_{12}; (2) 2 *N* NaOH, PTC; (3) 254 nm, NEt$_3$, C_6H_{12}; (4) biphenyl·2Li, THF, $-78°C$.

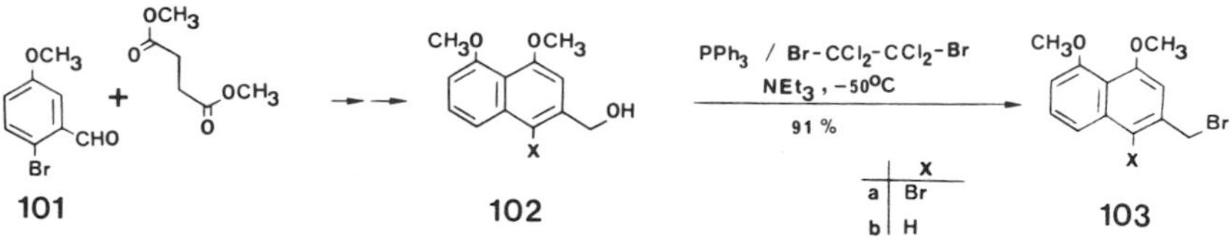

SCHEME 24. A nonbiomimetic pathway to methyl-functionalized napthalenes (66,67).

very smoothly be brought about using triphenylphosphine–1,2-dibromotetra-chloroethane (53). The desired aryl coupling (step 3, Scheme 23) could be achieved, now *intra*-molecularly, by irradiation of **98** to yield the cyclic ether **99** (15%). Beside formation of the six-membered ring **99**, hydrodehalogenation to **104** (15%) as well as ipso substitution to **105** (9%) were also found to occur (see Fig. 5).

The utility of the benzylether-type bridge in **98** became evident from the fact that the analogous *inter*molecular photolysis reaction led to quantitative hydro-dehalogenation of the naphthalene. In this case no aryl coupling was observed, not to mention the required ortho orientation.

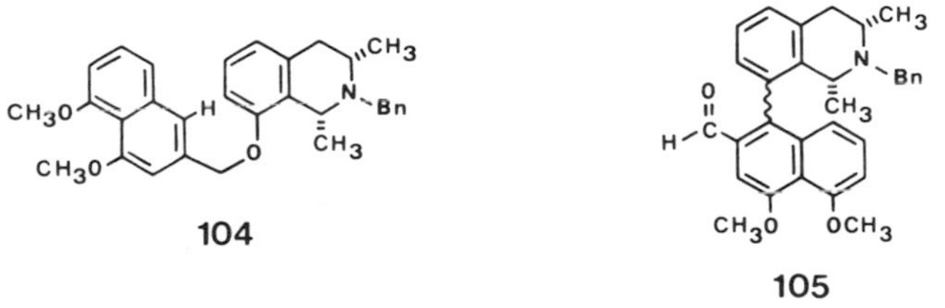

FIG. 5

After having facilitated the coupling entropically, the ether bridge in the cyclic ether **99** had become astonishingly stable to catalytic hydrogenation, as well as to Lewis acids (7). It could, however, be ring opened very smoothly (step 4, Scheme 23) by reduction with dilithium biphenyl (68), leading back to the methyl group and the hydroxy function, both now ortho to the new biphenyl linkage in **100** (7). Due to the restricted rotation at the 7–1' bond, the N-protected naphthyl isoquinoline **100** thus obtained occurs as stable atropisomers, which could be separated with some difficulty. Macroscopically, however, this atropisomerism did not become manifest until the reductive ring opening of the cyclic ether **99,** which was found, by NMR and UV evidence, to be planarized by the auxiliary bridge (7).

Catalytic dehydrogenation of **100** under simultaneous debenzylation (see Scheme 25) led to the known (33), fully aromatized naphthyl isoquinoline **106,**

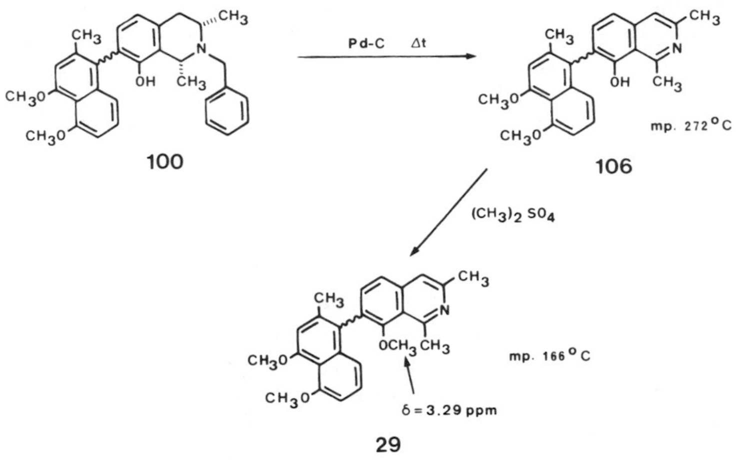

SCHEME 25. Completion of a first total synthesis of naphthylisoquinoline alkaloids (7).

identical in all respects with the reported dehydrogenation product from natural triphyophylline (**23**) and isotriphyophylline (**24**), thus confirming the constitution of these *Triphyophyllum* alkaloids (7). Subsequent O-methylation yielded *O*-methyltetradehydrotriphyophylline (**29**), a compound fully in agreement with the reported (*33*) data of the natural alkaloid that had been isolated in this racemic form from *Triphyophyllum peltatum*. The high-field shift of the 8-methoxy group indicated the correct position of the biphenyl linkage, as synthetically achieved by the ortho-specific coupling strategy.

A first total synthesis of the trans configurated tetrahydroisoquinoline alkaloid triphyophylline (**23**), *via* the lactone intermediate **108**, was accomplished (see Scheme 26) (*69*). The use of an ester type auxiliary bridge, as in **107**, still offered several crucial advantages over the above mentioned ether type prefixation of the molecular moieties.

1. The aryl coupling, now palladium catalyzed, proceeded in much higher yields, the resulting lactone **108** could be crystallized without further purification.
2. The ring opening of the planarized, not atropisomeric lactone **108** did no longer require dilithio biphenyl, but could be effected with complex hydrides, in some cases even with $NaBH_4$.
3. Due to the still functionalized methyl groups, the stable atropisomeric alcohols **109** could be separated from each other unusually well, before their transformation to the methyl compounds **23a** and **23b**.

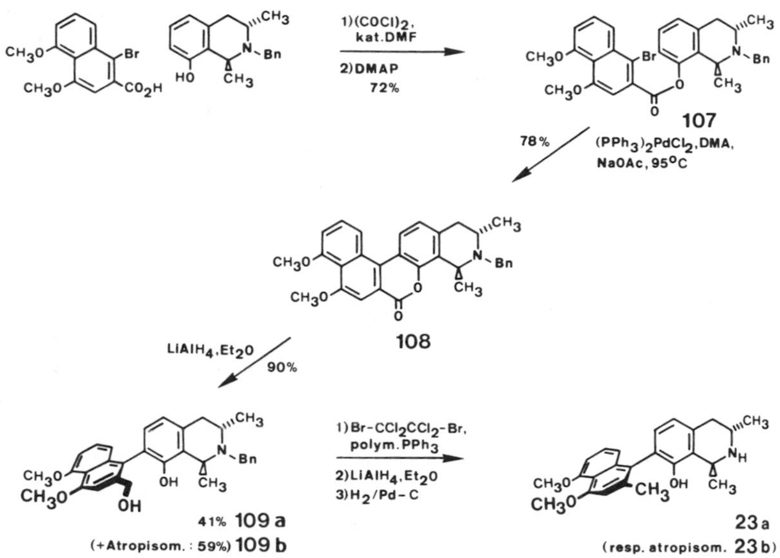

SCHEME 26. The total synthesis of racemic triphyophylline (23) (69).

4. The ring opening of 108 could, however, also be performed by base catalyzed ester hydrolysis. In this case, a recycling of the corresponding undesired atropisomer was possible by recyclization to 108.
5. Finally, this strategy will be perfectly suited for syntheses of the different
   possible stereo- and regioisomeric forms of the related, hydroxymethylated
   alkaloid triphyopeltine (27) (see Section III).

This first synthesis of triphyophylline (23) confirms its relative trans configuration of the two secondary methyl groups. A synthesis of this main *Triphyophyllum* alkaloid in optically active form, which is supposed to establish the chirality
at the biphenyl linkage as well as the entire absolute stereochemistry, is under
investigation in the author's laboratory.

E. ENANTIOSELECTIVE TOTAL SYNTHESES

After these described biomimetic syntheses of racemic naphthyl isoquinoline
alkaloids, optically active synthetic material had to be prepared in enantiomerically pure form, not only for a systematic screening of biological activity
of the minor alkaloids, but also for a thorough reinvestigation of several published, obviously uncertain structures that could not be determined by X ray. For
these purposes, an enantiomeric separation of the racemic material as obtained
from the biomimetic syntheses seemed uneconomical, given the fact that all

stereochemically established naphthyl isoquinoline alkaloids possess an S configuration at C-3 so that the $3R$ material would be lost, unless recycled by troublesome racemization and repeated enantiomer separation. For this reason, efficient methods were developed (70,71) for the reliable asymmetric synthesis of stereochemically defined naphthyl isoquinoline alkaloids.

1. By Reductive Amination of Aryl Propanones (70)

A directed formation of the important chiral center at C-3 by reductive amination of diketo precursors **A** (see Scheme 4, Section IV,A,4), selectively at the aliphatic keto function—as proposed for the biosynthesis—could not be achieved chemically in the presence of the aromatic keto function, owing to the fast ring closure, for example, of **75** (cp. Scheme 20, Section V,C,3) to isoquinolinium salts like **90.** As this aromatic keto function could not be protected selectively, the "biomimetic" nitrogen incorporation was performed on the *mono*ketone **110,** readily available according to Scheme 14 (Section V,B).

Close to techniques used in amphetamine chemistry (72), reductive amination of **110** with the inexpensive (S)-1-phenylethylamine (**111**) with NaCNBH$_3$ or, more effectively, with Raney nickel, stereoselectively (d.e. > 95%) gave the desired diastereomer **112.** Compound **112** could easily be purified by crystallization, debenzylated to the enantiomerically homogeneous aryl isopropyl amine

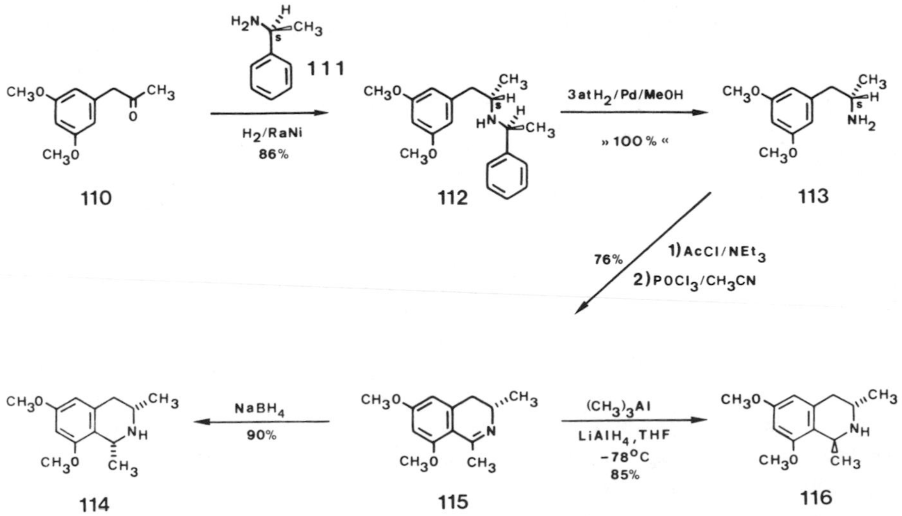

SCHEME 27. Asymmetric synthesis of optically active isoquinoline moieties of *Ancistrocladus* alkaloids (70).

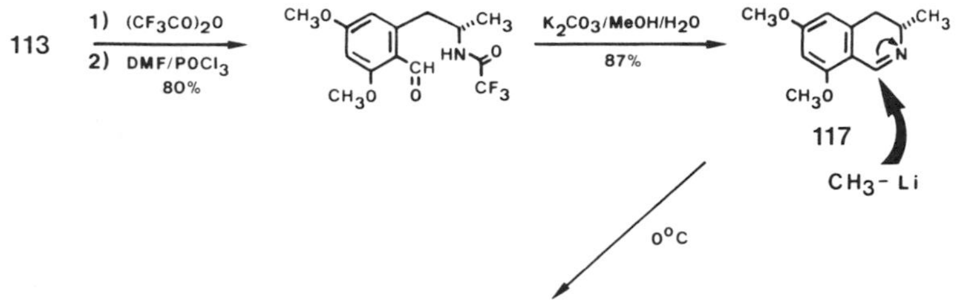

SCHEME 28. Stereocontrolled introduction of a methyl group at C-1 (70).

113, and subsequently cyclized by Bischler–Napieralsky reaction to the dihydro-isoquinoline **115,** the heterocyclic moiety of the *Ancistrocladus* alkaloids **18** and **21,** with the correct *S* configuration at C-3 (see Scheme 27) (70). Introduction of the chiral center at C-1 could also be manipulated to give either configuration. By reduction of **115** with complex hydrides or by catalytic hydrogenation, **114** could be obtained highly selectively, whereas the more common (cp. Section II) trans form **116** could very effectively be prepared from the same precursor **115** by alanate reduction in the presence of $Al(CH_3)_3$.

The much more important 1*S*,3*S* isomer **116** could also be obtained by nucleophilic alkylation of the 1-unsubstituted dihydroisoquinoline **117,** for example, with CH_3Li, stereoselectively occurring from the β face (70). Again, **117** was easily accessible from the chiral aryl isopropylamine **113** (see Scheme 27).

2. Syntheses from the Chiral Pool (71)

For a preparation of alkaloids 3*S* configurated beyond all doubt, synthetic pathways from the chiral pool were also elaborated (71). The C_3N unit destined to represent the crucial chiral center at C-3 turned out to be the amino acid alanine, in its natural L-form **118,** which should easily be incorporated into the isoquinoline framework by reductive amination with the acetophenone **119** and subsequent *intra*molecular acylation, as outlined in Scheme 29. The practical realization of such a non-biomimetic synthesis (see Scheme 30), however, differs from this very simple concept in three ways:

1. To minimize of racemization reactions, resulting from keto–enol tautomerism, the dimethyl acetal **120** was used (Pomerantz–Fritsch synthesis), instead of the acid **118.**

2. The lack of reactivity at the ring closure position in **119** was compensated by an additional methoxy group, as in **121,** which, after the cyclization of

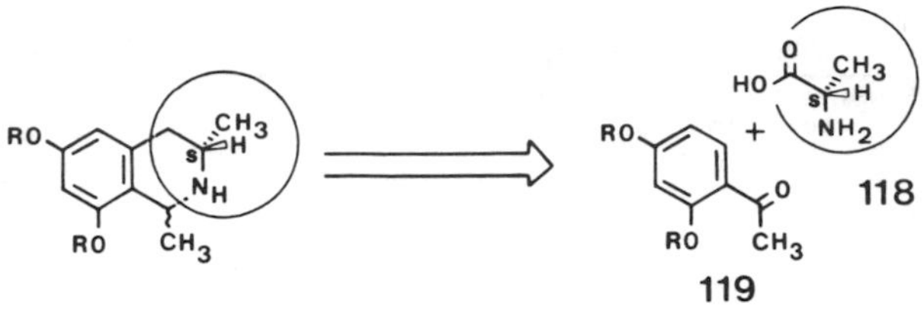

SCHEME 29. Retrosynthetic disconnection to the acetophenone **119** and L-alanine (**118**).

122 to **123**, could selectively be removed again, together with the benzylic oxygen function at C-4, by Birch reduction.

3. In order to avoid diastereomer formation in the amination reaction, the 1-methyl group was again introduced stereoselectively *after* the ring closure.

3. Atropisomer-Selective Synthesis of Ancistrocladine (**1**) (*73*)

The first naphthyl isoquinoline alkaloid to be synthesized in optically active form was ancistrocladine (**1**) itself, the most important and widespread of all the *Ancistrocladus* alkaloids (see Scheme 31). The effectivity of the ester bridge assisted aryl coupling method (as presented in Section V.D.) was demonstrated by the regioselective formation of the highly strained lactone **124**, which—different from the corresponding triphyophylline precursor **108** (see Scheme

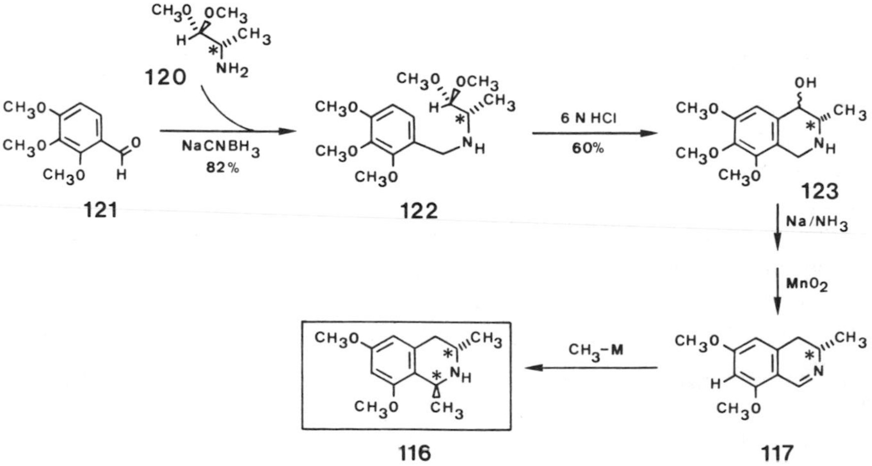

SCHEME 30. Synthesis of the isoquinoline moiety **116** of *O*-methylancistrocladine (**5**) from the chiral pool (*71*).

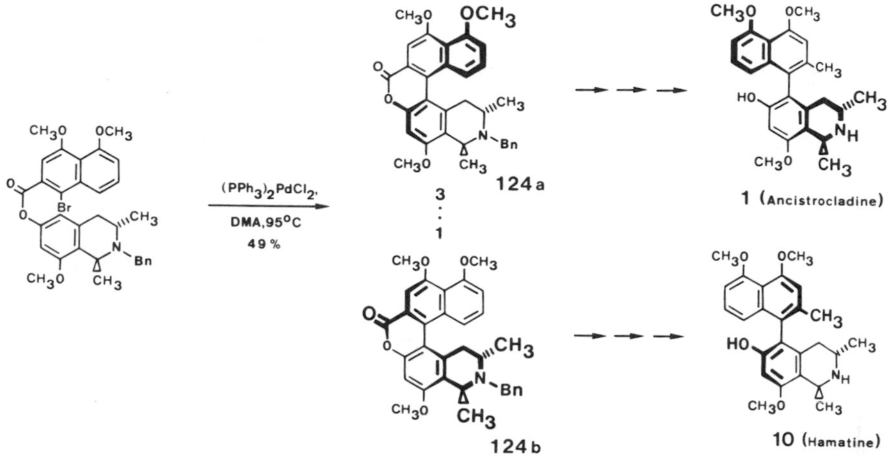

SCHEME 31. Total synthesis of (−)-ancistrocladine (**1**) and (+)-hamatine (**10**) (*73*).

26)—was now indeed split into a pair of helicene-like atropisomers, with a remarkable atropisomer excess (3:1) favouring **124a**, the "ancistrocladine type," which is predominant also in the plants. Using the procedure elaborated for the triphyophylline synthesis (see Section V.D.), the lactones **124a** and **124b** were separately transformed into ancistrocladine (**1**) and its natural atropisomer hamatine (**10**).

VI. Concluding Remarks

Naphthyl isoquinoline alkaloids have proved to be a novel group of structurally intriguing natural products. Their extraordinary substitution pattern seems to correspond to an unprecedented biosynthesis of isoquinolines from acetate units via β-polycarbonyl precursors.

This unusual biosynthetic pathway—the synthesis and regio-controlled cyclization of β-polyketones, optionally leading to tetrahydro-, dihydro-, or fully conjugated isoquinolines, or to naphthalenes and respective naphthoquinones—could be imitated efficiently *in vitro*. The smooth course of these chemical model reactions underlines the chemical plausibility of the proposed biogenetic scheme and simultaneously provides the basis for a first and variable total synthesis of naphthyl isoquinoline alkaloids.

A broad and systematic synthetic elaboration of this fascinating class of natural products seems urgent not only for a confirmation (or disproval) of several uncertain alkaloid structures but also for an intensive investigation of their biological activity. Though, for instance, the roots of *A. tectorius* have been used to treat dysentery and malaria (*24*) and extracts of *Ancistrocladus* plants have been shown to possess spasmolytic (*A. heyneanus,* Ref. *74*) or tumor-inhibiting (*A. tectorius,* Ref. *23*) activity, interesting pharmacological properties of purified naphthyl isoquinoline alkaloids have hitherto been found only for minor components such as ancistrocladidine (**21**), which shows a good spasmolytic activity, comparable to that of papaverine (*43,75*). Its extremely low content in the roots of *A. heyneanus* (*11,17*), which does not allow large-scale exploitation for pharmacological purposes, constitutes another strong motivation for a chemical synthesis of naphthyl isoquinoline alkaloids like **21**. New, efficient methods of isolation and structural elucidation, combined with modern synthetic strategies, will in the future give rise to further remarkable results in this interesting field.

Acknowledgment

I wish to express my thanks to the Deutsche Forschungsgemeinschaft (DFG) and to the Fonds der Chemischen Industrie for the financial support of our work in this field.

REFERENCES

1. M. Shamma, *in* "The Isoquinoline Alkaloids," Academic Press, New York, 1972.
2. A. Pictet and T. Spengler, *Chem. Ber.* **44,** 2030 (1911).
3. D. R. Dalton, *in* "The Alkaloids—a Biogenetic Approach." Dekker, New York, 1979.
4. T. R. Govindachari and P. C. Parthasarathy, *Indian J. Chem.* **8,** 567 (1970).
5. See Ref. *1,* p. 501.
6. T. R. Govindachari and P. C. Parthasarathy, *Tetrahedron* **27,** 1013 (1971).
7. G. Bringmann and J. R. Jansen, *Tetrahedron Lett.* **25,** 2537 (1984).
8. E. Gilg, *in* "Die Natürlichen Pflanzenfamilien" (A. Engler and K. Prantl, eds.), Vol. III/6, p. 274. W. Engelmann, Leipzig, 1895.
9. R. Hegnauer, *in* "Chemotaxonomie der Pflanzen," Vol. 3, p. 115. Birkhäuser, Basel, 1964.
10. J. Carrick, K. C. Chan, and H. T. Cheung, *Chem. Pharm. Bull.* **16,** 2436 (1968).
11. T. R. Govindachari and P. C. Parthasarathy, *Heterocycles* **7,** 661 (1977).
12. T. R. Govindachari, K. Nagarajan, P. C. Parthasarathy, T. G. Rajagopalan, H. K. Desai, G. Kartha, S. Lai Chen, and K. Nakanishi, *J. Chem. Soc., Perkin Trans. 1,* 1413 (1974).
13. N. Harada and K. Nakanishi, *Acc. Chem. Res.* **5,** 257 (1972).
14. T. R. Govindachari, P. C. Parthasarathy, and H. K. Desai, *Indian J. Chem.* **9,** 1421 (1971).
15. H. K. Desai, D. H. Gawad, T. R. Govindachari, B. S. Joshi, P. C. Parthasarathy, K. S. Ramachandran, K. R. Ravindranath, A. R. Sidhaye, and N. Viswanathan, *Indian J. Chem.* **14B,** 473 (1976).
16. T. R. Govindachari, P. C. Parthasarathy, and H. K. Desai, *Indian J. Chem.* **10,** 1117 (1972).
17. T. R. Govindachari, P. C. Parthasarathy, and H. K. Desai, *Indian J. Chem.* **11,** 1190 (1973).
18. T. R. Govindachari, P. C. Parthasarathy, T. G. Rajagopalan, H. K. Desai, K. S. Ramachandran, and E. Lee, *J. Chem. Soc., Perkin Trans. 1,* 2134 (1975).
19. P. C. Parthasarathy and G. Kartha, *Indian J. Chem.* **22B,** 590 (1983).
20. T. R. Govindachari, P. C. Parthasarathy, T. G. Rajagopalan, H. K. Desai, K. S. Ramachandran, and E. Lee, *Indian J. Chem.* **13,** 641 (1975).

21. T. R. Govindachari, P. C. Parthasarathy, H. K. Desai, and M. T. Saidane, *Indian J. Chem.* **15B,** 871 (1977).
22. J. P. Foucher, J. L. Pousset, A. Cavé, and R. R. Paris, *Plantes Méd. Phytothér.* **9,** 26 (1975).
23. Z. X. Chen, B. D. Wang, K. W. Quin, B. E. Zhang, Q. L. Su, and Q. C. Lin, *Yaoxue Xuebao* **16,** 519 (1981).
24. N. Ruangrungsi, V. Wongpanich, P. Tantivatana, H. J. Cowe, P. J. Cox, S. Funayama, and G. A. Cordell, *J. Nat. Prod.* **48,** 529 (1985).
25. J. P. Foucher, J. L. Pousset, A. Cavé, A. Bouquet, and R. Paris, *Plantes Méd. Phytothér.* **5,** 16 (1971).
26. J. P. Foucher, J. L. Pousset, Ad. Cavé, and An. Cavé, *Phytochemistry* **13,** 1253 (1974).
27. J. P. Foucher, J. L. Pousset, and A. Cavé, *Phytochemistry* **14,** 2699 (1975).
28. J. P. Foucher, J. L. Pousset, A. Cavé, A. Bouquet, and R. Paris, *Plantes Méd. Phytothér.* **9,** 87 (1975).
29. J. P. Foucher, Thèse Doct. Etat, Université Paris-Sud, 1975.
30. T. R. Govindachari, P. C. Parthasarathy, and H. K. Desai, *Indian J. Chem.* 9, 931 (1971).
31. J. Bruneton, A. Bouquet, A. Fournet, and A. Cavé, *Phytochemistry* **15,** 817 (1976).
32. M. Lavault, M. T. Kouhon, and J. Bruneton, *C. R. Seances Acad. Sci. (Paris)* **C285,** 167 (1977).
33. M. Lavault and J. Bruneton, *C. R. Seances Acad. Sci. (Paris)* **C287,** 129 (1978).
34. H. K. Airy Shaw, *Kew Bull.* **IV,** 341 (1952).
35. G. Bringmann, *in* "35 Jahre Fonds der Chemischen Industrie, 1950–1985," p. 151. Verband der Chemischen Industrie, Frankfurt, 1985.
36. G. Bringmann, *Liebigs Ann. Chem.,* 2126 (1985).
37. K. Mothes and H. R. Schütte, *in* "Biosynthese der Alkaloide," p. 41. VEB Deutscher Verlag der Wissenschaften, Berlin, 1969.
38. G. Bringmann and J. R. Jansen, *Heterocycles,* **24** (1986), in press.
39. E. Leete, *Acc. Chem. Res.* **4,** 100 (1971).
39a. M. F. Roberts, *Phytochemistry* **10,** 3057 (1971); *ibid.* **16,** 1381 (1977).
40. Ph. van Tieghem, *J. Bot.* **17,** 151 (1903).
41. R. Mehta, O. P. Arora, and M. Mehta, *Indian J. Chem.* **20B,** 834 (1981).
42. B. Z. Ahn and F. Zymalkowski, *Tetrahedron Lett.,* 821 (1976).
43. T. R. Govindachari, personal communication.
44. G. Bringmann and T. R. Govindachari, unpublished results.
45. G. Bringmann, J. R. Jansen, J. Reisch, G. Reisch, and A. A. L. Gunatilaka, unpublished results.
46. J. R. Jansen, Diploma Thesis, Münster, 1983.
47. R. Durand and M. H. Zenk, *Tetrahedron Lett.,* 3009 (1971).
48. K. H. Overton and D. J. Dicken, *Fortschr. Chem. Org. Naturst.* **34,** 283 (1977).
49. T. R. Govindachari, P. C. Parthasarathy, and J. D. Modi, *Indian J. Chem.* 9, 1042 (1971).
50. G. Bringmann and S. Schneider, *Liebigs Ann. Chem.,* 765 (1985).
51. G. Bringmann and S. Schneider, *Heterocycles* **19,** 1449 (1982).
52. P. J. Wittek, Dissertation, Vanderbilt University, 1975.
53. G. Bringmann and S. Schneider, *Synthesis,* 139 (1983).
54. G. Bringmann, *Angew. Chem.* **94,** 205 (1982); *Angew. Chem. Int. Ed. Engl.* **21,** 200 (1982).
55. G. Bringmann, *Tetrahedron Lett.* **23,** 2009 (1982).
56. G. Bringmann, *Liebigs Ann. Chem.,* **2105** (1985).
57. A. J. Birch, P. Fitton, D. C. C. Smith, D. E. Steere, and A. R. Stelfox, *J. Chem. Soc.,* 2209 (1963).
58. C. L. Kirkemo and J. D. White, *J. Org. Chem.* **50,** 1316 (1985).
59. G. Hesse and E. Bayer, *Z. Naturforsch.* **19B,** 875 (1964).

60. T. M. Harris and C. M. Harris, *Tetrahedron* **33,** 2159 (1977).
61. J. S. Hubbard and T. M. Harris, *J. Am. Chem. Soc.* **102,** 2110 (1980).
62. I. Iijima, M. Miyazaki, N. Taga, and T. Tanaka, *Heterocycles* **8,** 357 (1977).
63. G. Bringmann and J. R. Jansen, *Liebigs Ann. Chem.,* 2116 (1985).
64. S. Mongkolsuk and C. Sdarwonvivat, *J. Chem. Soc.,* 1533 (1965).
65. G. S. Sidhu, A. V. B. Sankaram, and S. Mahmood Ali, *Indian J. Chem.* **6,** 681 (1968).
65a. A. Mathey, B. Steffan, and W. Steglich, *Liebigs Ann. Chem.,* 779 (1980).
66. B. O. Handford and W. B. Whalley, *J. Chem. Soc.,* 3896 (1963).
67. G. Bringmann and J. R.Jansen, in preparation.
68. J. J. Eisch, *J. Org. Chem.* **28,** 707 (1963).
69. G. Bringmann, J. R. Jansen, and H. P. Rink, in preparation.
70. G. Bringmann and J. R. Jansen, Angew. Chem., submitted for publication.
71. G. Bringmann, J. R. Jansen, and J.-P. Geisler, presented on the "Chemiedozententagung," Würzburg, 1986.
72. D. E. Nichols, C. F. Barfknecht, D. B. Rusterholz, F. Benington, and R. D. Morin, *J. Med. Chem.* **16,** 480 (1973).
73. G. Bringmann, J. R. Jansen, and H. P. Rink, *Angew. Chem.,* submitted for publication.
74. S. C. Sharma, Y. N. Shukla, and J. S. Tandon, *Phytochemistry* **14,** 578 (1975).
75. A. A. L. Gunatilaka, *J. Nat. Sci. Council Sri Lanka* **6,** 39 (1978).

—— CHAPTER 4 ——

ALKALOIDAL SUBSTANCES FROM *Aspergillus* SPECIES

YUZURU YAMAMOTO AND KUNIZO ARAI

Faculty of Pharmaceutical Sciences
Kanazawa University
Kanazawa, Japan

I. Introduction

For a long time, it has been considered that fungi have poor ability to elaborate alkaloids because of their evolutionarily primitive status. The extensive investigations of the so-called antibiotics showed that this view is erroneous, and, particularly in last 20 years, many nitrogen-containing metabolites including true alkaloids have been isolated from fungi, especially the genera *Aspergillus* and *Penicillium*.

The fungal alkaloids seem to be characteristic in utilizing amino acid itself as its component in contrast to the higher plant alkaloids which are biosynthesized mainly from the amine produced by decarboxylation of amino acid (*1,2,3*). In this chapter, discussion is limited to fungal alkaloidal metabolites isolated from *Aspergillus* species, which are classified by biogenetic category.

II. Pyrazine Metabolites

Since aspergillic acid, a pioneer compound of this group, was isolated as an antibacterial substance of the culture broth of *Aspergillus flavus* by White and Hill (*4*) in 1943, numerous pyrazine metabolites have been found during the course of investigations of antiphage and antibacterial substances. Several reviews (*5,6,7*) of these pyrazine metabolites have been published. In the 1960s several metabolites were added during the screening of mycotoxins in Japanese industrial seed mold (*8,9*), and in the early 1970s some other metabolites were obtained in cultivation on a medium supplemented with methionine (*10*). Several metabolites complexed with iron (*11,12*) or zinc (*9a*) have been also reported. Pyrazine metabolites are listed in Table I (*13–23*).

THE ALKALOIDS, VOL. 29

TABLE I

Pyrazine Metabolites from Aspergillus Species

Tautomeric equilibrium (N-hydroxy pyrazinone ⇌ N-oxide): a pyrazine ring bearing R¹ and R² substituents, shown as the N-hydroxy/keto form (N–OH, C=O) in equilibrium (⇌) with the N-oxide/hydroxy form (OH, N→O).

Name	Structure R¹	Structure R²	mp (°C)	[α]D (conc, solvent)	Occurrence	Ref.
N-Oxide Group 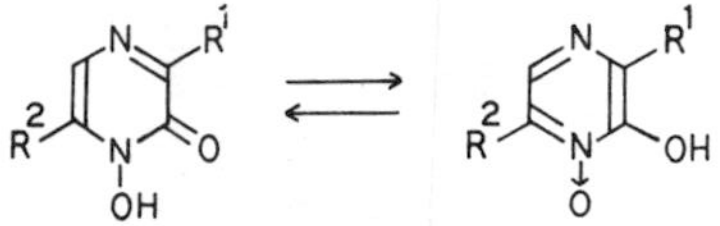						
Aspergillic acid (1)	Me_2CHCH_2-	$\underset{Me}{MeCH_2CH}-$	97–99	+13.3° (3.9, EtOH)	*Aspergillus flavus*, *Asp. sojae*	*4,13–17*
Hydroxyaspergillic acid (4)	Me_2CHCH_2-	$\underset{OH}{\overset{Me}{MeCH_2C}}-$	148–150; 152–153	+36° (0.1, EtOH); +35.85° (0.3, EtOH)	*Asp. flavus*, *Asp. oryzae*, *Asp. sojae*	*19,20*
Neoaspergillic acid (8)	Me_2CHCH_2-	Me_2CHCH_2-	125–126; 126–127.5	Inactive	*Asp. flavus*, *Asp. ochraceus*	*11,21*
Neohydroxyaspergillic acid (9)	Me_2CHCH_2-	$\underset{OH}{Me_2CHCH}-$	170–171; 164–166	−57° (0.64, EtOH); −58° (1.01, EtOH)	*Asp. sclerotiorum*, *Asp. sojae*	*18,23*
β-Hydroxyneoaspergillic acid	Me_2CHCH_2-	$\overset{OH}{Me_2CCH_2}-$	143–144	—	*Asp. ochraceus*	*11*
Mutaaspergillic acid	Me_2CHCH_2-	$\overset{OH}{Me_2C}-$	173–174	0 ± 0.3°	*Asp. oryzae*, *Asp. sojae*	*8*

Compound			mp (°C)		Organism	Ref.
1-Hydroxy-3,6-di-(1-methylpropyl)pyrazine-2-one	Me | MeCH₂CH —	Me | MeCH₂CH —	Oil	—	*Asp. flavus* *Asp. sojae*	*9a,9b*
1-Hydroxy-6-(1-hydroxy-1-methylpropyl)-3-(1-methylpropyl)-pyrazine-2-one	Me | MeCH₂CH —	Me | MeCH₂C— | OH	120–121	—	*Asp. sojae*	*9b*
1-Hydroxy-6-(1-methyl-ethyl)-3-(1-methylpropyl)-pyrazine-2-one	Me | MeCH₂CH —	Me₂CH —	205–207	—	*Asp. oryzae*	*22*
1-Hydroxy-6-(1-methyl-ethyl)-3-(2-methylpropyl)-pyrazine-2-one	Me₂CHCH₂ —	Me₂CH —	94–95	—	*Asp. sojae*	*9b,9c*
S-Containing Group						
1-Hydroxy-3,6-di[2-(methyl-thio)ethyl] pyrazine-2-one	MeSCH₂CH₂ —	MeSCH₂CH₂ —	106–107	—	*Asp. flavus*	*10*
1-Hydroxy-6-(1-methyl-ethyl)-3-[2-(methylthio)-ethyl]pyra-zine-2-one	MeSCH₂CH₂ —	Me₂CH —	69–75	—	*Asp. flavus*	*10*
1-Hydroxy-3-(2-methylpropyl)-	Me₂CHCH₂ —	MeSCH₂CH₂ —	114–116	—	*Asp. flavus*	*10*

(continued)

TABLE I (Continued)

Name	Structure		mp (°C)	$[\alpha]_D$ (conc, solvent)	Occurrence	Ref.
	R¹	R²				
6-[2-(methyl-thio)ethyl]pyra-zine-2-one						
1-Hydroxy-6-(2-methylpropyl)-3-[2-(methyl-thio)ethyl]pyra-zine-2-one	MeSCH₂CH₂—	Me₂CHCH₂—	114–116	—	*Asp. flavus*	*10*
1-Hydroxy-6-(1,1-dimethyl-ethyl)-3-[2-(methylthio)-ethyl]pyra-zine-2-one	MeSCH₂CH₂—	Me₃C—	74–76	—	*Asp. flavus*	*10*

Deoxy Group

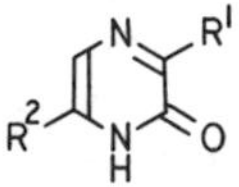

Name	Structure		mp (°C)	$[\alpha]_D$ (conc, solvent)	Occurrence	Ref.
	R¹	R²				
Deoxyaspergillic acid (**2**)	Me₂CHCH₂—	$\underset{}{\overset{\text{Me}}{\text{MeCH}_2\text{CH}}}—$	85	—	*Asp. sojae*	*9b*
Deoxyhydroxyas-pergillic acid	Me₂CHCH₂—	$\underset{\text{OH}}{\overset{\text{Me}}{\text{MeCH}_2\text{C}}}—$	107	—	*Asp. sojae*	*9b*

Compound	R^1	R^2	mp (°C)		Organism	Ref.
Flavacol (7) (deoxyneoaspergillic acid)	$\mathrm{Me_2CHCH_2-}$	$\mathrm{Me_2CHCH_2-}$	144.5–147, 144	Inactive	*Asp. sojae*, *Asp. ochraceus*, *Asp. flavus*	9,21
Deoxyneo-β-hydroxyaspergillic acid	$\mathrm{Me_2CHCH_2-}$	$\mathrm{Me_2C(OH)CH_2-}$	149–150, 122.5–123	—	*Asp. ochraceus*	21
Deoxydehydromutaaspergillic acid	$\mathrm{Me_2CHCH_2-}$	$\mathrm{Me_2CH-}$	111	—	*Asp. flavus*	9
Deoxymutaaspergillic acid	$\mathrm{Me_2CHCH_2-}$	$\mathrm{Me_2C(OH)-}$	133–134.5	—	*Asp. sojae*	9
3,6-Di(1-methylpropyl)pyrazine-2-one	$\mathrm{MeCH_2CH(Me)-}$	$\mathrm{MeCH_2CH(Me)-}$	129	—	*Asp. flavus*, *Asp. sojae*	9
6-(1-Hydroxy-1-methylpropyl)-3-(1-methylpropyl)pyrazine-2-one	$\mathrm{MeCH_2CH(Me)-}$	$\mathrm{MeCH_2C(OH)(Me)-}$	117	—	*Asp. sojae*	9

Metal Complexes

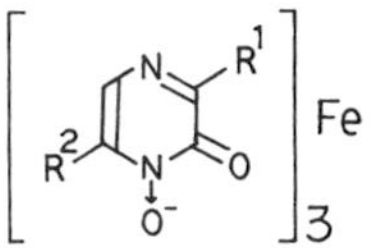

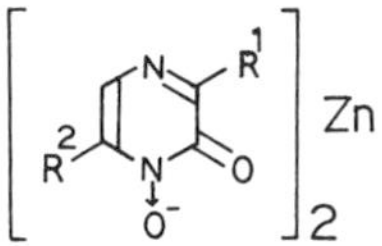

$$\left[\ \underset{R^2}{\overset{R^1}{\underset{\displaystyle}{\mathsf{N}\text{–pyrazin-2-one ring}\ (C{=}O,\ N{\to}O^-)}}}\ \right]_3 \mathrm{Fe} \qquad \left[\ \underset{R^2}{\overset{R^1}{\mathsf{N}\text{–pyrazin-2-one ring}\ (C{=}O,\ N{\to}O^-)}}\ \right]_2 \mathrm{Zn}$$

Compound	R^1	R^2	mp (°C)		Organism	Ref.
Ferriaspergillin	$\mathrm{MeCH_2CH(Me)-}$	$\mathrm{Me_2CHCH_2-}$	Oil	—	*Asp. flavus*	12
Ferrineoaspergillin	$\mathrm{Me_2CHCH_2-}$	$\mathrm{Me_2CHCH_2-}$	135–136	—	*Asp. flavus*	11,12,21
Zn–aspergillin	$\mathrm{Me_2CHCH_2-}$	$\mathrm{MeCH_2CH(Me)-}$	229–230	—	*Asp. sojae*	9a

189

1. Chemistry

Aspergillic acid (**1**), $C_{12}H_{20}N_2O_2$, is a monobasic acid that also possesses a weakly basic group. A deep red coloration with $FeCl_3$ and formation of a gteen cupric salt suggest that aspergillic acid is a hydroxamic acid derivative, and this is confirmed by the formation of deoxyaspergillic acid (**2**), $C_{12}H_{20}N_2O$, through dry distillation of **1** with copper chromite catalyst (Scheme 1). Bromination of **1** and **2** followed by reduction with Zn and AcOH yields a diketopiperazine (**3**) (*14*). Vigorous hydrolysis of **3** with HBr yields a mixture of DL-leucine and DL-isoleucine (*15,16*). The structure of aspergillic acid was finally concluded by the chemical synthesis of deoxyaspergillic acid (*17*). Aspergillic acid has been assigned the corresponding *N*-oxide or the tautomeric pyrazine hydroxamic acid structure.

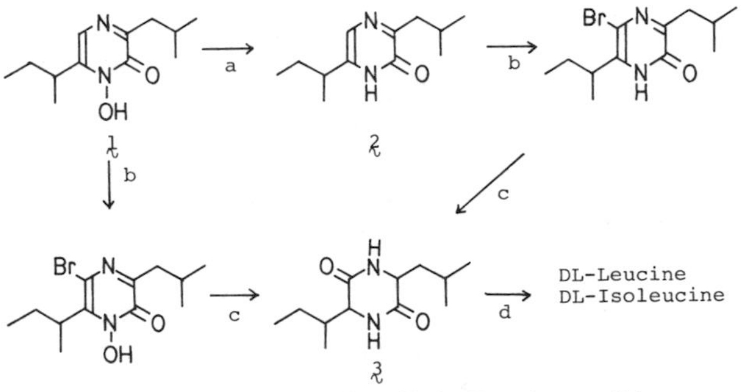

SCHEME 1. Chemical degradation of aspergillic acid (**1**). Reaction conditions: a, copper chromite; b, Br_2–H_2O; c, Zn–AcOH; d, HBr.

The location of the additional oxygen atom of hydroxyaspergillic acid (**4**), $C_{12}H_{20}N_2O_3$, was ascertained to be on the *sec*-butyl side chain from the following results (*19*) (Scheme 2). Reduction with HI in AcOH affords the dehydrodeoxy derivative (**5**) which becomes optically inactive. The UV spectrum indicates that a newly generated double bond is connected to the pyrazine ring. Ozonolysis of **5** yields acetaldehyde and **6**, which gives a positive iodoform reaction.

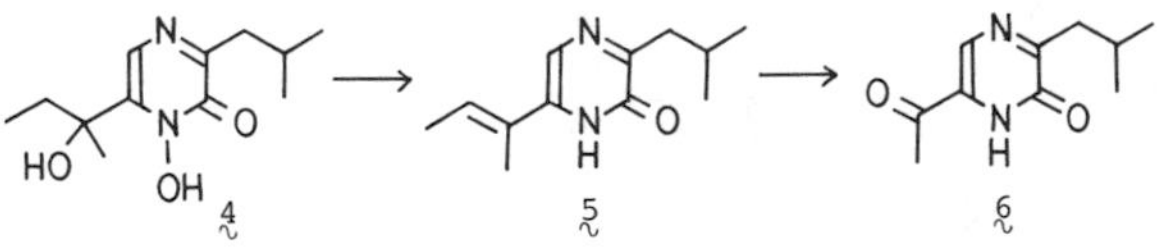

SCHEME 2. Degradation of hydroxyaspergillic acid (**4**).

2. Biosynthesis

MacDonald (*24*) found that labeled aspergillic acid (**1**) is produced by *Asp. flavus* grown on a medium supplemented with radioactive leucine and isoleucine. Hydrolysis of the labeled metabolite shows that it is biosynthesized from one molecule each of leucine and isoleucine. The same result is also recognized in hydroxyaspergillic acid (**4**).

In the early stages of fungal growth, the amount of **1** is greater than that of **4**. However, most of **1** in the later stages of growth has disappeared, whereas more **4** is formed. Appreciable ^{14}C-labeled **4** is obtained from labeled **1** provided in cultures, but the reverse reaction does not occur, indicating that aspergillic acid is a precursor of hydroxyaspergillic acid (*25*). Further, [^{14}C]deoxyaspergillic acid is oxidized to give **1** and **4** in *Asp. sclerotiorum* (*26*). This biosynthetic behavior is similar in the other related products (Scheme 3); [^{14}C]flavacol (**7**) (deoxyneoaspergillic acid) can serve as a substrate for neoaspergillic acid (**8**) and neohydroxyaspergillic acid (**9**) (*27*). These findings support the biosynthetic pathway shown in Scheme 3 for neohydroxyaspergillic acid. Though cyclo(L-leucyl-L-leucine) (**10**) is an effective precursor in biosynthesis of pulcherriminic acid (**11**) in *Candida pulcherrima* (*28*), evidence of **10** as an intermediate between leucine and flavacol is still not obtained.

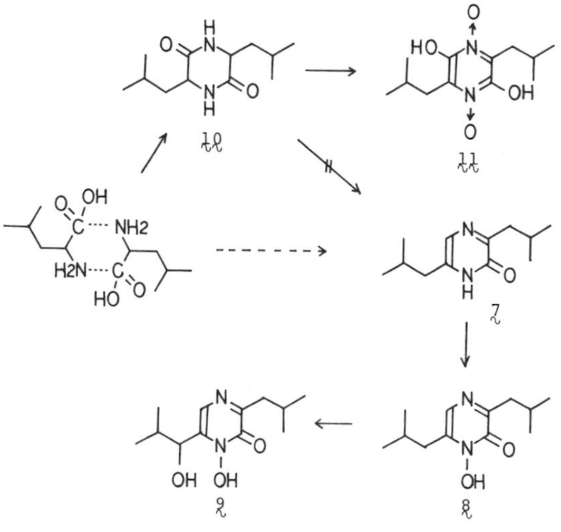

SCHEME 3. Biosynthetic pathway of neohydroxyaspergillic acid (**9**).

3. Biological Activity

Generally, N-oxide metabolites show an antibiotic activity in microorganisms and a convulsive effect in animals. On the other hand, deoxy compounds do not. The hydroxamic acid moiety in these metabolites seems to be essential for exhibition of their biological activities. Usually, deoxy compounds have a fluorescent violet color, whereas N-oxides are not fluorescent (9b). It seems useful to distinguish mycotoxin-producing fungi in seed.

III. Diketopiperazine Metabolites Derived from Tryptophan and Alanine

Since the discovery of echinulin in 1943, many fungal metabolites having a 2.5-diketopiperazine moiety have been isolated. Since 1970, this group has attracted interest, particularly due to the isolation of many biologically active metabolites such as austamide, fumitremorgen, and tryptoquivaline. In most cases, the diketopiperazine ring of fungal metabolites utilizes tryptophan as one of the components. In this section fungal metabolites composed of tryptophan and alanine are described.

A. ECHINULIN AND RELATED METABOLITES

1. Echinulin

Quilico and Panizzi (29) isolated a nitrogen-containing metabolite from the mycelium of *Aspergillus echinulatus* and named it echinulin (**12**). This compound has since been found in several molds of the *Asp. glaucus* group (*Asp. chevalieri, Asp. amstelodami, Asp. ruber, Asp. repens, Asp. niveo-glaucus, etc.*) (30,31). The yield of echinulin from *Asp. echinulatus* was high (5–6% of dry mycelium), but, owing to the considerable difficulty of combustion in elementary analysis, the formula was misassigned as $C_{28}H_{37}N_3O_2$ at an early stage. So in spite of the extensive degradation work of Quilico's group (32), research on the chemical structure has not proceeded smoothly. The formula was corrected to $C_{29}H_{39}N_3O_2$ on the basis of biosynthetic evidence by Birch *et al.* in 1961 (33).

Echinulin absorbs 3 mol of H_2 by catalytic hydrogenation on PtO_2 (Scheme 4). Pyrolysis of the potassium salt of hydrogenated echinulin affords a viscous liquid, $C_{23}H_{37}N$ (**13**), which is positive to Ehrlich reaction and classified as a 2,5,7-trisubstituted indole by comparing the UV and IR spectra. KOH hydrolysis of hydroechinulin gives hydroechinin (**14**). Oxidation of **14** with peracetic acid affords substituted N-acylanthranilic acid which is hydrolyzed with alcoholic KOH to give 1,1-dimethylbutyric acid and disubstituted anthranilic acid. The

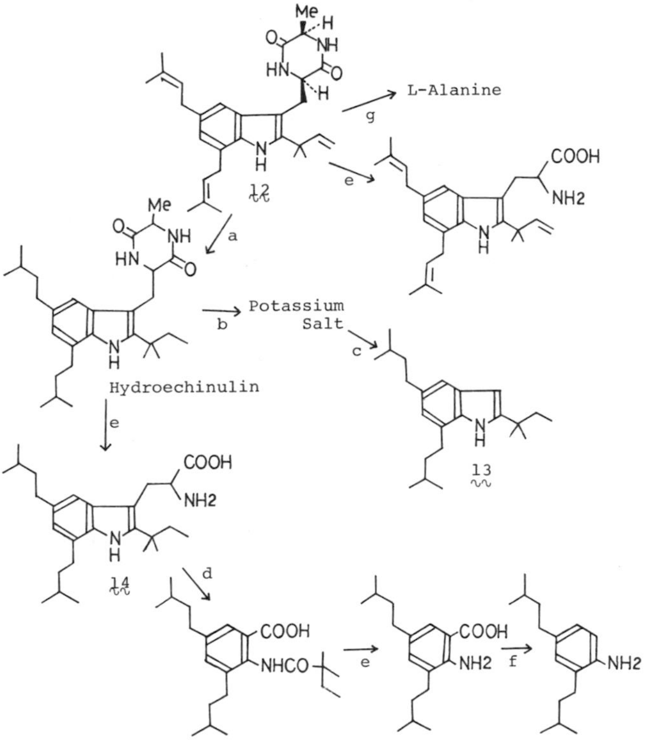

SCHEME 4. Chemical degradation of echinulin (12). Reaction conditions: a, H_2–PtO_2; b, KOH; c, pyrolysis; d, H_2O_2–AcOH; e, KOH–EtOH; f, heat; g, HBr.

latter was decarboxylated, and the product was identified with 2,4-diisoamyl aniline. In the IR spectrum of echinulin, two CO–NH absorption bands appear at 1670 and 1635 cm^{-1} which are practically superimposable on that of al-anylalanine anhydride; thus, the diketopiperazine ring was assigned.

The incorporation experiments by Birch and co-workers (33,34), using [2-^{14}C]mevalonic acid, L-[1-^{14}C]alanine, and L-tryptophan, provided valuable information for the structural elucidation. Echinulin possesses two asymmetric centers. L-Alanine is obtained by acid hydrolysis, but another chiral center on the tryptophan moiety is easily racemized. Later, it was determined as the L form by microbioassay of the aspartic acid obtained by ozonolysis (35). Finally, the chemical structure 12 was confirmed by the stereoselective total synthesis of optically active echinulin by Kishi and co-workers (36) (Scheme 5).

It took a rather long period to determine the novel structure of echinulin, but once upon it, chemical structures of successively isolated metabolites were deter-

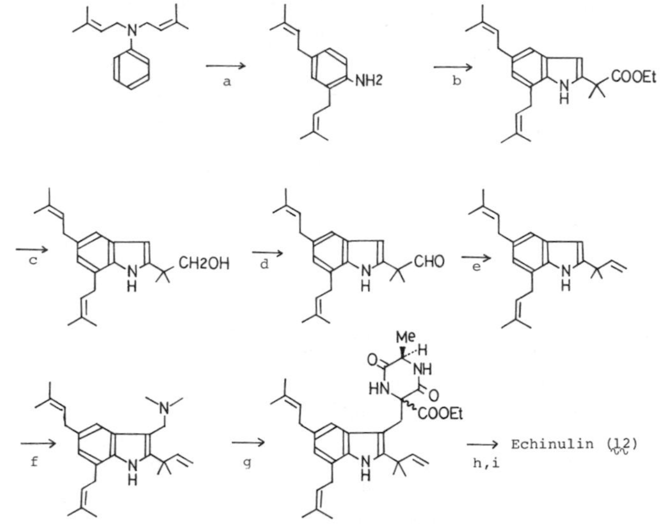

SCHEME 5. Total synthesis of echinulin (**12**) by Kishi and co-workers. Reaction conditions: a, $ZnCl_2$–Xylene–heat; b, $BrCH_2COC(CH_3)_2CO_2Et$; c, $LiAlH_4$; d, $DMSO–Ac_2O$; e, $CH_2{=}PPh_3$; f, $(CH_3)_2NH–CH_2O–AcOH$; g, ⟨structure⟩ –KOH powder; h, NaOH; i, heat.

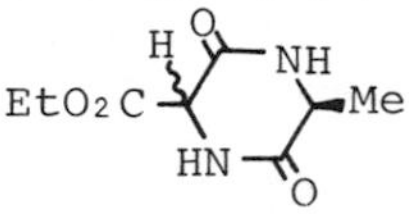

mined by comparison of physicochemical properties. The ^{13}C-NMR spectra of the echinulin and neoechinulin group were discussed in several reports (*37,38,39*).

In 1976, two key intermediates in the biosynthesis of **12** were isolated from filtrates of *Asp. chevalieri* (Mangin) Thom and Church grown on potato extract medium (*40,41*). One is cyclo(L-alanyl-L-tryptophan) (CAT, **15**), and the other is the 2-isoprenylated derivative (MICAT or preechinulin) (**16**). The latter was also obtained by using a partially purified enzyme which transfers the isoprene unit from 3-methyl-2-butenyl-1-phosphate to CAT (*42*).

B. NEOECHINULIN AND RELATED METABOLITES

Since 1969 many metabolites having an isoprenylated dehydrotryptophanyl moiety have been isolated from mycelia of *Aspergillus* species, which are classified into neoechinulin, isoechinulin, and cryptoechinulin groups (Table II). All metabolites are substituted by a *tert*-pentenyl (reversed isoprenyl) group at the C-2 position of indole ring as in echinulin. But a further isoprenyl group (if present) is located at C-6 in neoechinulins, at C-5 in isoechinulins, and at C-4 and C-5 in cryptoechinulin G. In many metabolites, the methyl group derived from alanine is modified to $=CH_2$ or $=O$ by higher oxidation. The metabolites of the dehydro series can be distinguished from echinulin by UV spectra (echi-

TABLE II

Neoechinulin and Related Metabolites

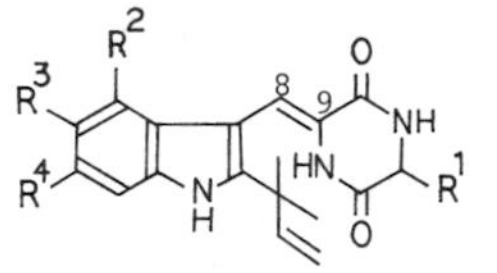

	Structure					
Name	R¹	R²	R³	R⁴	mp (°C)	Occurrence
Neoechinulin A (**18**)	Me	H	H	H	285 264–265	*Aspergillus amstelodami, Asp. niger, Asp. ruber*
Neoechinulin B (**19**) (cryptoechinulin C)	=CH₂	H	H	H	234–236	*Asp. amstelodami*
Neoechinulin C (**20**) (cryptoechinulin A)	=CH₂	H	H	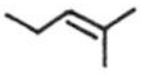 [prenyl]	205–207	*Asp. amstelodami*
Neoechinulin D (**21**)	Me	H	H	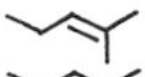 [prenyl]	223–225	*Asp. amstelodami*
Neoechinulin E	=O	H	H	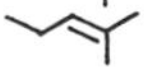 [prenyl]	275	*Asp. amstelodami*
Neoechinulin (**17**)	=O	H	H	H	245–246	*Asp. amstelodami*
Isoechinulin A (**22**)	Me	H	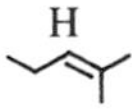 [prenyl]	H	—	*Asp. ruber*
Isoechinulin B (**23**)	=CH₂	H	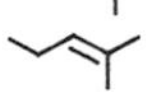 [prenyl]	H	—	*Asp. ruber*
Isoechinulin C (**24**)	=CH₂	H	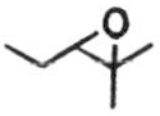 [prenyl with O]	H	—	*Asp. ruber*
Cryptoechinulin G (**25**)	=CH₂	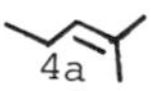 [prenyl, 4a]	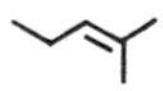 [prenyl]	H	—	*Asp. ruber*

nulin, λ_{max} 228, 285, and 295 nm; neoechinulin, λ_{max} 230, 287, and 420 nm; isoechinulin A, λ_{max} 227, 287, and 341 nm).

Stereochemistry of Z around the 8,9 double bond was established by ¹H- and ¹³C-NMR analysis (*37,50*) and synthetic study (*52*). Neoechinulin and related metabolites afford substituted indoles together with the corresponding 3-formyl derivatives by hydrolysis with KOH in alcohol. Oxidation of hydro derivatives with H_2O_2 in cold AcOH yields the corresponding N-2,2-dimethylbutyrylanthranilic acid derivatives.

1. Neoechinulins

Neoechinulin (**17**) and neoechinulins A–E have been isolated from ether extracts of mycelia of *Asp. amstelodami* (*44–46*) and independently from *Asp. ruber* (*47,48*) by many groups. Neoechinulins C (**20**) and B (**19**) were also isolated by Cardillo and co-workers, who called them cryptoechinulins A and C, respectively (*49*).

Oxidation of neoechinulin with $KMnO_4$ in alkaline medium and subsequent

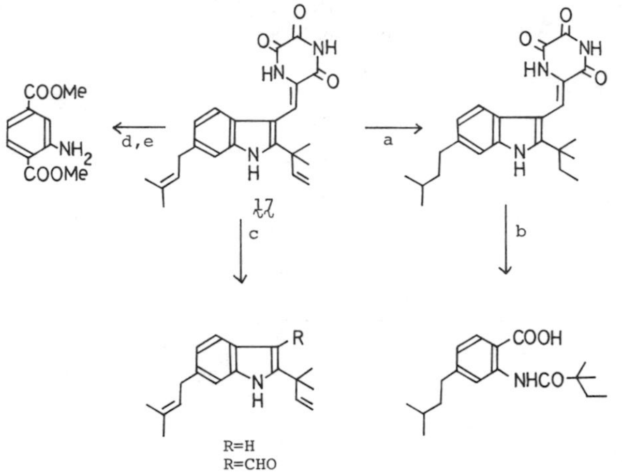

SCHEME 6. Chemical degradation of neoechinulin (17). Reaction conditions: a, H_2–PtO_2; b, H_2O_2–AcOH; C, KOH; d, $KMnO_4$; e, CH_2N_2.

methylation with CH_2N_2 after acidification affords 2-aminoterphthalic acid di-methyl ester, confirming the presence of the 6-substituted indole nucleus (50) (Scheme 6). Neoechinulin was synthesized by Kishi and co-workers (51), and neoechinulins A (18) and D (21) were synthesized by Goto and co-workers (52). The essential difference in NMR spectra between neoechinulins A and B consists of the absence of a $-CH_3$ doublet at δ1.42 for 18 and the appearance of $=CH_2$ at δ4.48 and 5.09 for neoechinulin B.

2. Isoechinulins

Isoechinulins A, B, and C (22–24) have been isolated from the mycelium of *Asp. ruber* by Tamura and co-workers (37,48,53), in the course of screening for insecticidal metabolites of fungi. Among them, isoechinulin A causes growth inhibition of silkworm larvae (*Bombyx mori* L.) by oral administration at a concentration of 1000 ppm.

Isoechinulin A (22) was considered to have an isoprenyl group at position 5 or 6 of the indole nucleus, judging from the coupling constant in the NMR spectrum. Hexahydroisoechinulin A is oxidized with H_2O_2 and then methylated to give *N*-2,2-dimethylbutyrylisoamylanthranilic acid methylester. Hydrolysis of the ester with HCl in MeOH gives isoamylanthranilic acid methylester. Conversion to the *N*-acetyl compound causes a remarkable downshift (from δ6.62 to 8.60) in only one of three aromatic protons in the NMR spectrum. Accordingly, the location of the prenyl group was determined as C-5.

3. Cryptoechinulin G

The optically inactive alkaloid cryptoechinulin G (**25**), $C_{29}H_{35}N_3O_2$, was isolated from the mycelium of *Asp. ruber* grown on sugar beet molasses (*54*). Spectroscopic evidence allowed assignment of the structural formula. Particularly, the presence of two adjacent aromatic hydrogens at δ7.05 and 7.15 (J = 8.2 Hz) in the ^{1}H-NMR spectrum and the signal of C-3 at 101 ppm in the ^{13}C-NMR spectrum determined that two adjacent isopentenyl groups are located at C-4 and C-5 instead of C-6 and C-7. In similar metabolites that lack substitution at C-4, the signal of C-3 appears at 103–104 ppm. The observed shift may be due to the shielding γ-effect of CH_2 at position C-4a.

4. Cryptoechinulins B and D

Another two isoprenylated dehydrotryptophanyl derivatives named cryptoechinulins B (aurechinulin, **26**) and D (**27**) were isolated from the mycelium of *Asp. amstelodami* by Fuganti and co-workers (*55*). The mass and elementary analysis data of these optically inactive compunds show the formula $C_{43}H_{49}N_3O_5$ and $C_{38}H_{41}N_3O_5$, respectively. The former is a mono-3,3-dimethylallyl derivative of the latter. Alkaline hydrolysis of cryptoechinulin B affords   2-(1′,1′-dimethylallyl)-6-(3′,3′-dimethylallyl)-3-indolecarbaldehyde, which is a degradative product of neoechinulin C resulting from alkaline hydrolysis.

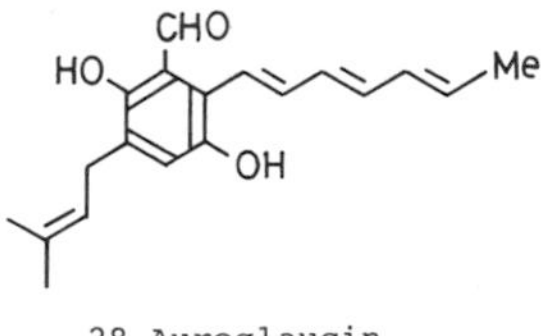

28 Auroglaucin

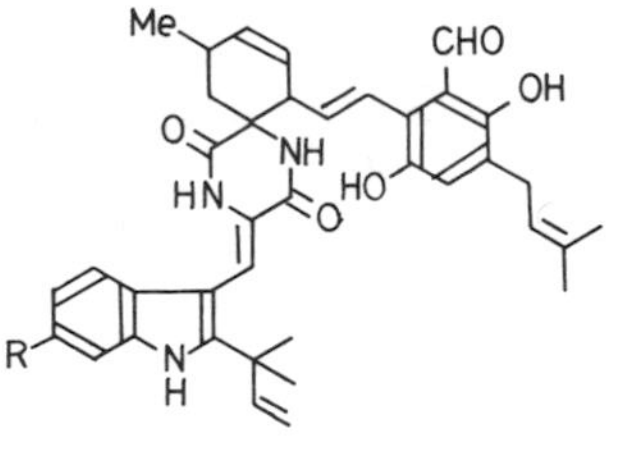

26 R=H

27 R= -CH₂CH=C⟨Me Me⟩

The strong MS peaks of **26** at m/z 298 and 389 correspond to auroglaucin (**28**), $C_{19}H_{22}O_3$, and neoechinulin C, $C_{24}H_{27}N_3O_2$, respectively. The formula of **26** is equal to the sum of these compounds, the both of which are metabolites of the *Asp. amstelodami* strain. Analysis of ^{1}H- and ^{13}C-NMR spectra indicates the structure (**26**) for cryptoechinulin B. Compounds **26** and **27** were synthesized by Diels–Alder reaction of auroglaucin and neoechinulins C or B, respectively (*56*).

C. Biosynthesis

Initially, Birch and co-workers (33,34) found that L-tryptophan, L-alanine, and mevalonic acid are incorporated into echinulin (Scheme 7). Slater (57) reported that chemically prepared cyclo(L-alanyl-[3'-¹⁴C]-L-tryptophan) (CAT), when added to the culture medium of Asp. amstelodami, is incorporated efficiently into echinulin (16.2%), whereas incorporation of L,D isomer is poor (0.2%). Further, the feeding of L,L-CAT together with 5 mol of unlabeled L-tryptophan did not significantly lessen the incorporation ratio (15.5%). This means that the L,L-cyclopeptide is incorporated intact without extensive prior hydrolysis to L-tryptophan.

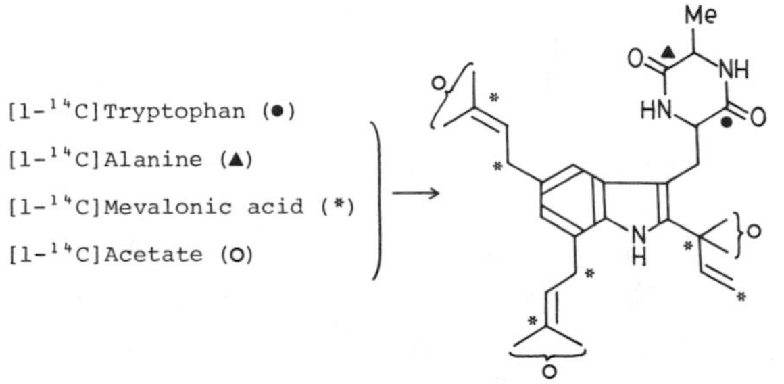

[1-¹⁴C]Tryptophan (●)

[1-¹⁴C]Alanine (▲)

[1-¹⁴C]Mevalonic acid (*)

[1-¹⁴C]Acetate (O)

SCHEME 7. Incorporation of tryptophan, alanine, mevalonic acid, and acetate into echinulin (12).

Allen (43) showed that enzymatically prepared ¹⁴C,³H-labeled MICAT (16) is also incorporated into echinulin well (14%) without change of the ¹⁴C/³H ratio *in vivo*. However, neither tryptophan nor linear dipeptide was isoprenated by this enzyme. These results suggest that cyclization proceeds prenylation at C-2. With administration of cyclo(L-[U-¹⁴C]alanyl-L-[5,7-³H₂]tryptophan (58), all isolated neoechinulins keep constant the ³H/¹⁴C ratio, which means the modified groups (at C-12) are produced after ring formation.

The incorporation of MICAT into neoechinulin shows that Δ⁸-unsaturation occurs after prenylation at C-2. Feeding experiments of (3R)- or (3S)-L-[3-³H,3-¹⁴C]tryptophan established the stereospecific removal of a *pro-S* hydrogen atom from the β-methylene position to effect the Δ⁸-unsaturation (59). The prenylation site in the benzene unit seems to depend on the presence or absence of Δ⁸-unsaturation in the substrate. In the saturated compound the C-5 and C-7 positions of indole ring are available for direct electrophilic attack, whereas in the Δ⁸-unsaturated compound C-4 and C-6 become relatively active, which may be caused by the contribution of resonance structure of B (56) (Fig. 1).

Another interesting structural feature of echinulin family is the presence of reversed isoprene unit at position 2. Many mechanisms (60–67) involving either a direct attack at position 2 of the indole system or a primary attack at position 3

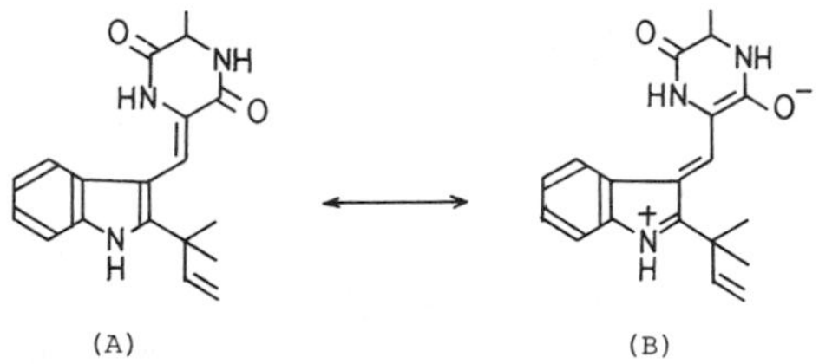

FIG. 1

or *N* followed by rearrangement have been discussed, but direct substitution seems more likely [cf. asterriquinone (**101**)]. In conclusion, the biosynthetic network of the echinulin group may be fitted as shown in Scheme 8, but the postulated relationships still remain to be verified.

D. ASTECHROME

An iron-containing metabolite named astechrome, mp 188–189°C (dec.), was isolated by Yamamoto and co-workers (68) as dark-red needles from the my-

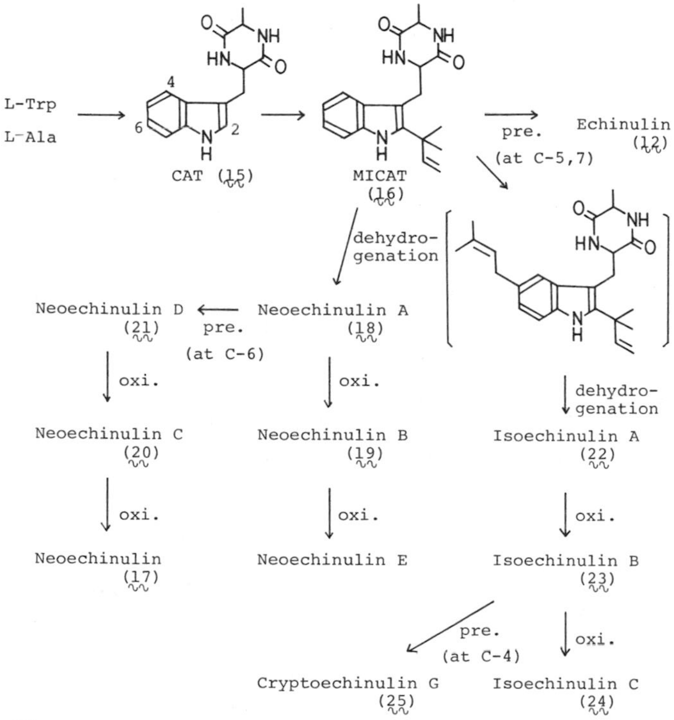

SCHEME 8. Biosynthetic network of the echinulin group. Key: pre., prenylation; oxi., oxidation.

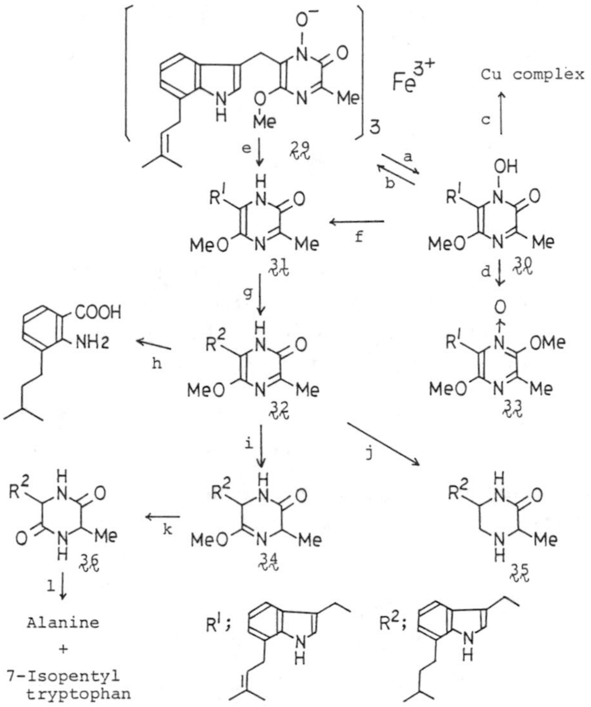

SCHEME 9. Chemical degradation of astechrome (29). Reaction conditions: a, NaOH–acetone; b, FeCl$_3$; c, CuSO$_4$; d, CH$_2$N$_2$; e, PBr$_3$; f, (CH$_3$O)$_3$P; g, H$_2$–Pd/C; h, H$_2$O$_2$–NaOH; i, H$_2$–PtO$_2$–EtOH; j, H$_2$–PtO$_2$–AcOH; k, HCl–AcOH; l, 6 N HCl.

celium of *Asp. terreus* IFO 6123 from shake cultures on Czapek–Dox medium supplemented with polypeptone and ferrous ion. Astechrome is optically inactive, and the presence of iron atom in the molecule was suggested by NMR measurement and analyzed quantitatively (∼5.3%). The molecular weight (*m/z* 1112) was obtained from FD–MS, whereas in EI–MS the highest fragment peak appeared at *m/z* 351. Alkaline treatment of astechrome (29) gives an iron-free colorless powder (30), C$_{20}$H$_{22}$N$_3$O$_3$, which is regenerated to 29 with FeCl$_3$ and which, with CuSO$_4$, yields a greenish copper complex, (C$_{20}$H$_{22}$N$_3$O$_3$)$_2$Cu, mp 162°C, *m/z* 767 (Scheme 9). The measured Magnetic moment (5.60 bohr magnetons) corresponds to a paramagnetic Fe^{3+}. From this evidence the chemical formula (C$_{20}$H$_{22}$N$_3$O$_3$)$_3$Fe is assigned to astechrome. Astechrome has weak antibiotic activity.

Treatment of 29 with PBr$_3$ [or 30 with (CH$_3$O)$_3$P] affords a deoxy compound (31), C$_{20}$H$_{24}$N$_3$O$_3$, which possesses a 3,3-dimethylallyl group in the molecule (see Scheme 9). It is hydrogenated to give 32, C$_{22}$H$_{24}$N$_3$O$_3$, which in the NMR spectrum shows one each of methyl, methoxy, and methylene groups and four ring protons together with a new isopentyl group. Three of the ring protons are vicinal and another proton is coupled with NH ($J = 2.5$ Hz). The presence of the indole ring is suggested by positive Ehrlich reaction and by the UV spectrum.

Oxidation of **32** with H_2O_2 in NaOH gives 3-isopentylanthranilic acid, which determines isopentyl group at position 7 of the indole ring. The presence of NH–CO bands (3400, 1540, and 1510 cm^{-1}) in the IR spectrum of **32** is supported by the shift of UV absorption in alkaline solution. These bands disappear in the IR spectrum of acetyl or methyl derivatives, in which −N=C−OAc or −N=C−OCH$_3$ groups have been produced. In parallel, in the methyl ether of **30** (**33**), the CO–N(OH) bands (3400 and 1620 cm^{-1}) in **30** disappear and the presence of the new *N*-oxide group is recognized by IR analysis.

The change of the UV absorption maximum (450 nm at pH 7; 520 nm at pH 2) also supports the theory that astechrome is a complex of iron and hydroxamate. DL-[2-^{14}C]Tryptophan and DL-[1-^{14}C]alanine are effectively incorporated into astechrome (14.0 and 5.6%, respectively). The arrangement of methoxy and hydroxamate groups in the pyrazine ring was determined by decoupling experiments of further hydrogenated compounds (**34** and **35**). Demethylation of **34** with HCl affords diketopiperazine **36** which is hydrolyzed to alanine and 7-isopentyltryptophan.

IV. Diketopiperazines Derived from Tryptophan and Proline

The fundamental metabolite of the tryptophan–proline derived series possessing no substituent in the skeleton is brevianamide F (L-prolyl-L-tryptophanyldiketopiperazine, **37**, which was isolated from *Pen. brevi-compactum* (*69*). Though **37** has not been isolated from *Aspergillus* species yet, many biologically active derivatives such as austamides, fumitremorgens, and verruculogen have been found in *Aspergillus* species. All these metabolites have a C$_5$ side chain at position 2 of the indole ring. Based on differences in the substituted form (normal or reversed) of the isoprene unit, metabolites are classified in this chapter into the austamide and fumitremorgen groups.

A. AUSTAMIDE GROUP

Metabolites of the austamide group have a 1,1-dimethylallyl (reversed isopentenyl) moiety at position 2 of the indole ring. Steyn (*70*) separated the crude toxic extracts (50 g) from dried moldy maize (6.3 kg) of *Asp. ustus* (Bainier) Thom and Church, by chromatography on formamide-impregnated cellulose powder in systematic fractionation guided with bioassay in ducklings. Prolyl-2-(1′,1′-di-

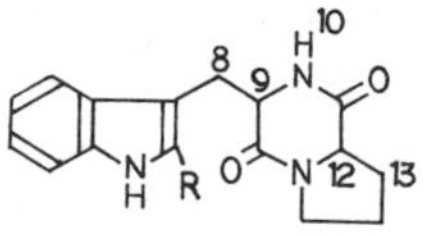

37 R=H
38 R= ⊢
39 R= ⊢ (12,13-dehydro)

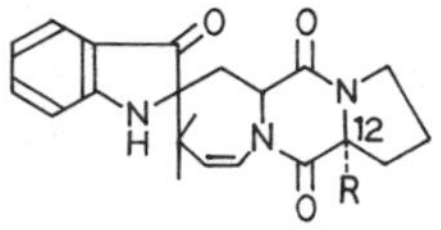

41 R=H
43 R=OH

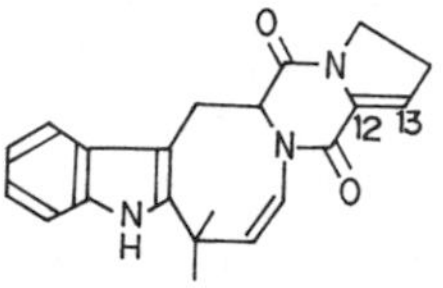

42

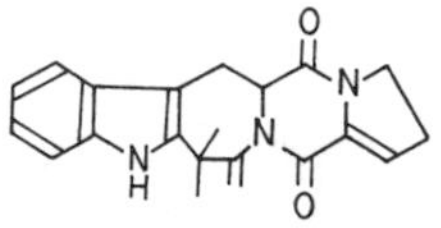

46

methylallyltryptophanyl)diketopiperazine (deoxybrevianamide E, **38;** 25 mg), the 12,13-dehydro derivative (**39;** 120 mg), austamide (**40;** 2.7g), 12,13-di-hydroaustamide (**41;** 40 mg), and 10,20-dehydro-[12,13-dehydroprolyl-2-(1′,1′-dimethylallyltryptophanyl)diketopiperazine] (**42;** 8 mg) were isolated. Later, 12,13-dihydro-12-hydroxyaustamide (**43;** 10 mg/10 kg maize) was also isolated from the same strain (*71*).

1. Deoxybrevianamide E and Related Metabolites

The indoles **38** and **39** give a negative Ehrlich color reaction, indicating substitution at position 2 and 3 of the indole ring. The mass spectra of these compounds show a base peak at m/z 198, $C_{14}H_{16}N$, which originated via cleavage of the C-8–C-9 bond. On the basis of spectral data, **38** seemed to be identical to deoxybrevianamide E which had been obtained by Birch and Wright (*72*) upon treatment of brevianamide E (**44**) with Zn and AcOH. Later, **38** and **44** were

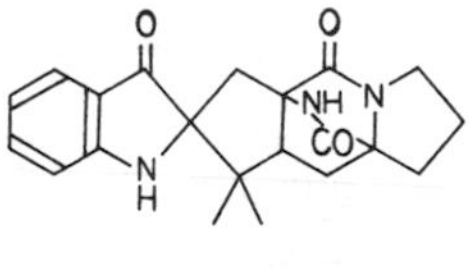

45

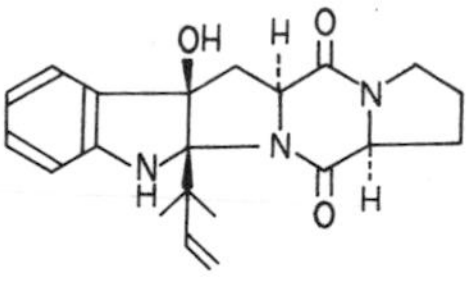

44

synthesized by several workers (*73,74*). Allen (*75*) also prepared **38** from brevianamide F, using a partially purified enzyme possessing a prenylation ability (cf. echinulin). Proline obtained by acid hydrolysis of **38** shows a peak at 224 nm [φ] +225° (0.1 N HCl) which is identical in the Cotton effect to that of L-proline. Then, **38** was determined to have the 12*S* conformation. Further, from a comparison of the CD spectra of L-tryptophanyl-L-prolyldiketopiperazine and

D-tryptophanyl-L-prolyldiketopiperazine, the L,L configuration of **38** was confirmed. The Dreiding model of this compound indicates a boat form for the diketopiperazine ring and a half-chair conformation for the proline ring. 12, 13-Dehydroprolyl-2-(1′, 1′-dimethylallyltryptophanyl)diketopiperazine, $C_{21}H_{23}N_3O_2$, shows a similar IR and UV spectra to that of **38** with the exception of an additional sharp band at 1648 cm^{-1} attributable to the enamide moiety in the IR spectrum.

2. Austamide and Related Metabolites

Austamide (**40**), $C_{21}H_{21}N_3O_3$, $[\alpha]_D^{20}$ +152°, is an amorphous yellow compound with an intense green fluorescence and is toxic in 1-day-old ducklings. It shows UV absorptions at 234, 256, 268(sh), 282, and 392 nm, among which the absorptions at 234, 256, and 392 nm are typical of the ψ-indoxyl chromophore which occurs in brevianamide A (**45**) and other ψ-indoxyl alkaloids (*76*). The UV absorptions at 268 and 282 nm are absent in tetrahydroaustamide, so these are associated with the cyclic enamide chromophore.

In the IR spectrum of austamide, bands arising from an NH (3420 cm^{-1}), a ψ-indoxyl CO (1700 cm^{-1}), a diketopiperazine unit (1680 cm^{-1}), and an enamide group(1650 cm^{-1}) are recognized. Two geminal methyl groups (δ0.80 and 1.50) and cis olefinic protins (δ4.80 and 6.66, J = 10 Hz) are observed in the NMR spectrum, but the reversed isoprene group is not recognized. Also only one exchangeable NH (br, δ5.26) is present, and the preparation of *N*-nitroso derivative of tetrahydroaustamide establishes the secondary nature of the ψ-indoxyl nitrogen. These results suggest that the endo methylene of the reversed isoprene unit bonds to a nitrogen atom of the diketopiperazine ring. The other protons appearing as A_2M_2X in the proline ring correspond to that of **39**. The splitting pattern in the aromatic region is identical in austamide and brevianamide A. These assignments are confirmed by decoupling experiments.

The most important fragmentation of austamide originates from cleavage of the spiran ring leading to the allicyclic fragment *m/z* 218 (Fig. 2). Catalytic hydrogenation over Pd/C gives the two pairs each of stereoisomers of dihydro- and tetrahydroderivatives. The major product of dihydro derivative, mp 235–238°C, $[\alpha]_D^{22}$ +55°, is identical to naturally occurring 12,13-dihydroaustamide

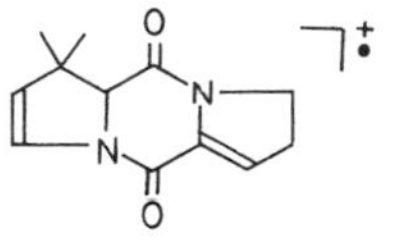

m/z 218

FIG. 2

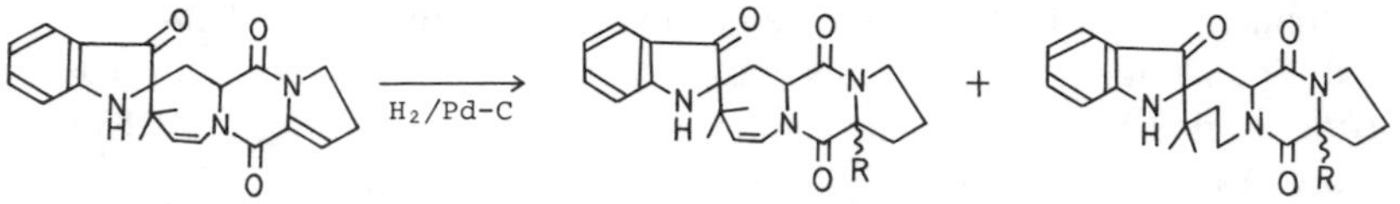

FIG. 3

(41) (Fig. 3). Acid hydrolysis of **41** liberates approximately 1 mol of proline, [ϕ] +460° (224 nm, 0.1 N HCl), which establishes the S conformation at C-12. The stereochemistry at position 9 is based on the $9S,12S$ configuration of deoxybrevianamide E, an apparent bioprecursor. The relative configuration of the spiro atom (position 2) to that of position 9 is based on NMR-shift data. The proposed structure and absolute configuration of (12S)-tetrahydroaustamide was subsequently confirmed by X-ray crystallography (77). dl-Austamide was synthesized by Hutchison and Kishi (78).

The structure of 12,13-dihydro-12-hydroxyaustamide (**43**), mp 164–165°C, was determined by comparison with the spectral data of austamide. The lack of a proton at C-12 ($\sim\delta 4.1$) strongly supports the presence of an OH group at C-12. The mass spectrum of **43** shows a strong peak at m/z 363, $C_{21}H_{21}N_3O_2$, due to loss of H_2O and subsequent fragmentation, virtually identical to that of austamide, with the base peak at m/z 218.

10,20-Dehydro-[12,13-dehydroprolyl-2-(1',1'-dimethylallyltryptophanyl)diketopiperazine] (**42**), $C_{21}H_{21}N_3O_2$, was found in *Asp. ustus* irregularly and in poor yield. The UV and IR spectra are virtually identical to those of **39**, differing in the NH region where only two peaks occur at 3485 and 3350 cm^{-1} (3480, 3460, and 3360 cm^{-1} in **39**). The NMR spectrum is also virtually identical to austamide, though a considerable difference is recognized in the chemical shift. From the above data two structures, **42** and **46,** are possible. However, the down-field shift of two olefinic protons is identical to that observed for these protons in austamide, and a secondary methyl is not exhibited by hydrogenation; thus **46** is excluded. The chemical structure **42** is in agreement with biosynthetic considerations and can be regarded as the probable precursor of austamide. Therefore, closure of the diketopiperazine onto the terpene unit may occur prior to the indole oxidation which furnishes the ψ-indoxyl moiety as in austamide and brevianamide A.

B. FUMITREMORGEN GROUP

Another group of tryptophan–proline-derived fungal metabolites having a 3,3-dimethylallyl (normal isopentenyl) unit at position 2 of the indole ring is the fumitremorgen group. This group includes many tremorgenic metabolites such as fumitremorgens A and B and verruculogen. Structural characteristics include a methoxyl group at position 6 of the indole ring and a fused five-ring skeleton obtained by condensation of a substituted isoprene residue onto a nitrogen atom of a diketopiperazine ring.

The tremorgenic toxin was first isolated from *Asp. flavus* by Wilson and Wilson (*79*). Though this metabolite itself has not been further investigated owing to its natural occurrence in very small amounts, the tremorgenic compounds known were so rare at the time that this work excited interest in this field. Tremorgenic toxins comprise a group of about 20 metabolites which cause vigorous tremor and convulsions in experimental animals.

1. Fumitremorgens A and B

During investigation of toxigenic food-borne fungi, Yamazaki and co-workers (*80*) found that a certain strain of *Asp. fumigatus* Fres. has a strong tremorgenic action on mice by intraperito-neal (ip) injection of crude extract. From the AcOEt extracts of the toxic strain, 10 metabolites were isolated by column chromatography on silica gel. Two tremorgenic principles were named fumitremorgens A (**47**) and B (**48**) (FTA and FTB), and other nontremorgenic metabolites were initially designated as "FTC–FTJ" (*81*). Metabolites of the latter group were demonstrated to be tryptoquivaline-related compounds, and their names were changed to "tryptoquivalines C–J" (*82*). FTA and FTB show neurotoxic activities. The ED_{50} values for causing tremor in mice by ip administration are 0.177 mg/kg for FTA and 3.5 mg/kg for FTB. The LD_{50} of FTA in mice by intravenous (iv) administration is 0.185 mg/kg (*83*).

48 Fumitremorgen B
(FTB)

The UV spectral patterns of FTA and FTB indicate the presence of an indole nucleus in the molecules. The NMR spectral data suggest a 5- or 6-methoxyindole, and comparison with the ^{1}H-NMR data of 2,3-dimethyl-5- and -6-methoxyindole indicates that the methoxyl group is present at position 6. The diketopiperazine ring in the molecule is indicated by strong IR absorption bands at 1685 and 1645 cm^{-1} and the absence of amide II-bands. In feeding experiments (*80*), [3-^{14}C]tryptophan, [U-^{14}C]proline, and [2-^{14}C]mevalonic acid are effectively incorporated into FTA as well as into FTB. The diketopiperazine ring is assumed to be biosynthesized from two amino acids, 6-methoxytryptophan and proline.

Fumitremorgen B (FTB) (meaning, fumigatus toxin B) (*84*), $C_{27}H_{33}N_3O_5$, *m/z* 479, colorless needles (MeOH), mp 211–212°C, $[\alpha]_D^{25}$ +24°, shows two isoprenyl units in the NMR spectrum. Comparison with 3-methyl-*N*-isopentenylindole indicates that one of the two isoprenyl groups is located at position N-1 of the indole ring. A signal at δ4.52 (d, 2*H*) is assigned to a methylene adjacent to the indole nitrogen atom which is coupled with an olefinic proton at δ5.04. In the mass spectrum, the base peak appears at *m/z* 311 ($C_{20}H_{25}NO_2$) owing to the elimination of the diketopiperazine fragment which shows two isopentenyl groups in the indole ring. A negative color reaction to Ehrlich's reagent supports the presence of another C_5 unit at position 2. Then, a signal at δ5.97 (d, 1*H*) coupling with an olefinic proton at δ4.67 in a second isoprenyl chain is assigned as a methine binding to N-10. The remaining two oxygen atoms are considered as a secondary OH at position 8 and a tertiary OH at 9, on the basis of spectral data. From this evidence the chemical structure of fumitremorgen B (FTB; **48**) was elucidated. The absolute configuration of FTB was determined on the basis of X-ray analysis data, and the configuration of proline was obtained through acid hydrolysis (*85*). Comparison between FTB and lanosulin (a metabolite from *Pen. lanosum*) revealed that they are identical (*86*).

Fumitremorgen A (FTA), $C_{32}H_{41}N_3O_7$ (colorless needles), mp 206–209°C, $[\alpha]_D^{20}$ +61° (acetone), shows UV and IR spectra similar to those of FTB, but the presence of only one hydroxyl group is expected from the NMR spectrum. The difference between the formulas of FTA and FTB is $C_5H_8O_2$, a fact which suggests an additional isopentenyl group compared with FTB. By refluxing with 0.1% H_2SO_4 in MeOH, FTA gives a degradation product (*49*), $C_{29}H_{37}N_3O_7$, mp 222–224°C (Scheme 10). The UV, IR, and NMR spectra indicate elimination of one isoprenyl group and formation of two new methoxyl groups in **49**. On hydrogenation of **49** over Adam's catalyst in AcOEt, a second product (**50**), $C_{24}H_{31}N_3O_6$, mp 218–219°C, is obtained. The disappearance of the second isopentenyl group and the appearance of a tertiary OH (δ2.50) and an NH of the indole (δ10.00) in **50** are concluded from the NMR analysis. Thus, it is presumed that the isopentenyl group attached to N-1 of the indole ring may be removed. It is known that a secondary–tertiary dialkyl peroxide gives a ketone and a tertiary alcohol by treatment with Lewis base or electron donors. Accordingly, it is assumed (*87*) that the two extra oxygen atoms in FTA belong to the secondary–tertiary dialkyl peroxide which forms a secondary and a tertiary alcohol upon catalytic hydrogenation with subsequently removal of the N-substituted secondary alcohol. On refluxing of FTA with 0.1% H_2SO_4 in acetone, verruculogen is obtained as a minor product. In conclusion, the structure of fumitremorgen A (FTA) as shown in **47** was also confirmed later by X-ray analysis by Clardy and co-workers (*88*).

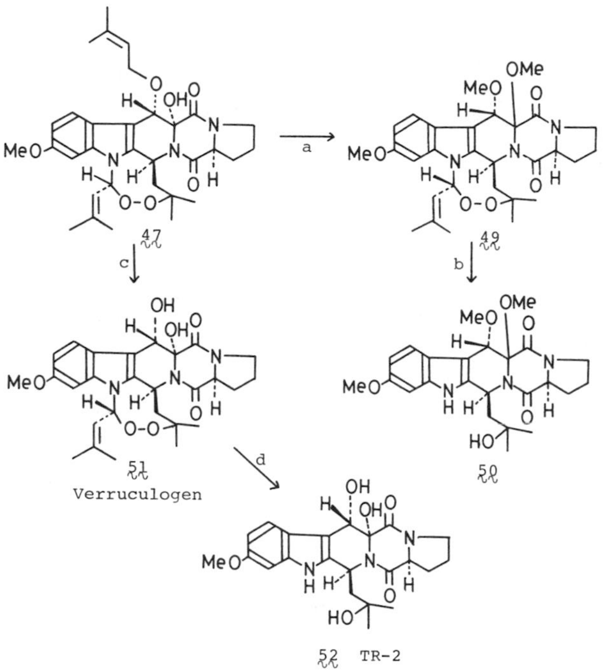

SCHEME 10. Chemical degradation of fumitremorgen A (**47**). Reaction conditions: a, 0.1% H_2SO_4–MeOH; b, H_2–PtO_2–AcOEt; c, 0.1% H_2SO_4–acetone; d, H_2–Pd/C–EtOH.

2. Verruculogen and Related Metabolites

In 1972 (*89*), another strong tremorgenic mycotoxin named verruculogen (TR-1, **51**), $C_{27}H_{33}N_3O_7$, mp 233–235°C (dec.), *m/z* 511.236, was obtained from a strain of *Pen. verruculosum* isolated from peanuts in the United States. Later, this toxin was found in *Asp. caespitosus* (*90*) and other *Penicillium* strains (*91,92*). The LD_{50} (ip) of verruculogen is 2.4 mg/kg, and the ED_{50} for tremor response is 0.39 mg/kg (ip) in Swiss mice. Oral doses are 40 times less effective (*89*). In surface cultures of *Pen. simplicissimum,* the diketopiperazine ring of verruculogen is known to be biosynthesized from L-tryptophan and L-proline (*93*). The structure was determined by X-ray diffraction experiments (*91*).

The indole portion of the verruculogen molecule is planar, the diketopiperazine ring is folded into a boat conformation, and the hydroxyl groups at C-8 and C-9 are cis as those of FTA and FTB (Fig. 4). In contrast, all the nontremorgenic compounds have a different stereostructure (chair and trans) in which there is no possibility of the presence of the strong hydrogen bonds among the CO and two OH groups. The stereochemistry around these groups determines

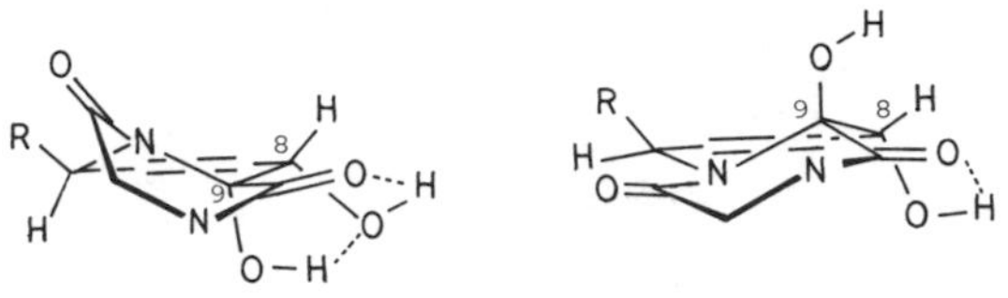

(A)

cis

(B)

trans

FIG. 4

the tremorgenic activity; in particular, the cis conformation of the two hydroxy groups is essential (*94*).

Another tremorgenic metabolite, TR-2 (**52**), $C_{22}H_{27}N_3O_7$, originally obtained by catalytic hydrogenation of verruculogen with Pd/C in EtOH (*95*) was later isolated from *Asp. caepitosus* Raper and Thom, together with FTB and cyclopiamine B (*96*), and from *Asp. fumigatus* (*97*). Upon administration to *Pen. raistrickii* (*98*), ^{14}C-labeled TR-2 is incorporated in above 27% yield into verruculogen, which suggests that TR-2 is an important intermediate in biosynthesis of verruculogen. It is interesting that verruculogen is metabolized to the less potent tremorgens (TR-2 and others) in the rat (*99*).

In 1982, the related tremorgenic metabolite acetylverruculogen, mp 221–224°C(dec.), was isolated from *Pen. verruculosum*. It affords 4-hydroxyproline as a hydrolyzed product upon acid treatment, and the structure **53** was determined by the X-ray analysis (*100*).

Fumitremorgen C (**54**), the simplest member of this group of mycotoxins, and epoxyfumitremorgen C have been isolated from culture extracts of *Asp.*

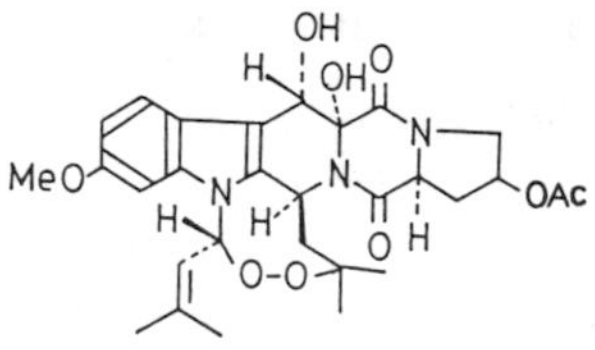

53

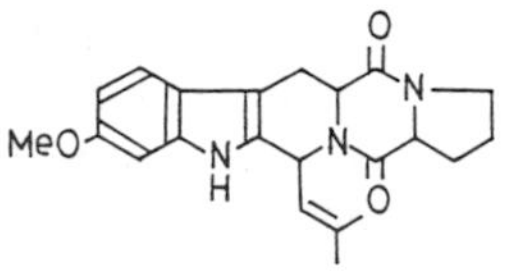

54

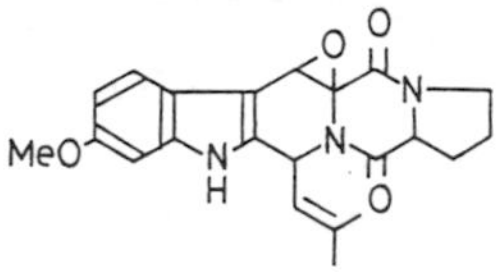

Epoxyfumitremorgen C

fumigatus collected from silage by Cole (*101*). The structure of fumitremorgen C was determined by X-ray analysis.

3. Cyclopiamine B

Cyclopiamine B, mp 243–245°C, is another example of the type of metabolite derived from tryptophan and proline. It was previously known to be produced by *Pen. cyclopium* Westling (*102*) and *Pen. urticae* Bainier (*103*), and it was also isolated from the CHCl$_3$–MeOH extracts of mold-infected material of *Asp. caespitosus* Raper and Thom (*96*). By solvent partition and subsequent partition chromatography on formamide-impregnated cellulose powder, Steyn *et al.* (*96*) isolated cyclopiamine B as a minor cometabolite (0.00025%) of fumitremorgen B and TR-2. The structure of cyclopiamine B (**55**) was elucidated by chemical degradation and derivation together with application of physicochemical tech-

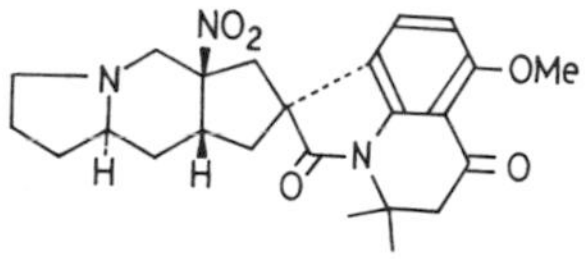

55

niques. The final structure and relative stereochemistry were established by single-crystal X-ray analysis (*102*). The co-occurrence of these metabolites confirms their biosynthetic relationships. Cyclopiamine B seems to be biogenetically derived from tryptophan, proline, and two units of dimethylallylpyrophosphate. The different linkage of the two C$_5$ units leads to the novel structure.

V. Other Diketopiperazines Derived from Tryptophan

Few fungal metabolites consist of a diketopiperazine ring derived from tryptophan and an amino acid other than alanine and proline. Those known in *Aspergillus* species include ditryptophenaline and D-valyl-L-tryptophan anhydride.

1. Ditryptophenaline

Ditryptophenaline (**56**), mp 204–205°C, was isolated from the mycelium of *Asp. flavus* during the investigation for food-borne mycotoxin by Büchi and coworkers (*104*). The chemical formula, C$_{42}$H$_{40}$N$_6$O$_4$, is assigned on the basis of the high-resolution mass spectrum (*m/z* 692.3095), and the NMR spectrum indicates a dimeric structure. It possesses a diketopiperazine ring derived from tryptophan and phenylalanine and has UV absorptions at 244 and 303 nm. The chemical structure was determined by single-crystal X-ray experiments. Later, Hino and co-

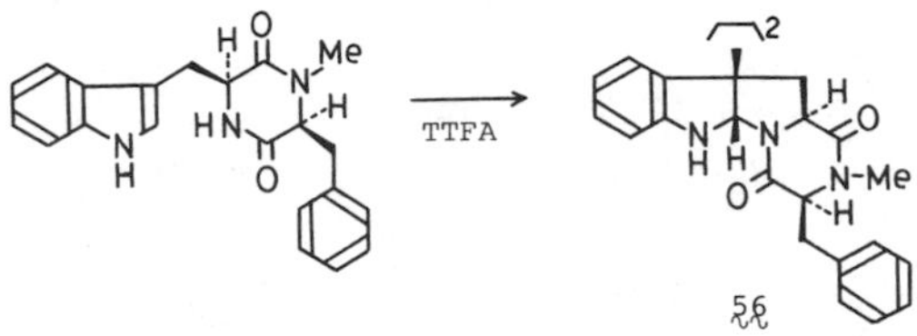

56

workers (*105*) determined the absolute configuration by oxidative coupling of cyclo(L-*N*-methylphenylalanyl-L-tryptophan) with thallium(III) trifluoroacetate (TTFA) in 3% yield. Ditryptophenaline does not possess significant toxic or antibiotic properties.

2. D-Valyl-L-tryptophan Anhydride

D-Valyl-L-tryptophan anhydride (**57**) was isolated by Stipanovic *et al.* (*106*) from cultures of *Asp. chevalieri* (Margin) Thom and Church together with pre-echinulin (MICAT). The rhombic crystals change to long needles between 235 and 255°C, and **57** melts at 277–279°C. High-resolution mass measurement

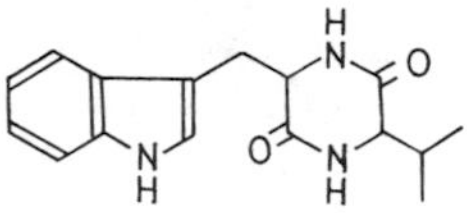

57

indicates the formula $C_{16}H_{19}N_3O_2$. The IR spectrum shows the presence of at least one amide group (1660 cm^{-1}), and the UV spectrum (λ_{max} 220, 274.5, 281, and 290 nm) suggests the presence of an indole nucleus. In the ^{1}H-NMR spectrum the presence of two methyl protons at δ0.85 and 0.88 coupling to a methine (m, δ2.23) and a doublet proton (d, δ3.23) indicates that a portion of the molecule is derived from valine. The metabolite was tentatively assumed to be valyltryptophan anhydride, and this is confirmed by following synthesis. L-Tryptophan methylester is reacted with carbobenzoxyvaline, the product hydrolyzed with NaOH, and then the protected carbobenzoxy group is removed by hydrogenation with Pd/C. The product is refluxed for 48 hr in toluene and yields L-valyl-L-tryptophan anhydride, mp 298.5–303°C, isolated by TLC. The mass spectrum of this compound is similar to that of the metabolite, however, their IR and NMR spectra and R_f values are different. Then, DL-valyl-L-tryptophan anhydride was synthesized as described above and separated on silica gel. The product from the upper band is rechromatographed, and the purified compound, mp 278–279°C, proved to be identical to the naturally occurring metabolite. The lower band corresponds to that of L,L compound.

3. Neoxaline

A weakly neurotoxic metabolite, neoxaline (**58**) (*107*), mp 202°C (dec.), $[\alpha]_D^{24}$ −16.3° (CHCl$_3$), is described here even though it does not possess a diketopiperazine system in the molecule. It was isolated from cultures of *Asp. japonicus* (yield, 230 mg/18 liters) (see this treatise, Vol. 22 (1984), p. 317) (*108*). Its physicochemical properties are very similar to oxaline (**59**), isolated from *Pen. oxalicum,* and the unique structure was determined by correlation with oxaline (*109*). The efficient conversion of roquefortine (**60**) to oxaline (24.3%) is known (*110*), and neoxaline is probably also biosynthesized from **60**, which has been isolated from *Pen. roquefortii* (*111*) as a tremorgenic toxin and has a tryptophan–histidine-derived diketopiperazine system in the molecule.

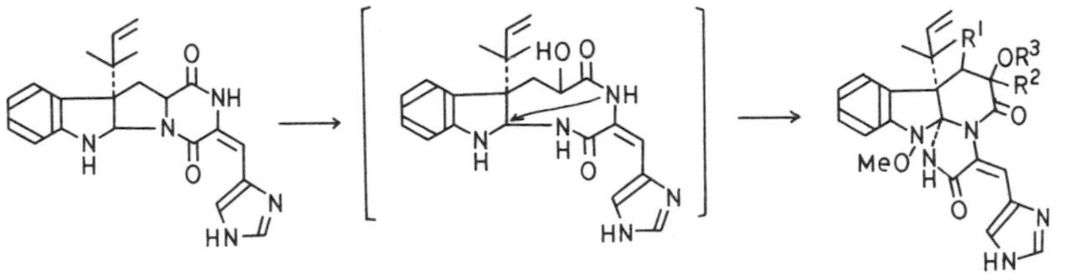

60

58 R^1=R^2=R^3=H

59 R^1=R^2 (dehydro)

R^3=Me

VI. Diketopiperazines Derived
from Phenylalanine

Several metabolites in which phenylalanine participates as a component of the diketopiperazine ring have been isolated from *Aspergillus* species. Other possible components of the diketopiperazine skeleton are serine (gliotoxin), phenylalanine (aranotin and emestrin), and glycine (aspirochlorine). It is interesting that all of these metabolites possess two–four sulfur atoms in the molecule. These inhibit bacterial growth and are rather toxic to mammalian cells.

A. GLIOTOXIN GROUP

1. Gliotoxin

Gliotoxin (**61**), C$_{13}$H$_{14}$N$_2$O$_4$S$_2$, mp 194–195°C, $[\alpha]_D^{22}$ −289° (MeOH), a highly antifungal and antibacterial metabolite, was first isolated from a *Trichoderma* species in 1936 by Weindling and Emerson (*112*). This substance has since been isolated as a metabolite of a variety of microorganisms, including *Asp. fumigatus* (*113*), *Asp. chevalieri* (*114*), *Gliocladium* species (*115*), and some *Penicillium* species (*116*).

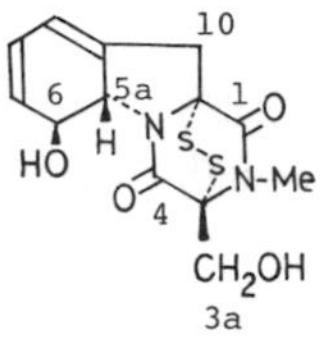

The chemistry of gliotoxin was extensively studied by Johnson and co-workers (*115,117*). The correct structure was finally proposed in 1958 and confirmed by X-ray analysis (1966) (*118*) which also revealed the absolute configuration. Many reviews have been published on the structure, chemistry, biosynthesis, and biological activity of gliotoxin (*119–123*). A total synthesis of gliotoxin, using the Michael reaction as a key step, was achieved by Kishi and co-workers (*124*) (Scheme 11).

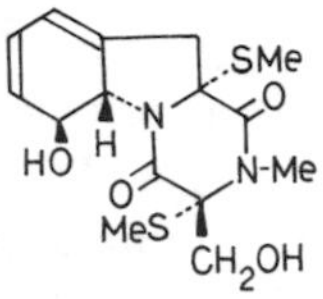

In 1980 bisdethiobis(methylthio)gliotoxin (**62**) was isolated from *Gliocladium deliquescens* (yield, ~6% of gliotoxin). Obtained as amorphous, it was characterized via the crystalline, bis-4-bromobenzoate of the didehydro derivative and identified by synthesis from gliotoxin by reduction and methylation (*125*).

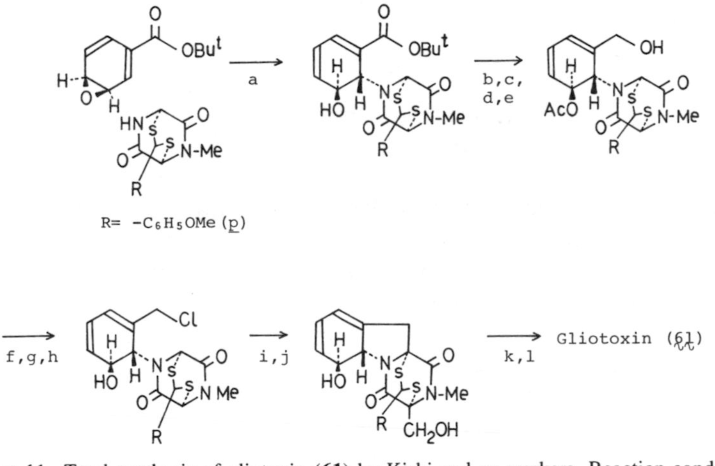

SCHEME 11. Total synthesis of gliotoxin (**61**) by Kishi and co-workers. Reaction conditions: a, Triton B; b, Ac$_2$O–Py; c, TFA; d, ClCO$_2$Et–Py; e, NaBH$_4$; f, MsCl–Et$_3$N; g, LiCl–DMF; h, NaOMe–MeOH/CH$_2$Cl$_2$; i, C$_6$H$_5$Li–ClCH$_2$OCH$_2$C$_6$H$_5$; j, BCl$_3$; k, *m*–ClPBA; l, HClO$_4$.

2. Biosynthesis

DL-[1-^{14}C]Phenylalanine is incorporated into gliotoxin (**61**) in 4.2% yield, and degradation of labeled gliotoxin shows that essentially all of the radioactivity (82%) resides at position 1, suggesting the intact incorporation of phenylalanine (*126*). DL-[2-^{14}C]Phenylalanine incorporates in 12.4% yield and DL-[1-^{14}C]serine in 1.9% (*127*), but labeled DL-tryptophan and *m*-tyrosine are not (*128,129*). Further, in other feeding experiments with the four stereoisomers, only cyclo(L-phenylalanyl-L-serine) is incorporated (48%) into gliotoxin (*130*). Cyclo(L-[4'-^{3}H]phenylalanyl-L-[3-^{14}C]serine) gives gliotoxin with no detectable change of the ^{3}H/^{14}C ratio (*129*). However, no intermediates are detected, and the sequence of the major reactions, namely, N-methylation, oxidative cyclization on the phenyl ring, and introduction of sulfur, still remains undefined. In feeding experiments of the unnatural compound cyclo(L-2-aminobutanoyl-L-phenylalanine) (**63**), a new "primitive" type of sulfur-containing metabolite, 3-benzyl-6-ethyl-3,6-bis(methylthio)piperazine-2,5-dione (**64**), was isolated (*131*) (Scheme 12). This result suggests that biosynthetic introduction of sulfur atom precedes N-methylation or oxidative modification of the phenyl ring via an epoxy intermediate.

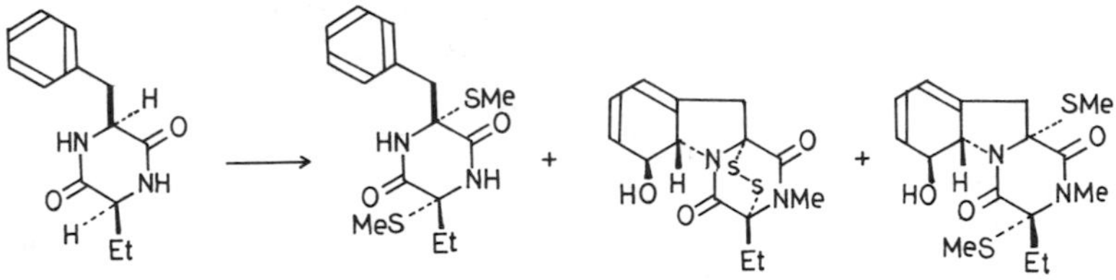

SCHEME 12. Bioconversion of **63** in *Gliocladium deliquescens*.

Bisdethiobis(methylthio)gliotoxin (**63**) is formed irreversibly from gliotoxin (**61**) (8.65%) in *Gliocladium deliquescens*. Several gliotoxin-related metabolites including sirodesmins A, B, C, and G from *Sirodesmium* species (*132*) and phomamide from *Phoma lingam* (*133*) have been isolated. The isolation of the tetrasulfide analogue of gliotoxin, gliotoxin G, from *Asp. fumigatus* was reported (*133a*).

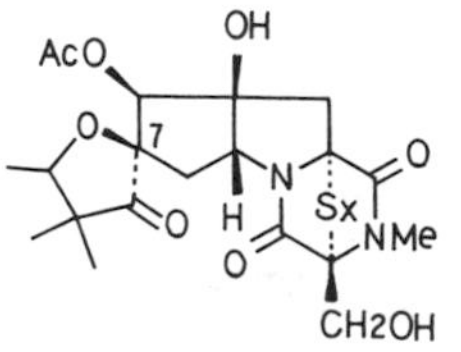

X=2 Sirodesmin A
     Sirodesmin G
     (epimer at C-7)
X=3 Sirodesmin C
X=4 Sirodesmin B

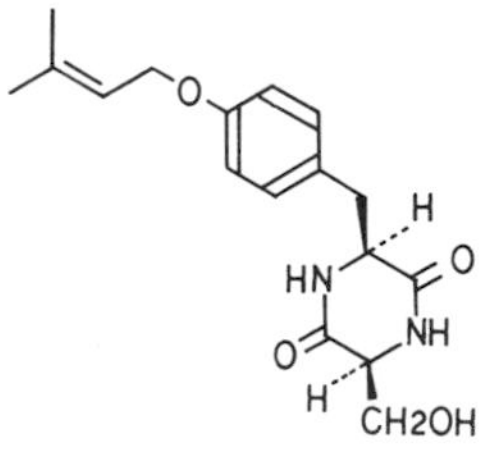

Phomamide

B. ARANOTIN GROUP

The aranotin group of sulfur-containing diketopiperazines includes alkaloids composed of two phenylalanines. In certain cases the benzene rings are deformed to oxepin and other rings. From *Arachniotus aureus,* aranotin (**65**), acetylaranotin (**66**), bisdethio(dimethylthio)acetylaranotin (BDA, **67**; major metabolite), and, successively, apoaranotin (**68**) and bisdethio(dimethylthio)acetylapoaranotin (BDAA, **69**) have been isolated by Nagarajan and co-workers (*134,135*). Acetylaranotin and BDA have been isolated independently from *Asp. terreus,* as antibiotics named LL-S88α and LL-S88β (*136*), respectively, and these compounds were also isolated as plant-growth inhibitors (*137*).

1. Aranotin and Related Metabolites

Acetylation of aranotin (**65**), $C_{20}H_{18}N_2O_7S_2$, mp 198–200°C, affords another metabolite, acetylaranotin (**66**), $C_{22}H_{20}N_2O_8S_2$, mp 201–215°C, which has a symmetrical structure. Raney-nickel desulfurization of acetylaranotin or BDA, $C_{24}H_{26}N_2O_8S_2$, mp 213–217°C, affords the same compound, bisdethioacetylaranotin (**70**), $C_{22}H_{22}N_2O_8$, mp 201–202°C (Scheme 13). Therefore, acetylaranotin and BDA have the same carbon skeleton and differ only with respect to the sulfur atom function. From the CD curve, the same configuration at the asymmetric centers of the diketopiperazine moiety in both gliotoxin and acetylaranotin was confirmed. Data from the UV, IR, mass, and NMR spectra suggest that acetylaranotin bears a disulfide bridge and that BDA is the dithiomethyl derivative. This was confirmed by conversion of acetylaranotin to BDA by reduction with $NaBH_4$ followed by methylation with CH_3I in $CHCl_3$–MeOH, in which the retention of the configuration is recognized (Scheme 13).

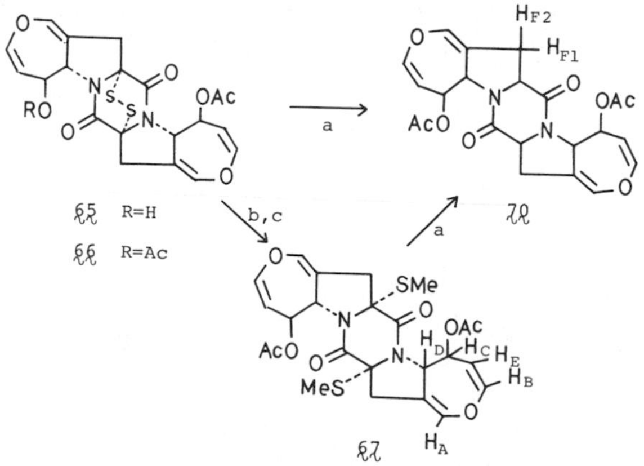

SCHEME 13. Conversion of aranotin (**65**) to acetylaranotin (**66**), BDA (**67**), and **70**. Reaction conditions: a, Raney Ni; b, NaBH₄; c, CH₃I–CHCl₃/MeOH.

The NMR spectrum of BDA reveals the presence of SCH_3 ($\delta2.25$), $OCOCH_3$ ($\delta2.06$), CH_2 ($\delta3.05$) groups and shows five CH signals. Chemical shifts of protons at $\delta6.59$ (dd, H_A) and $\delta6.31$ (dd, H_B) are in good agreement for the α position in an enol ether, and $\delta4.70$ (dd, H_E) for the β position, respectively. The coupling constant ($J_{BE} = 8.0$ Hz) is in excellent agreement with a seven-membered cyclic enol ether ring. Another proton at $\delta5.83$ (m) moves upfield to $\delta4.7$ upon deacetylation. Therefore, H_C is on the same carbon atom bearing the acetyl group. The other proton (H_D) at $\delta5.09$ is on a carbon atom adjacent to oxygen or nitrogen atom. The new CH (q, $\delta4.36$) introduced during desulfurization is coupled to methylene (H_{F1}, H_{F2}, each d, $\delta2.80$ and 2.98 in **70**), so the position of the sulfur-containing group is determined.

The structures of acetylaranotin and BDA, including the absolute configuration, were studied (*138*) and confirmed by independent X-ray analyses (*139,140*). The spectral characteristics of apoaranotin (**68**), $C_{20}H_{18}N_2O_6S_2$, mp 200–205°C, shows its similarity to gliotoxin and acetylaranotin. The highest molecular weight fragment in the mass spectrum appears at m/z 382, with an accompanying loss of a 64-unit mass which suggests the presence of a disulfide bridge in diketopiperazine as in acetylaranotin. The identity of half of this molecule with that of acetylaranotin is obvious from the coupling pattern and chemical shifts in the NMR spectrum. The structure of the other half of the molecule is determined by oxidation of the acetate of apoaranotin with MnO_2. The phenol obtained (dehydroapoaranotin, mp 183–186°C) has a UV spectrum very similar to that of dehydrogliotoxin, and it also shows good agreement in the NMR spectrum with the corresponding aromatic protons of the product afforded by MnO_2 oxidation of gliotoxin. The relationship of apoaranotin (**68**) and BDAA (**69**), $C_{24}H_{26}N_2O_7S_2$, mp 105–107°C, corresponds to that of aranotin and BDA (*113,135*).

68   69

2. Biosynthesis

Cyclo(L-[U-^{14}C]phenylalanyl-L-phenylalanine) is incorporated into BDA in 19.9% yield, in cultivation for 48 hr in *Asp. terreus* (*141*). In contrast, the cyclo L,D and D,D isomers are not incorporated. Further, the isolation of *cis*-3,6-bibenzyl-3,6-bis(methylthio)piperazine-2,5-dione (**71**), mp 291–293°C, $[\alpha]_{549}$ $-122°$, from *Asp. terreus* (*142*) suggests that the sulfur atoms are introduced in an early stage of aranotin biosynthesis, as in gliotoxin biosynthesis.

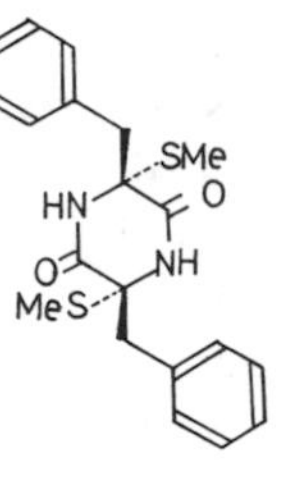

22

From the structures of aranotin and related metabolites, Neuss *et al.* (*135*) suggested that oxepin oxide might be involved in the biosynthesis of naturally occurring dihydrooxepins. The generation of oxepin oxide is suggested to involve monooxygenation of phenylalanine to give arene oxide (A) (Scheme 14). This system is in equilibrium with the isomeric oxepin (B). An intramolecular nucleophilic attack by the diketopiperazine amide group on A produces a substituted cyclohexadienol (C; gliotoxin type). Alternatively, further oxidation of the oxepin B to D, followed by a similar nucleophilic process, might yield the aranotin-type system (E). The complete biosynthesis of alkaloids of the aranotin group remains undefined, particularly concerning the introduction of sulfur bridge.

A few other aranotin-related metabolites have been noted. Epicorazine A has been isolated from *Epicoccum nigrum* (*143*). Bipolaramide was found in *Bipolaris* species (*144*).

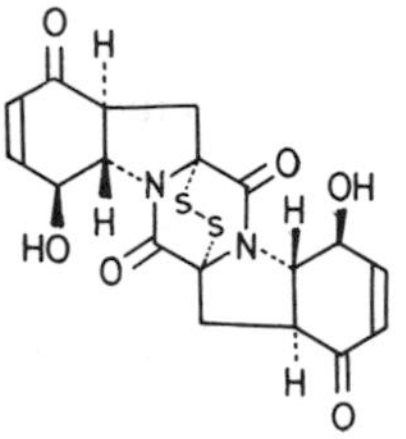

Epicorazine A 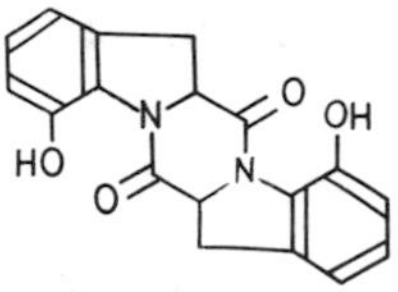   Bipolaramide

C. EMESTRIN AND ASPIROCHLORINE

1. Emestrin

A macrocyclic epithiodiketopiperazine derivative, emestrin (**72**), $C_{27}H_{22}N_2O_{10}S_2$, mp 233–236°C(dec.), was isolated by Seya *et al.* (*145*) from the mycelium of *Emericella* (=*Aspergillus*) *striata* collected in Nepal. It was identical to a mycotoxin, EQ-1, which had been obtained from *Emericella quadrilineata* collected at a date plantation in Iraq by Yamazaki and co-workers

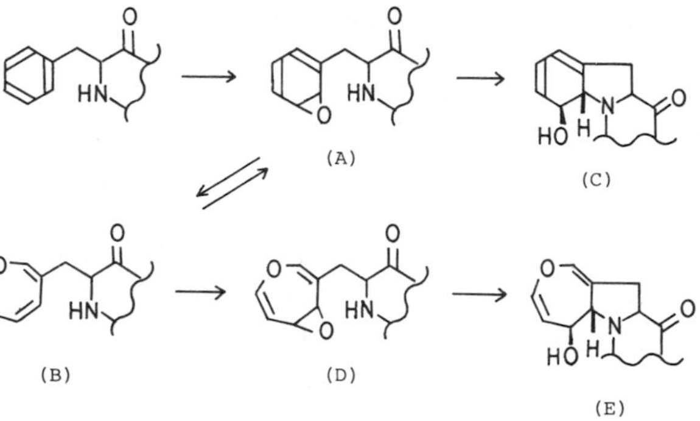

SCHEME 14.

(*146*). By NMR decoupling experiments, the partial structures of the diphenyl ether and oxepine moiety were established. Additional evidence is provided by X-ray crystallography. Peaks in the CD spectrum (λ_{max}) at 235, 270, and 340 nm show that this metabolite has the same absolute configurations (3*R*,11a*R*) as alkaloids of aranotin group. The diketopiperazine ring of **72** is derived biogenetically from two molecules of phenylalanine, and the macrocyclic ring is constructed with one benzoate. It has a strong antifungal activity.

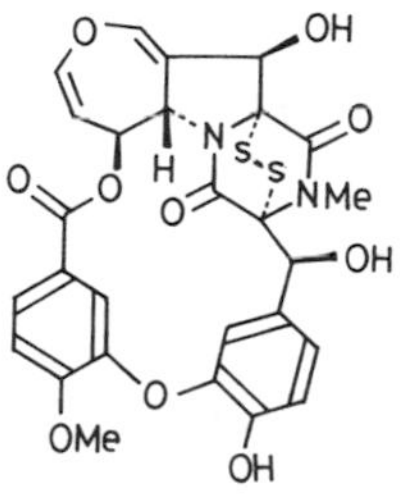

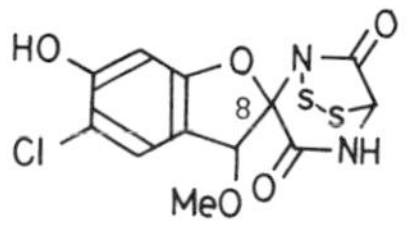

2. Aspirochlorine

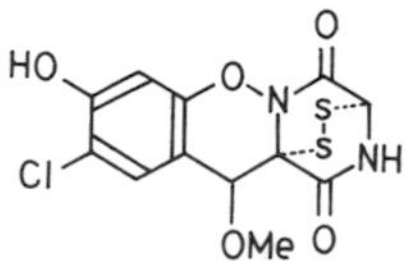

Aspirochlorine (**73**), $C_{12}H_{19}ClN_2O_5S_2$, mp 160–172°C, $[\alpha]_D^{21}$ +66.7°, possesses a diketopiperazine ring that is derived from phenylalanine and glycine and also contains sulfur atoms. It was isolated from culture filtrates of *Asp. flavus* in 1982 by Sakata *et al.* (*147*) and determined to be identical to antibiotic A-30641 from *Asp. tamarii* (*148, 149*) and also to oryzachlorin from *Asp. oryzae* (*150,151*), by comparison of their spectroscopic data. The chemical structure **73a** assigned to antibiotic A-30641 as deduced from chemical results was altered by Sakata *et al.* (*147*) to **73** to explain the ^{13}C signal appearing at 102.6 ppm which must be assigned to a teritary bridgehead carbon (C-8). Recently, incorporation of [U-^{14}C]phenylalanine and [U-^{14}C]glycine into this metabolite was reported (*152*). Compound **73** has activities against phytopathogenic fungi and yeasts (*151*).

VII. Metabolites Derived from Tryptophan and Anthranilic Acid

A. TRYPTOQUIVALINES

Tryptoquivalines are group of novel metabolites having a unique hexacyclic skeleton including a spiro-γ-lactone and have been isolated from two *Aspergillus* species. Clardy *et al. (153)* and Büchi *et al. (154)* isolated six compounds from two toxigenic strains (NRRL-5890 and MIT-M-18) of *Asp. clavatus* collected from mold-damaged rice. Independently, Yamazaki *et al. (81,82,155,156)* reported many metabolites, named initially as fumigatus toxins C–N and later designated as tryptoquivalines C–N (FTC–N), from an *Asp. fumigatus* strain which simultaneously produces fumitremorgen A (**47**) and B (**48**) (FTA and -B). To date, 14 naturally occurring tryptoquivalines have been reported (Table III).

TABLE III
Tryptoquivalines

Name	Structure			mp (°C)	$[\alpha]_D$ (conc, solvent)
	R^1	R^2	R^3		
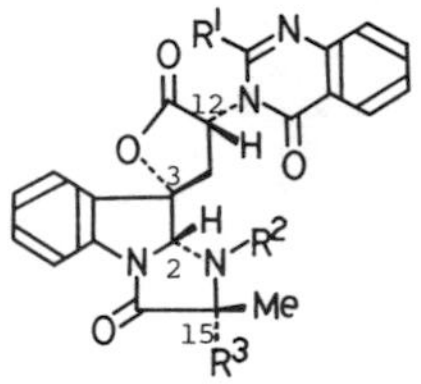					
Tryptoquivaline (FTC) (**74**)	H	OH	Me	155–157	+130 (0.22, CHCl₃)
				215–217	+168 (0.23, CHCl₃)
Nortryptoquivaline (FTD) (**77**)	H	OH	H	256–258	+170 (0.06, CHCl₃)
				224–225 (dec.)	+115 (0.23, CHCl₃)
Nortryptoquivalone (FTB) (**78**)	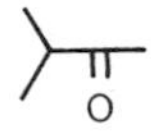	OH	H	208–209	+255 (1.05, CHCl₃)
Deoxytryptoquivaline (**79**)	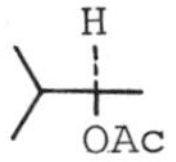 H	H	Me	150–152	+56.8 (0.78, CHCl₃)
Deoxynortryptoquivaline (**80**)	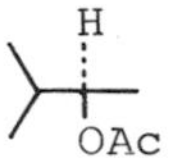 H	H	H	158–160	+69.5 (0.82, CHCl₃)

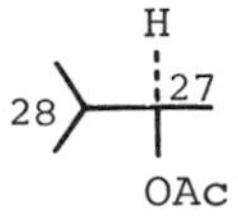

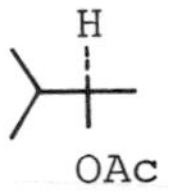

TABLE III   (*Continued*)

Name	R¹	R²	R³	mp (°C)	[α]D (conc, solvent)
Deoxynortryptoquivalone (FTN) (**81**)	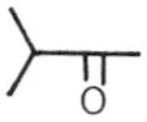	H	H	192–193 193–197	+171 (0.79, CHCl₃) +127 (0.03, DMSO)
FTI (**82**)	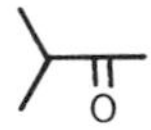	OH	Me	232–235.5 (dec.)	+239 (0.16, CHCl₃)
FTE (**83**)	H	OH	H	~257 (dec.)	+257 (0.009, CHCl₃)
FTG (**85**)	H	OH	Me	240–241.5 (dec.)	+215 (0.011, acetone)
FTJ (**87**)	H	H	H	254–258	+135 (0.024, acetone)

$$R^1, R^2, R^3, Me \text{ structure}$$

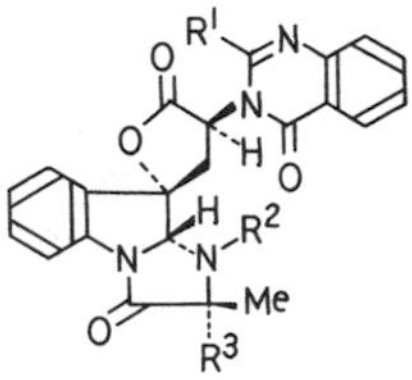 Name	R¹	R²	R³	mp (°C)	[α]D (conc, solvent)
FTF (**84**)	H	H	H	~277 (dec.)	−109 (0.006, CHCl₃)
FTH (**86**)	H	OH	H	~274 (dec.)	−155 (0.021, acetone)
FTL (**88**)	H	OH	H	265–268	−229 (0.03, DMSO)
FTM (**89**)		H	OH	H	157–164

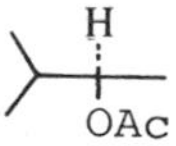

(For FTM: −154 (0.50, CHCl₃), with OAc substituent shown)

1. Tryptoquivaline and Related Metabolites

Tryptoquivaline (FTC, **74**), $C_{29}H_{30}N_4O_7$, mp 155–157°C, $[\alpha]_D^{25}$ +130°, exhibits absorption maxima at 228, 275, 305, and 317 nm in the UV spectrum. From the IR spectrum the presence of hydroxy, γ-lactone, ester, and amide groups in the molecule is expected. The NMR spectrum indicates the presence of one each of

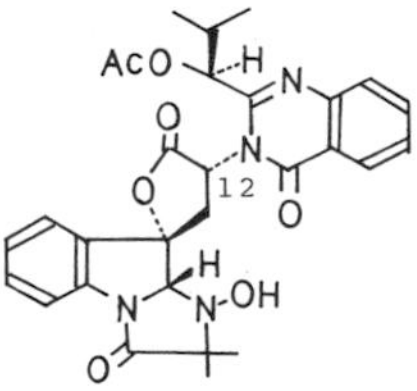

74

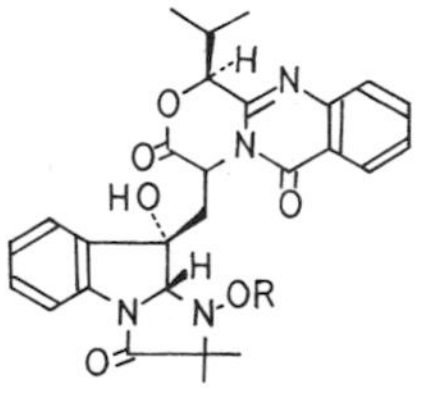

75   R=H
R= p-BrC₆H₄NHCO-

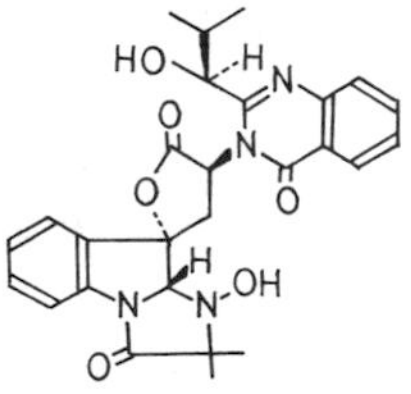

76

$(CH_3)_2CH(C-)H-$, $-CH_2-$, gem-dimethyl, methine, acetyl, and hydroxyl groups as well as eight aromatic hydrogens.

FTC tests positive with triphenyltetrazonium chloride (TTC), suggesting the presence of a hydroxylamine group. Methanolysis (0.5% HCl) gives a deacetyl derivative (75), mp 252–253°C, for which the TTC test is still positive and the UV spectrum is unchanged. On the other hand, a carboxyl band of the lactone is shifted from 1790 to 1765 cm^{-1}, which suggests tryptoquivaline is not the simple acetate of 75.

The intermediate hydroxy-γ-lactone probably has undergone acyl transfer to the more stable trans disubstituted hydroxy-δ-lactone. The p-bromophenyl-urethane derivative of 75, mp 196–198°C, (TTC negative) was utilized in determination of the structure and of the relative configuration by X-ray analysis (153).

Hydrolysis of tryptoquivaline in methanolic alkali (0.1% KOH) produces an uncharacterized carboxylate which on acidification is transformed to the hydroxy-γ-lactone (CO, 1800 cm^{-1}), mp 175–176°C, $[\alpha]_D^{25}$ −142°. The new γ-lactone appears to be C-12-epitryptoquivaline (76) and shows that epimerization occurs prior lactone ring opening.

Nortryptoquivaline (FTD, 77), $C_{28}H_{28}N_4O_7$, mp 256–258°C, $[\alpha]_D^{25}$ +170°, produces a UV spectrum similar to that of tryptoquivaline (FTC), and all IR absorptions associated with functional groups are identical in these two compounds. The NMR spectrum of 77 is identical to that of FTC except that a doublet methyl (δ1.58, $J = 7$ Hz) appears instead of two singlet methyls (δ1.47 and 1.49) in FTC. Therefore 77 is nor derivative, and its CD spectrum indicates the stereochemistry relative to FTC. Yamazaki's FTD was identified as nortryptoquivaline by direct comparison (156). A single-crystal X-ray diffraction experiment was performed on nortryptoquivaline, and the absolute configuration was established as 2-S,3-S,12-R,15-S,27-S. The existence of the spiro-γ-lactone was also confirmed (157).

Nortryptoquivalone, $C_{26}H_{24}N_4O_6$, mp 208–209°C, $[\alpha]_D^{25}$ +255°, displays UV and IR spectra similar to those of tryptoquivaline. It is also positive to the TTC test, and the NMR spectrum shows that it is a nor-type compound, like FTD. The production of 2,4-dinitrophenylhydrazone shows the existence of the carbonyl group responsible for shifting of a neighboring CH signal (m, δ4.13) to lower field (cf. δ2.57 in tryptoquivaline). Based on the above data the structure 78 is suggested for nortryptoquivalone.

Three other TTC-negative metabolites isolated from Strain MIT-M-18 (154) were separated by HPLC and TLC. These compounds are oxidized with m-chloroperbenzoic acid to the corresponding hydroxyamine derivatives, namely, tryptoquivaline, nortryptoquivaline, and nortryptoquivalone, respectively. These three metabolites possessing a secondary amine in the molecule were designated as deoxytryptoquivaline (79), deoxynortryptoquivaline (80), and deoxynortryptoquivalone (81), respectively.

Among the other related metabolites (FTE–N) isolated by Yamazaki *et al.* (*81,82*), FTI and FTN were supposed to belong to the tryptoquivalone group, judging from the NMR data in which a multiplet methine (C-28) is appearing at abnormally lower field (δ4.07 and 4.08, respectively). In combination with the other spectral results, FTI was determined as **82** (*155*), and FTN was identified as deoxynortryptoquivalone (*156*).

The other metabolites FTE, FTF, FTG, FTH, FTJ, and FTL (**83–88**) represent a different group from the above metabolites, though the spectral data resemble those of other tryptoquivalines. A single aromatic proton (δ~8.5, 1*H*) is characteristically observed in their NMR spectra instead of the acetoxyisobutyl group which appears in the spectra of FTC and FTD, which means lack of a side chain.

It is of interest that the optical rotations of FTF, FTH, and FTL are opposite (levorotary) to those of FTJ, FTE, and FTG (dextrorotary), respectively, though their chemical structures are identical except for the stereochemistry at C-12. Yamazaki *et al.* (*156*) found that FTF, FTH, and FTL are obtained upon treatment of FTJ, FTE, and FTG, respectively, with 0.1% KOH in MeOH. This fact indicates that the metabolites of the former group are the C-12 stereoisomers of the latter group (*S* configuration), respectively. The possibility that the compounds identified as FTF, FTH, and FTL are artifacts remain. FTM (**89**) is also obtained from nortryptoquivaline by the KOH–MeOH reaction, demonstrating their stereoisomerism.

2. Synthesis

The first total synthesis of tryptoquivaline G (FTG) was achieved by Büchi *et al.* (*158*) (Scheme 15), which confirmed the proposed structure and established the relative and absolute configurations. Subsequently, Ban's group (*159*) effectively prepared Büchi's intermediate (**90**) through thallium(III) trinitrate (TTN) oxidation at a crucial step of spirolactonization. More recently, Hino's group achieved a short-step total synthesis of tryptoquivaline (*160,161*) and tryptoquivaline G (*162*) through oxidative double cyclization of the *N*-acyltryptophan precursor with *N*-iodosuccinimide (NIS) in CF₃COOH (Scheme 16).

3. Biosynthesis

Tryptoquivalines appear to be tetrapeptides derived from tryptophan, anthranilic acid, valine, and alanine (or methylalanine). Deoxynortryptoquivalone (FTN) may be the first metabolite formed in the pathway of tryptoquivaline biosynthesis. Oxidation of the secondary amine to the hydroxyamine would form nortryptoquivalone. If the isobutyl side chain from the above tryptoquivalones is lost by further oxidation, FTJ and FTE would result. On the other hand, if reduction of the carbonyl group following acetylation occurs on the side chain, nortryptoquivaline (FTD) or deoxynortryptoquivaline respectively, would be

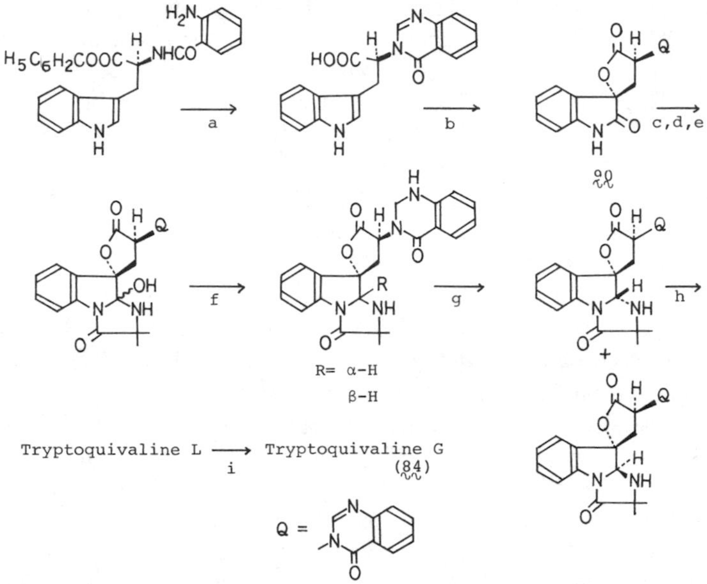

SCHEME 15. Total synthesis of tryptoquivaline G (**84**) by Büchi *et al.* Reaction conditions: a, TsOH–xylene–reflux; b, (CH$_3$SO$_2$)$_2$O–DMSO; c, bis(trimethylsilyl)acetamide; d, NO$_2$C$_6$H$_4$CO$_2$CMe$_2$NHCO$_2$CH$_2$C$_6$H$_4$OMe–DMF–(CH$_3$)$_4$N$^+$Cl$^-$; e, TFA–anisole; f, NaBH$_3$CN; g, DDQ; h, *m*-ClPBA; i, KH/THF–DMF.

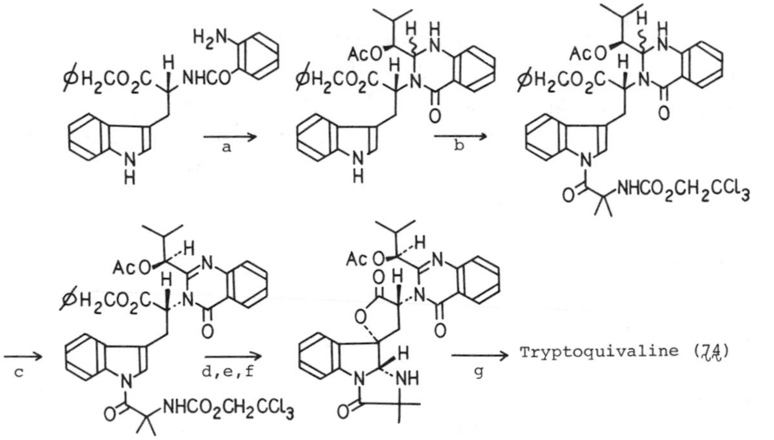

SCHEME 16. Total synthesis of tryptoquivaline (**74**). Reaction conditions:

a, ![structure with CHO, OAc] –molecular sieves (4A)–TsOH–CH$_2$Cl$_2$;

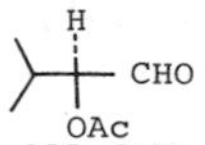

b, *p*-NO$_2$C$_6$H$_4$CO$_2$C(Me)$_2$NHCO$_2$CH$_2$CCl$_3$–KF–MeCN–18-crown-6–EtN-*i*-Pr$_2$; c, DDQ; d, H$_2$–Pd/C; e, NIS–TFA–reflux; f, Zn–AcOH; g, *m*-ClPAB.

formed. The geminal dimethyl group at position C-15 appears to be formed by incorporation of a C_1 unit into deoxynortryptoquivalone, or more likely by the direct participation of methylalanine. The assumed intermediate, deoxytrypto-quivalone, has not been isolated yet, but it might be changed to other metabolites according to the same transformation patterns. A hypothetical biosynthetic pathway of these tryptoquivalines was proposed (*155*) (Scheme 17).

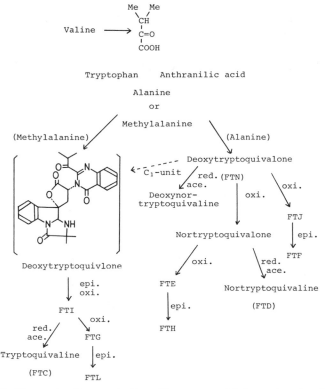

SCHEME 17. Biogenesis of tryptoquivalines. Key: ace., acetylation; epi., epimerization; oxi., oxidation; red., reduction.

B. OTHER TYPES OF TRYPTOPHAN–ANTHRANILIC ACID METABOLITES

Two metabolites, LL-S490β (*163*) and aszonalenin (*164*), possessing a 1,4-benzoazepine-2,5-dione moiety have been isolated from an *Aspergillus* strain. This ring system is supposed to be derived by condensation of tryptophan and anthranilic acid. LL-S490β, $C_{25}H_{25}N_3O_3$, mp 238–240°C, $[\alpha]_D$ +445° (MeOH), was isolated from an unidentified *Aspergillus* strain in 1973. LL-

S490β and aszonalenin, $C_{23}H_{23}N_3O_2$, mp 244–247°C, $[\alpha]_D^{20}$ +53°, were also found in the acetone extracts of dry mycelial mats of *Asp. zonatus,* using chromatography on a silica gel column and elution with benzene–acetone (19:1).

The bands at 3400, 1689, and 1647 cm^{-1} in the IR spectrum of LL-S490β (**91**) show the presence of NH and amide groups, respectively. The latter two absorptions suggest the presence of a dipeptide system. In the ^{1}H-NMR spectrum the presence of eight aromatic proton signals, a reversed isopentenyl group, an acetyl group (δ2.60), and an exchangeable NH (δ8.50) are recognized. The high-intensity peak at *m/z* 130 in the mass spectrum corresponds to the indoline-3-methylene ion, and the loss of 69 mass unit supports the presence of isopentenyl group.

From the molecular formula and the above evidence, a 3,4-dihydro-4-methyl-1*H*-1,4-benzodiazepine-2,5-dione moiety prepared by the combination of a tryptophan portion with anthranilic acid is inferred. The UV spectrum of this azepine compound corresponds to that of LL-S490β (λ_{max} 210, 245, and 284 nm). The remaining three of four protons that appeared as an ABX pattern at δ2.46 (d) and δ3.42 (d) are assigned to a geminal methylene. The geminal protons couple only to the C-5 methine proton (δ3.90, t), and another sharp signal at δ6.00 is reasonably assigned to the C-2 methine proton. Thus the inverted isopentyl group was placed at C-3.

The NMR spectrum of aszonalenin (**92**) is very similar to that of LL-S490β, although an acetyl signal in the latter is not observed and an additional singlet assignable to NH (δ~7.0) is present. Acetylation of **92** affords two acetyl derivatives, of which the monoacetate, mp 242–244°C, was identified as LL-S490β by comparison with spectroscopic data. The occurrence of an inverted isoprene unit at C-3 of the indole nucleus seemed unusual, but many examples have since been recognized in the roquefortine (**60**) and oxaline (**59**) groups. Aszonalenin apparently induces an abnormal second cleavage of sea urchin embryos at a concentration of 50 μg/ml.

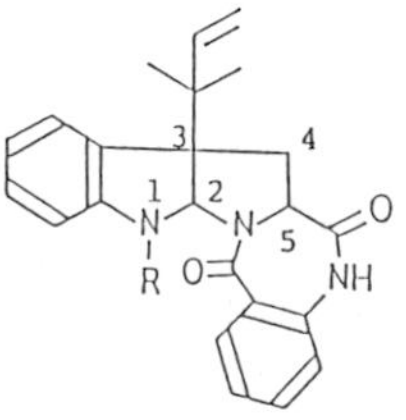

91 R=Ac
92 R=H

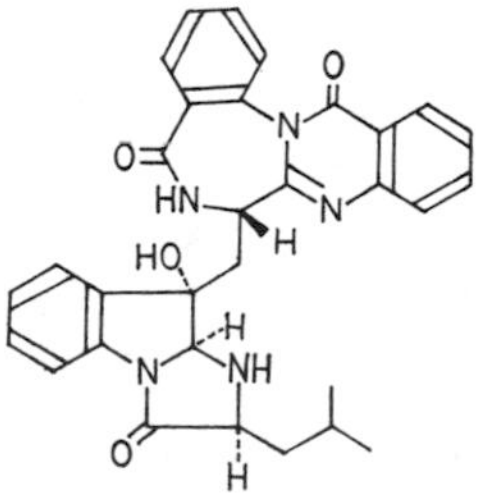

Asperlicin

Another type's structurally related metabolite, asperlicin, was isolated from *Asp. alliaceus,* which is a cholecystokinin antagonist (164a).

VIII. Indole–Mevalonate Metabolites

A. PASPALININE AND AFLATREM GROUP

Another group of fungal metabolites, biosynthesized from tryptophan and mevalonate, is increasing the size rapidly; many tremorgenic mycotoxins belonging to the paspalinine and aflatrem group are recognized. Among them, a subgroup having a unique hexacyclic array may be represented by paspaline (**93**).

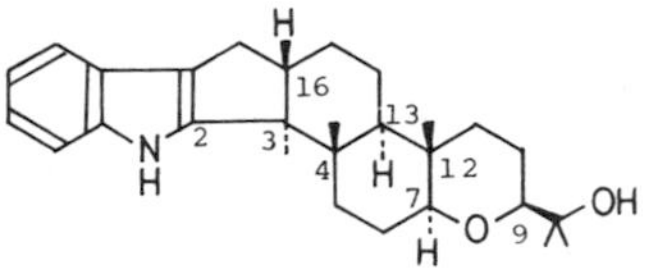

Most members were isolated from sclerotia of ergot fungus, *Claviceps paspali* (*165*), which is toxic to grazing cattle. Meanwhile, two tremorgenic metabolites, paspalinine (**94**) and aflatrem (**95**), were isolated from cultures of *Asp. flavus* (*166,167*), and another toxic metabolite, paxilline (**96**), was found in *Pen. paxilli* (*168*). In 1985, (−)-paspaline, the simplest member of this group, was total synthesized by Smith and Mewshaw (*171*).

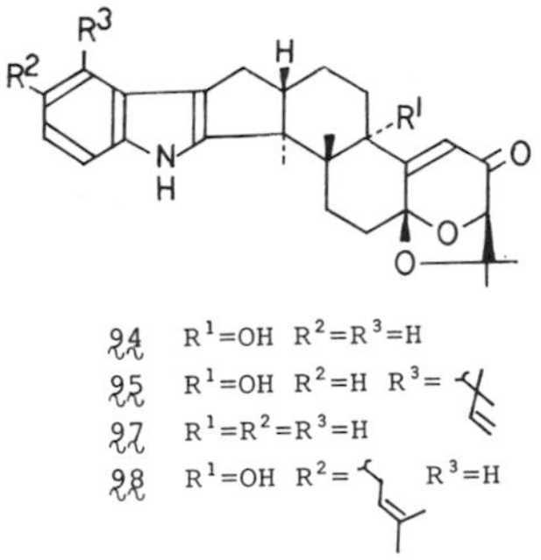

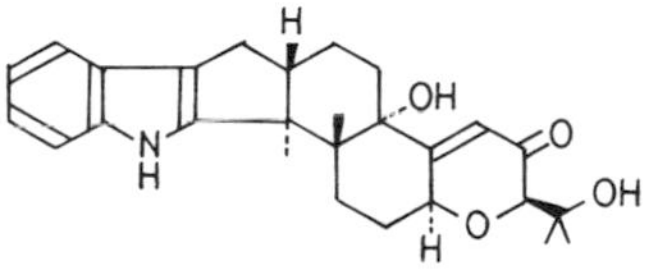

1. Paspalinine

Paspalinine has only been reported as a minor metabolite of *Claviceps paspali* (*169,170*), and the chemical structure was elucidated by comparison to the main metabolites, paspaline, $C_{28}H_{39}NO_2$, mp 264°C, and paspalicine (**97**), $C_{27}H_{31}NO_3$, mp ~260°C. Paspalinine possesses the composition, $C_{27}H_{31}NO_4$ (M^+ 433), accounting for one more oxygen atom than paspalicine. The UV (λ_{max} at 232, 25O, and 270 nm) and ^{13}C-NMR spectra match very closely those reported for paspalicine. However, the absence of the doublet at 39.2 ppm for

C-13 in paspalicine and its replacement by a new signal at 76.6 ppm indicates a hydroxyl group at C-13. The assumed structure has been checked in a single-crystal X-ray experiment (*170*).

2. Aflatrem

Aflatrem, $C_{32}H_{39}NO_4$, mp 233–235°C (dec.), was isolated as a tremorgenic metabolite in 1980 (*167*) and identified as the same tremorgen which had been reported by Wilson and Wilson (*79*). Aflatrem shows a remarkable similarity to paspalinine in the UV and IR spectra. The mass spectrum of aflatrem shows a parent peak at m/z 501, while paspalinine's parent ion occurs at m/z 433. In paspalinine as well as other related compounds (paspaline and paspalicine) a diagnostic peak appears at m/z 182 (dotted line in paspalinine), while the corresponding ion shows at m/z 250 in aflatrem, which suggests that the isoprene unit is attached to the aromatic portion (Fig. 5). The substituted position of this reversed isoprene unit is elucidated by the ^{1}H- and ^{13}C-NMR spectra. In indole nucleus, the ^{1}H-NMR signals for H-4′ (δ7.55) and H-7′ (δ7.40) appear further downfield than those for H-6′ (δ7.08) and H-5′ (δ7.00). In aflatrem, one of the more downfield resonances disappears. In the ^{13}C-NMR spectrum, the presence of a diagnostic peak at 111.2 ppm corresponding to C-7′ establishes that this carbon bears no substituent, and thus the alkyl substituent is placed at C-4′.

3. Pharmacology and Biosynthesis

The metabolites in this family show the same pattern in their CD spectra, so the existence of hydroxyl group attached C-13 seems essential to possess the tremorgenic activity (*169*). The ED_{50} of paxilline in mice (oral) is 25 mg/kg, and the ED_{50} values of aflatrem and paspalinine in 1-day-old chickens are <14 mg/kg. The metabolites possessing no hydroxyl group at C-13, namely, paspaline and paspalicine, are not toxic.

Biosynthesis studies with ^{13}C- and ^{14}C-labeled precursors were conducted by

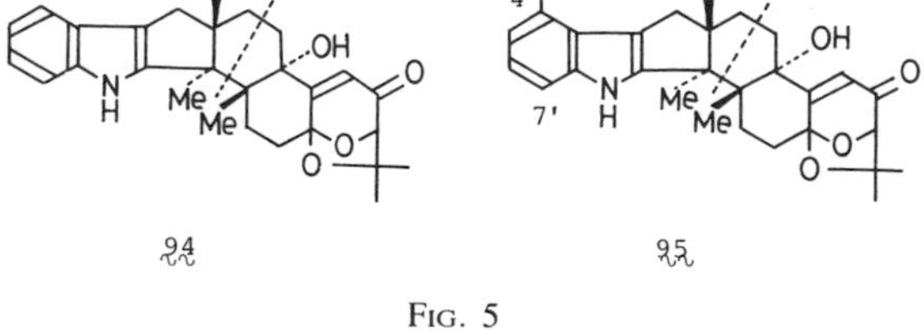

FIG. 5

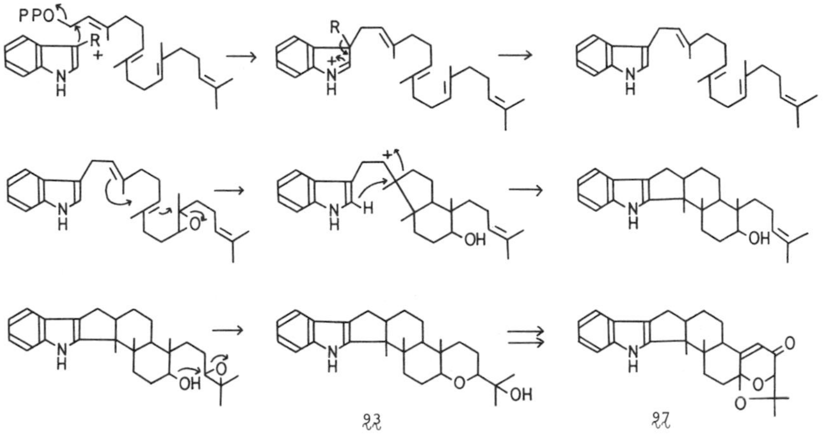

SCHEME 18. Biosynthesis of paspalicine (**97**).

Tanabe (*172*) and Acklin *et al.* (*173*) independently, and results were in good agreement. They showed that these related metabolites are derived from tryptophan and geranylgeraniol. Acklin *et al.* (*173*) reported that the probable biosynthetic sequence is paspaline (**93**) → paspalicine (**97**) → paspalinine (**94**) (Scheme 18). It might be further speculated that paspalinine is converted to aflatrem by addition of an isoprene unit at the 4′ position of the indole aromatic ring and that paspalitrem A (**98**) is derived via isoprenylation at the 5′ position.

From other fungal species more complex tremorgens, such as the janthitrems (A–G) from *Pen. janthinellum* (*174*), the penitrems (A–F) (*175,176*) from *Pen. crustosum* and other species, and the lolitrems (A–D) (*177*) from *Acremonium*

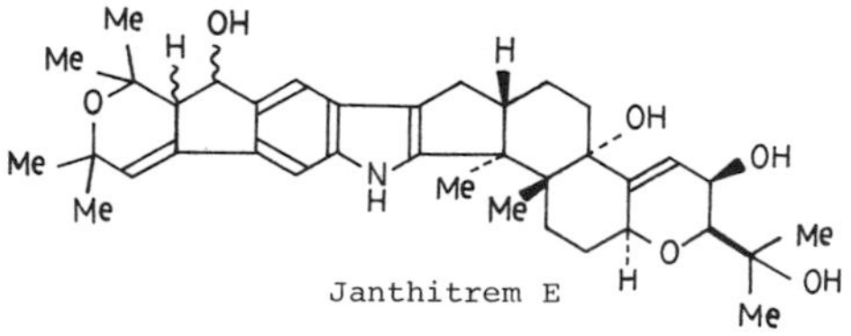

Janthitrem E

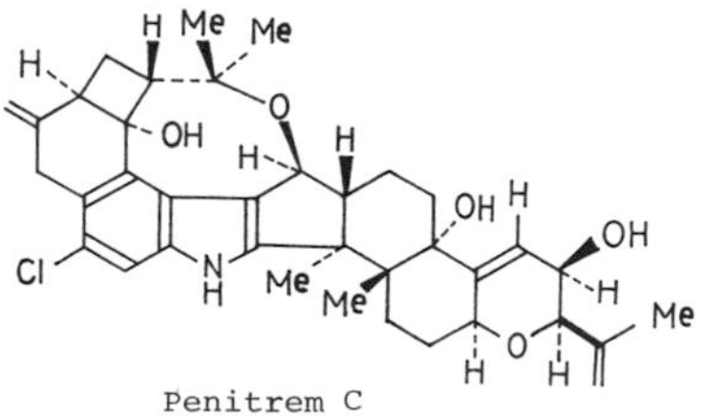

Penitrem C

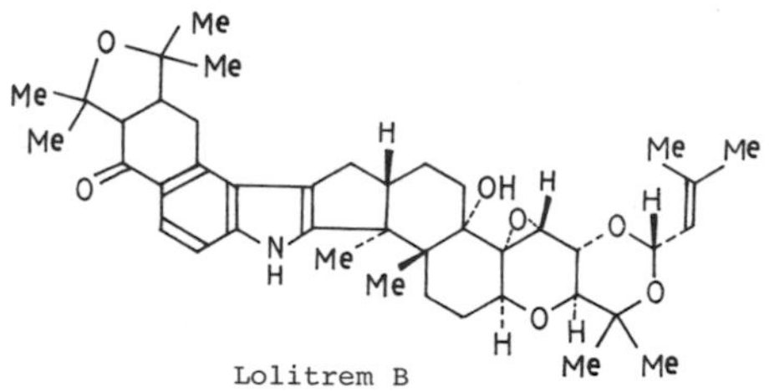

Lolitrem B

species, have been isolated. The structures, determined by X-ray analysis, are related to those of the paspaline group and also have an essential hydroxyl group in the corresponding position (C-13 in paspalinine). The distribution of ^{13}C labels in the diterpene part of penitrems indicates an origin similar to that proposed for paxilline and paspalicine (*178*).

B. Aflavinine and 20,26-Dihydroxyaflavinine

Strains of *Asp. flavus* producing aflatrem (**95**) also yield two nontremorgenic diterpene indoles, designated as aflavinine and 20,26-dihydroxyaflavinine. From culture extracts, aflavinine (**99**), mp 102–104°C, was isolated by Gallagher *et al.* (*179*), and dihydroxyaflavinine (**100**), mp 254–256°C, was obtained by Cole's group (*167*). The metabolites were assigned the molecular formulas, $C_{28}H_{39}NO$ and $C_{28}H_{39}NO_3$, respectively, by high-resolution MS. The UV spectra (λ_{max} 225, 284, and 292 nm) are typical of a 3-alkyl indole chromophore. Both structures were determined by X-ray analysis.

The stereochemistry of these compounds is virtually identical, and the configuration of the substituents is 15(S*), 18(S*), 19(S*), 22(R*), and 23(S*). The two cyclohexane rings are chair conformations, whereas the cyclohexene ring has a distorted half-chair conformation. The presence of two more hydroxyl groups in dihydroxyaflavinine generates two additional asymmetric centers, 20(R*) and 24(S*).

The coproduction of these metabolites with aflatrem and paspalinine suggests that all of these compounds may come from a common intermediate, 3-gera-

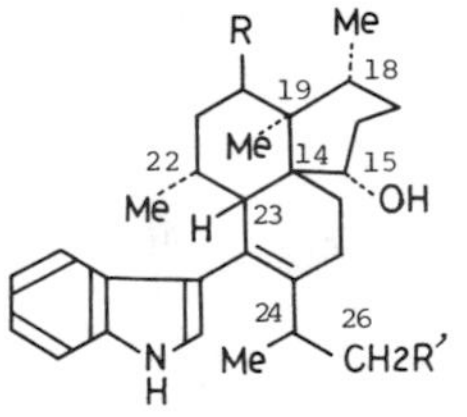

99   R=H   R'=H
100   R=OH   R'=OH

nylgeranylindole. Aflavinine may be produced by a different mode of cyclization and subsequent migration of a methyl group on C-14 to on C-19. A synthetic study toward aflavinine was reported by Danishefsky *et al.* (*180*).

IX. Bisindolylbenzoquinone Metabolites

Numerous metabolites possessing a bisindolylbenzoquinone moiety have been isolated from fungi in the 1980s. Among them, only metabolites of asterriquinone group were found in *Aspergillus* species. Other metabolites, such as cochliodinol (**102**), isocochliodinol (**103**), and neocochliodinol (**104**), have been isolated from several species of *Chaetomium* (*181,182*), and hinnuliquinone (**105**) was obtained from *Nodulisporium hinnuleum* (*183*) (Table IV).

TABLE IV
Metabolites Possessing a Bisindolylbenzoquinone Moiety

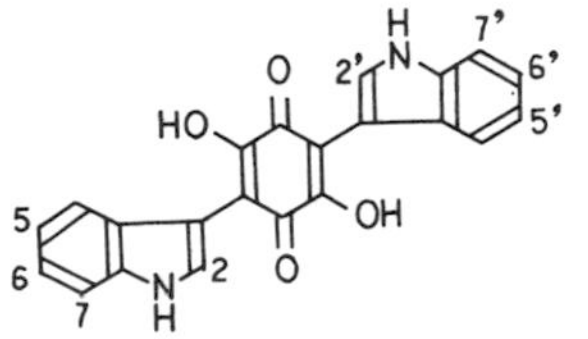

Name	Prenylated position 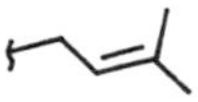	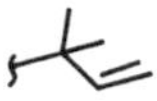	mp (°C)
	⌐⌐⌐	⌐⌐⌐	
Asterriquinone (**101**)	—	N,N'	218–220 (dec.)
Cochliodinol (**102**)	5,5'	—	206
Isocochliodinol (**103**)	6,6'	—	262 (dec.)
Neocochliodinol (**104**)	7,7'	—	285 (dec.)
Hinnuliquinone (**105**)	—	2,2'	243

A. ASTERRIQUINONE GROUP

1. Asterriquinone

Yamamoto *et al.* isolated a deep purple metabolite (λ_{max} 282, 290, and 482 nm) from stationary cultured mycelia of *Asp. terreus* IFO 6123 (*184*) and named it asterriquinone (**101**), $C_{32}H_{30}N_2O_4$. The structure was proposed on the basis of the following spectroscopic and chemical evidences.

The ^{1}H- and ^{13}C-NMR spectra of **101** account for half of the expected number of hydrogens and carbons. Also, in the ^{1}H-NMR spectrum of the tetraacetate that is afforded by reductive acetylation, four methyl signals of acetyl group appear as a single peak at δ1.86. This evidence indicates a symmetrical structure for this molecule.

Reduction with Zn–AcOH to give a colorless compound and recovery of the parent compound by air oxidation reveal a 1,4-quinone moiety in the molecule. The IR spectrum also supports the presence of a hydroxybenzoquinone structure [3300 (OH), 1640, and 1620 cm^{-1} (CO)]. In the ^{13}C-NMR spectrum measured at room temperature, the chemical shifts of carbonyl carbons and hydroxyl-bearing carbons are not clearly observed, which can be attributed to rapid inter-conversion between two equivalent tautomeric forms of the 2,5-dihydroxybenzoquinone ring moiety [in cochliodinol (*185*) an overlapping peak of CO and C—OH appears at average value, 168.5 ppm]. When measured at −75°C, this interconversion is repressed, and reasonable signals appear at 182.6 ppm (CO) and 149.7 ppm (C—OH). The dimethylether (**106**) shows the corresponding signals at 184.0 ppm (CO) and 153.2 ppm (C—OCH$_3$) at room temperature. The mass fragment peaks at m/z 69 (C$_5$H$_9$), 488 (M$^+$ −68), and 370 (M$^+$ −136) suggest the presence of two isopentenyl groups, which are both assigned as 1,1-dimethylallyl groups by the ^{1}H-NMR analysis of asterriquinone and its dihydro derivative.

Asterriquinone is oxidized with H$_2$O$_2$ in 0.1 N NaOH solution at room temperature to give N-(1',1'-dimethylallyl)indole-3-carboxylic acid. Asterriquinone shows a strong antitumor activity against Ehrlich ascites carcinoma and lymphatic leukemia L1210 in mice and against ascites hepatoma AH13 in rat. The methylether of asterriquinone has remarkably decreased activity (*186*).

2. Other Asterriquinones

During the course of surveys to find fungi possessing the ability to produce greater amounts of asterriquinone, it was discovered that *Asp. terreus* var. *africanus* IFO 8835 produces many similar metabolites (*187*). Cultivation in malt-extract medium supplemented with polypeptone (5 g/liter) greatly increases the yield of pigments (1.3 g/liter).

The pigments were roughly separated into four groups A–D according to their polarity on a silica gel column eluted with a mixture of benzene and AcOEt, and each group was rechromatographed on silica gel and/or alumina columns. These isolated pigments contain an indole-*p*-benzoquinone moiety, and the physical properties are very similar to those of asterriquinone, though two methoxyl groups substitute instead of two hydroxyl groups in asterriquinone. The general name "asterriquinone (AQ)" was proposed for this group, and the eleven iso-lated pigments were designated as AQ-A-1–A-4, B-1–B-4, C-1–C-2, and D (*106–116*) (Table V).

TABLE V

Other Asterriquinones

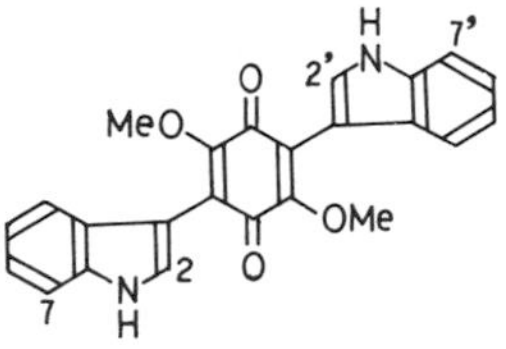

Name	Prenylated position 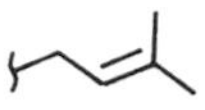	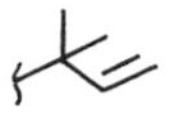	mp (°C)
AQ-A-1 (**106**)	—	N,N'	167–168
AQ-A-2 (**107**)	7,7'	2	189–192
AQ-A-3 (**108**)	—	N,2'	112–114
AQ-A-4 (**109**)	7	2,2'	252–253
AQ-B-1 (**110**)	7	2'	208–209
AQ-B-2 (**111**)	7	2	199–200
AQ-B-3 (**112**)	—	N'	148–150
AQ-B-4 (**113**)	—	2,2'	>300
AQ-C-1 (**114**)	—	2	213–214
AQ-C-2 (**115**)	7	—	202–203
AQ-D (**116**)	—	—	>300

The existence of two types of dimethylallyl groups (1,1 and 3,3) is recognized in the ^{1}H-NMR spectra of these pigments. The number of isoprene units in the molecule can be easily determined from the mass and ^{1}H-NMR spectra. The difference in molecular weight among these pigments is equal to that of an isoprene unit (68), so these compounds are analogs that differ in the number, form, and position of the dimethylallyl group in the indole ring. It is interesting that 1,1-dimethylallyl (reversed isoprenyl) groups are located at the nitrogen or position 2 of indole ring, whereas other positions of indole ring are substituted by 3,3-dimethylallyl (normal isoprenyl) group. N-Substituted 1,1-dimethylallyl groups (CH$_3$, δ1.75–1.86) are distinguishable from C-substituted ones (CH$_3$, δ1.37–1.54) by the chemical shifts of their methyl protons.

Chemically, oxidative cleavage of the benzoquinone ring with H$_2$O$_2$ under basic or acidic conditions was effective to determine the structure as shown in the example of AQ-C-1 (**114**) (Scheme 19). In this reaction, the presence of methoxyl and isopentenyl groups causes a decrease in yield of the degradative products. Therefore, it is desirable to use pigments which have been hydrogenated in the side chain and demethylated prior to oxidation.

Demethylation (*188*) of metabolites to dihydroxy compounds proceeds with

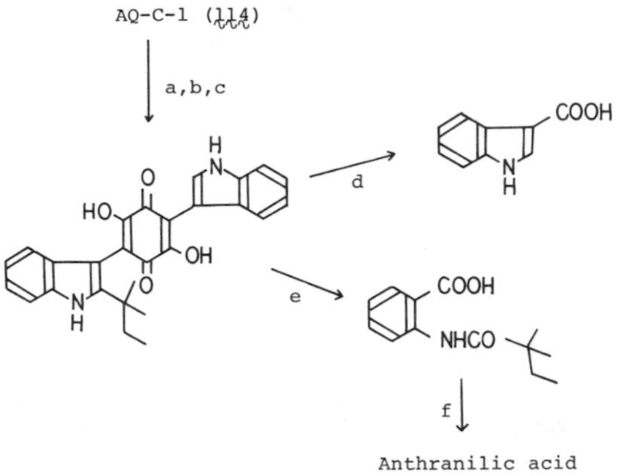

AQ-C-1 (114)

Anthranilic acid

SCHEME 19. Chemical degradation of AQ-C-1 (**114**). Reaction conditions: a, H_2–Pd/C; b, $FeCl_3$; c, HBr; d, H_2O_2–NaOH; e, H_2O_2–AcOH; f, HCl.

KOH in boiling EtOH for 1 to 2.5 hr in fairly good yield (56–85%). On the other hand, various kinds of side reaction occur in demethylation under acidic conditions. 1,1-Dimethylallyl groups at position 2 are fairly stable to acidic reagents, but those at N (AQ-A-1 and B-3) are subject to reactions such as elimination or rearrangement to position 2 with inversion. 3,3-Dimethylallyl groups located at position 7 (AQ-B-1 and C-2) promote cyclization between the N of indole ring via the *tert*-hydroxyisopentenyl derivative. In demethylation with BBr_3, partial demethylation is observed in the compounds AQ-C-1 and B-1 which possess a bulky 1,1-dimethylallyl group at position 2 (Fig. 6).

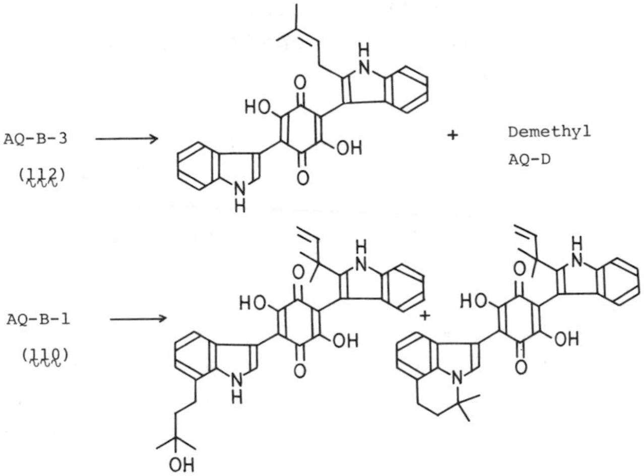

FIG. 6

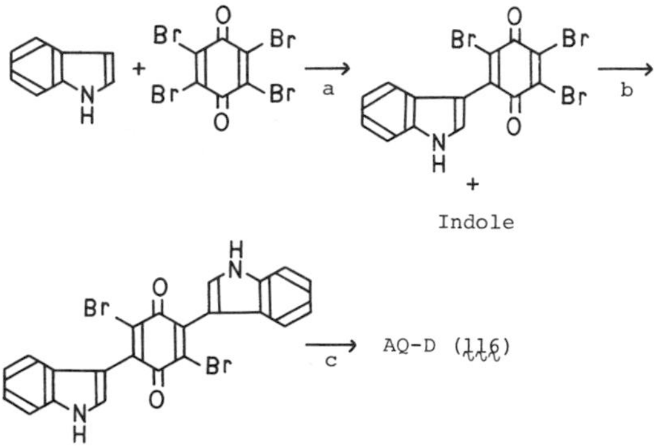

SCHEME 20. Total synthesis of AQ-D (**116**). Reaction conditions: a, K_2CO_3–CH_2Cl_2–silica gel; b, K_2CO_3–$C_2H_4Cl_2$–silica gel; c, NaOH–MeOH.

The simplest metabolite, AQ-D, was synthesized, through condensation of bromanil and indole catalyzed with silica gel, in 5% overall yield by Schwenner (*189*) (Scheme 20). The structures of asterriquinone D and cochliodinol were confirmed by X-ray analysis (*189,190*). Cochliodinol is reported to inhibit the growth of bacteria and fungi (*181*).

3. Related Colorless Metabolites

A number of related colorless metabolites were also isolated from strain IFO 8835 (*191*). Hydroquinones corresponding to AQ-A-1, C-1, and D (**117–119**) and dimethyl derivatives of quinol (**120–122**) corresponding to AQ-A-1, B-3, and C-2 were isolated as minor components. Further, three optically active metabolites (**123–125**) were isolated by extensive fractionation (30–50% AcOEt content) (yield, 1.8, 0.23, and 0.2 g/18 liters, respectively) together with a tripeptide determined as methyl nicotinylanthranilyltrimethoxyanthranilate (**126**), mp 142–143°C.

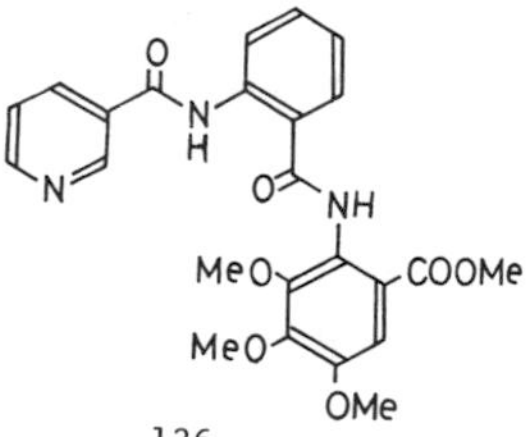

The UV spectrum (λ_{max} 223, 284, and 291 nm) of **123**, $C_{25}H_{22}N_2O_5$, mp 134–138°C, indicates the presence of an indole ring in the molecule, and the NMR spectrum shows the presence of an alcoholic hydroxyl ($\delta 5.30$) and an aliphatic NH group ($\delta 6.45$) which is coupling with a methine ($\delta 6.22$). An indole

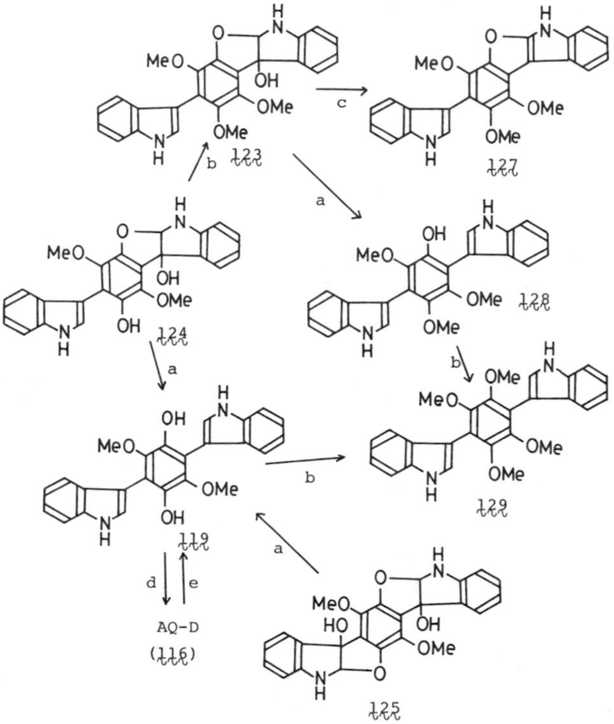

SCHEME 21. Chemical correlations among **123**, **124**, and **125**. Reaction conditions: a, LiAlH$_4$; b, CH$_2$N$_2$; c, *p*-toluenesulfonic acid; d, FeCl$_3$; e, Na$_2$S$_2$O$_4$.

NH (δ10.34) is also recognized in the ^{1}H-NMR spectrum. Dehydration of **123** with *p*-toluenesulfonic acid (Scheme 21) affords a compound (**127**), C$_{25}$H$_{20}$N$_2$O$_4$, mp 272–274°C, in which signals of the above OH, aliphatic NH, and methine have disappeared and another indole NH (δ11.19, s) has appeared in the ^{1}H-NMR spectrum. Reduction of **123** with LiAlH$_4$ affords **128**, which possesses two indole rings, three methoxyls, and one phenolic hydroxyl. The methylether (**129**) is identical to the compound derived from AQ-D by reduction and following methylation.

Metabolite **124**, C$_{24}$H$_{20}$N$_2$O$_5$, mp 131–140°C, is the monodemethylether of **123**. Metabolite **125**, C$_{24}$H$_{20}$N$_2$O$_5$, mp 221–223°C, has a symmetrical structure, and the UV spectrum (λ_{max} 235, 305, and 320 nm) shows the presence of an indole moiety. Both **124** and **125** afford the quinol of AQ-D with LiAlH$_4$.

B. BIOSYNTHESIS

The biosynthesis of bisindolylquinone metabolites has been studied in asterriquinone (*184,192*), cochliodinol (*185*), and hinnuliquinone (*183*) independently. Labeled DL-tryptophan is incorporated into these metabolites in 13.6, ~4, and

0.8% yield, respectively, whereas the incorporation of mevalonate is unexpectedly low in all metabolites (0.07, 0.15, and 0.19%, respectively). From degradative experiments of labeled asterriquinone, it is confirmed that the intact carbon skeleton of tryptophan is transported in the molecule.

In the IFO 8835 strain of *Asp. terreus,* the issue of variation in the yield of each AQ pigment has been pursued; AQ-D is initially recognized at the fourth day of cultivation, and the more isoprenylated metabolites appear later. When DL -[3-^{14}C]tryptophan (1.17×10^8 dpm) is added to 6-day-old cultures and cultivated for 11 days further, all metabolites are labeled, and the radiospecific activity is clearly different between the groups (D, 5.73; C group, 2.27–2.38; B group, 1.66–1.79; A group, 1.14–1.44; indoline group, **123,** 2.50, **124,** 2.49, **125,** 1.26 $\times 10^4$ dpm/mmol, respectively). The data suggest that the fundamental metabolite (AQ-D) is first built from tryptophan, and then isoprenylation proceeds one by one according to the pathway of AQ-D $\rightarrow$ AQ-C $\rightarrow$ AQ-B $\rightarrow$ AQ-A. The radioactivity of demethyl-AQ-D, added to 4-day-old culture medium, is incorporated into AQ-D, C-1, and indoline metabolites (**123–125**), but is not recognized in other pigments, even after a further 10 days' cultivation, which also shows that the formation of the indoline ring occurs at an early stage.

For further elucidation, partially purifired enzyme was prepared from homogenates of 7-day-old cultures of strain IFO 8835 by centrifugation, precipitation with 40% saturated $(NH_4)_2SO_4$, and Sephadex G-25 chromatography. The crude enzyme possesses the ability to form bisindolylquinone in the presence of ATP and Mg^{2+}, and prenylate in the presence of dimethylallylpyrophosphate (DMAPP). Incubation of [1-^{14}C]indolylpyruvate (or its methylester) with the crude enzyme and ATP in Tris–buffer solution (pH 8.2) gives radioactive demethyl AQ-D. However, the radioactivity of administered demethyl AQ-D is scarcely incorporated into AQ-C-1 and C-2, and no radioactivity is recognized in any demethyl AQs upon treatment with the crude enzyme solution in the presence of DMAPP. Also, none of the demethyl analogs of AQ-C-1, C-2, and B-3 become more prenylated in the same procedures. In contrast, labeled AQ-D changes to AQ-C-1, C-2, and B-3, simultaneously. These results show that prenylation proceeds only in the metabolites possessing a methoxyl group. AQ-C-1 is changed exclusively to B-4, whereas AQ-C-2 is prenylated to give B-1 and B-2 and, similarly, AQ-B-3 is further metabolized to A-1 and A-3. Moreover, AQ-B-1 is metabolized to A-2 and A-4; AQ-8-2 is not changed to other metabolites.

From the above evidence, the biosynthetic relationships of asterriquinones are deduced as presented in Scheme 22. Since the fungus has the ability to reduce the added quinones to quinols, biosynthesis may proceed via the equivalent quinol form. In the asterriquinone group, the 1,1-dimethylallyl (reversed isoprene) moiety seems to substitute directly at the N or 2 position of the indole ring. This might exclude the other hypothesis of rearrangement of a 3,3-dimethylallyl group introduced previously in other positions, and it might be true in echinulin group biosynthesis also.

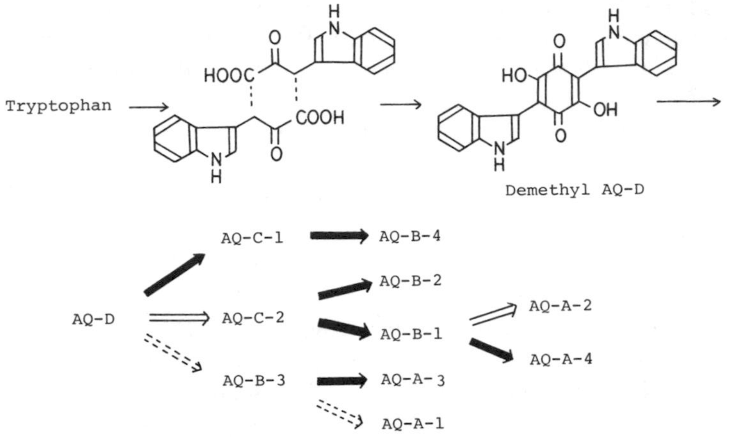

Demethyl AQ-D

SCHEME 22. Biosynthetic relationships of asterriquinones in *Aspergillus terreus* var. *africanus* IFO 8835. Key: ➡, 2-prenylation; ⇒, 7-prenylation; ⇢,N-prenylation.

X. Miscellaneous Metabolites

1. Cyclopiazonic Acid

In 1973, Abe and co-workers (*193*) isolated cyclopiazonic acid (**130,** α-CA), mp 245–246°C, cyclopiazonic acid imine (**131**), mp 277–278°C, and bissecode-hydrocyclopiazonic acid (**132,** β-CA), mp 168–169°C, from the combined extracts of mycelia and culture filtrates of *Asp. versicolor.* Cyclopiazonic acid was also isolated from *Asp. flavus* (*194*) and *Asp. oryzae* (*195*) by other groups. *Penicillium cyclopium* Westling (*196*) also produced the above metabolites as well as the intermediate, α-acetyl-γ-(β-indolyl)methyltetramic acid (**133**), mp 164–165°C. Cyclopiazonic acid is known to cause acute toxicosis in ducklings and rats (*197*).

The structure was deduced on the basis of chemical and physicochemical investigations. Synthetic studies on α-cyclopiazonic acid have been performed by several groups, independently (*198,199*), and biosynthetic studies have also been performed (*200,201*), as summarized by Holzapfel (*202*). These metabolites are derived by condensation of tryptophan, an isoprene unit, and two molecules of acetic acid. Bissecodehydrocyclopiazonic acid (β-CA) is a direct precursor of cyclopiazonic acid (α-CA), and the conversion may occur after dehydrogenation catalyzed by an oxidocyclase containing a flavin prosthetic group (*201*) (Scheme 23). ^{1}H- and ^{13}C-NMR spectroscopy and single-crystal X-ray crystallography indicate that both α-CA and its imine exist in CDCl$_3$ solution as an equilibrium mixture of four possible enolic tautomers, with structures **130** and **131** pre-

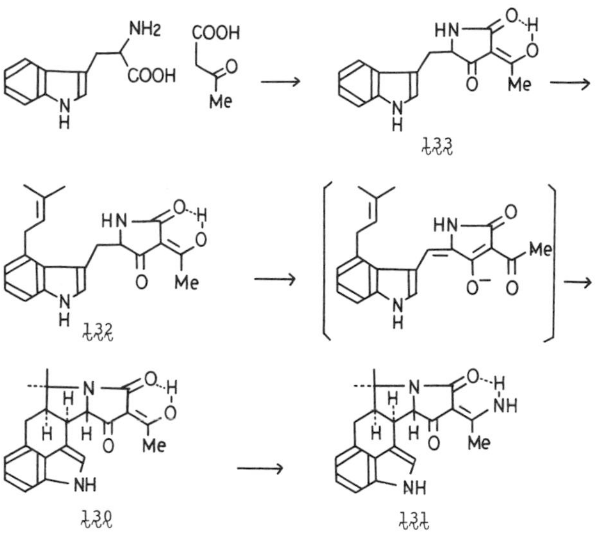

SCHEME 23. Biosynthesis of cyclopiazonic acid.

dominating, respectively (*203*) (Fig. 7). The ability of cyclopiazonic acid to complex with trace metals may be an important factor in toxicity *in vivo* (*204*).

2. Cytochalasins

Over 40 metabolites belonging to cytochalasin family have been isolated from various species of fungi, such as *Metarrhizium, Phoma, Phomopsis, Zygosporium, Rosellinia, Helminthosporium, Chaetomium,* and *Aspergillus*. The chemistry, structure, and biogenesis of this group has been reviewed by Binder and Tamm (*205*). Metabolites of this group have a characteristic structure of a highly substituted perhydroisoindolone group to which is fused a middle-membered (11–14) macrocyclic ring which is either carbocyclic, a lactone, or a cyclic carbonate (Table VI). A substituent at 3-position of isoindolone ring is variable, namely, a benzyl group in cytochalasins (with the exception of cytochalasin G), an indolylmethyl residue in chaetoglobosins (and cytochalasin G), and an isobutyl group in aspochalasins.

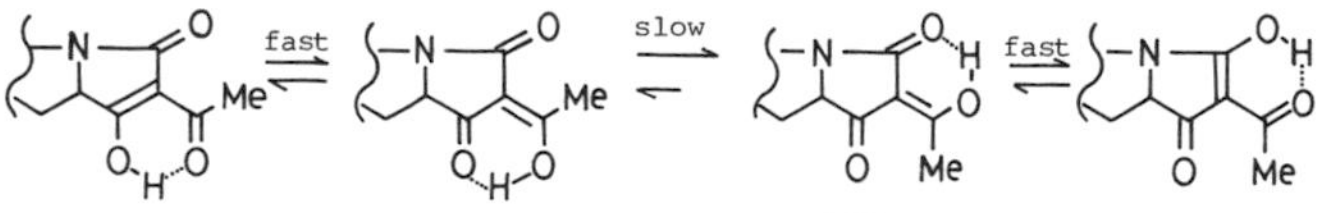

FIG. 7

TABLE VI
Cytochalasins

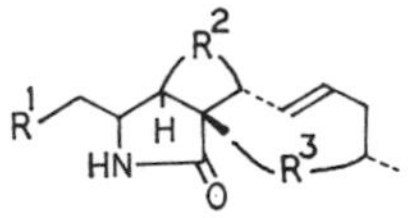

Compound	R[1]	R[2]	R[3a]
Cytochalasin	Ph	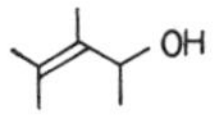	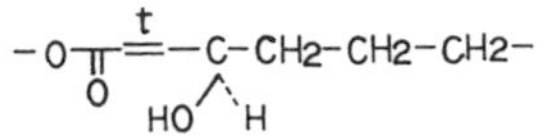
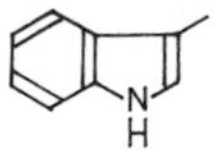		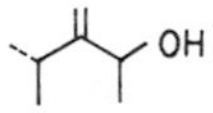	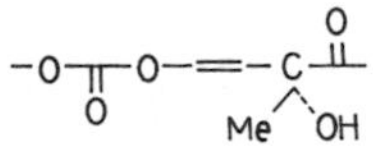
Chaetoglobosin		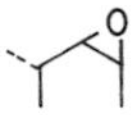	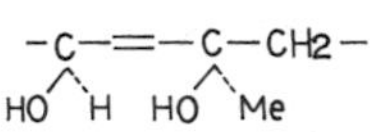
Aspochalasin	i-Bu	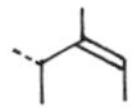	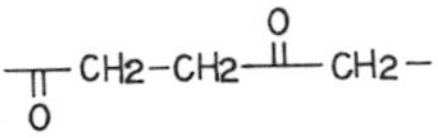

[a] Only a few examples are shown.

The structures of cytochalasins A and B were determined by Aldridge *et al.* (206) and Tamm and co-workers (207), independently, and structures of other metabolites were determined by correlations. The absolute configurations of many metabolites have been established by X-ray analysis. Approaches to synthesis of these complex compounds have also been reported by several groups (208), and successes in cytochalasin B and F have been reported by Stork's group (209).

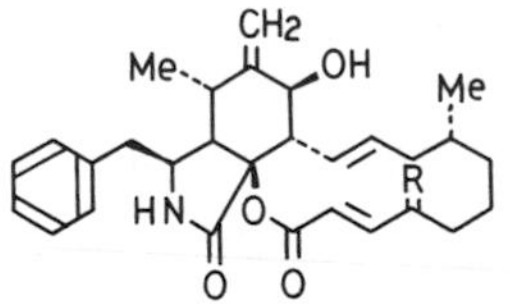

R=O Cytochalasin A
R⟨H, OH Cytochalasin B

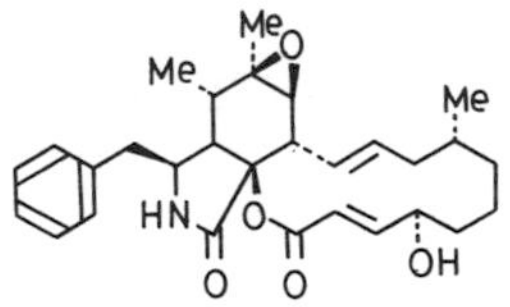

Cytochalasin F

The cytochalasins show unique effects on cell growth, movement, secretion, and phagocytosis. Moreover, some cytochalasins were reported as inhibitors of glucose transport in the plasma membrane of *Arthrobacter sialophilus,* in which the presence of an α,β-unsaturated carbonyl group in the macrolide moiety with appropriate bioisosteric placement are requisite for biological activity (*210*). On the other hand, it is probable that the effect of cytochalasins can be attributed to the interaction between the drug and the target protein, actin (*211,212*). Biological effects are reviewed by Carter (*213*).

a. Cytochalasins E and K.   From *Aspergillus* species cytochalasins E and K and aspochalasins A–D have been isolated. Cytochalasin E was isolated from highly toxic extracts of *Asp. clavatus* MIT-M16 (*214*), collected from mold-damaged rice in a Thai household where a child had died of an unidentified toxicosis. Cytochalasin E was previously known to be produced only by *Rosellinia* species (*215*), and later it was also isolated from *Asp. terreus* (*216*).

The structure of cytochalasin E (**134**), $C_{28}H_{33}NO_7$, mp 206–208°C (dec.), was elucidated by Aldridge *et al.* (*215,217*) and Büchi *et al.* (*218*), independently. The presence of intense mass peaks at m/z 91 (benzyl group) and m/z 190 ($C_{11}H_{12}NO_2$) in cytochalasin E and K (Fig. 8) strongly suggest that these compounds belong to the cytochalasin group, but their UV absorptions at 252, 257, and 263 nm are different from other cytochalasins. These have a unique structure in possessing an unusual vinyl carbonate moiety in a macrocyclic ring. As the above UV absorptions are absent in the 19,20-dihydro derivative, they can be ascribed to the vinyl carbonate unit.

The NMR spectrum of cytochalasin E was compared to that of cytochalasin F, which is isomeric with cytochalasin B. The partial structures are identical up to C-16, and the remaining $C_6H_6O_5$ unit is shown to contain a tertiary methyl group, a *tert*-hydroxyl group, and a cis-substituted double bond. Otherwise, the IR spectrum shows the presence of a carbonyl group (1729 cm^{-1}) and a vinyl carbonate (1762 and 1660 cm^{-1}). Hydrogenation over Pd/C yields only the 19,20-dihydro compound. Methanolysis in the presence of sodium bicarbonate gives a neutral substance (**135**) in which disappearance of the vinyl carbonate moiety (1765 and 1660 cm^{-1}) and appearance of a new carbonyl band at 1750 cm^{-1} are recognized. It shows the transformation of (A) to the CH_2CHO group [δ3.25 (dd) and 9.5] and a methyl ester (δ3.90) (B) (Fig. 9). Treatment of

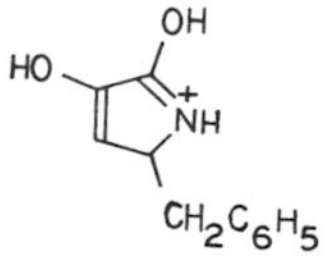

m/z 190

FIG. 8

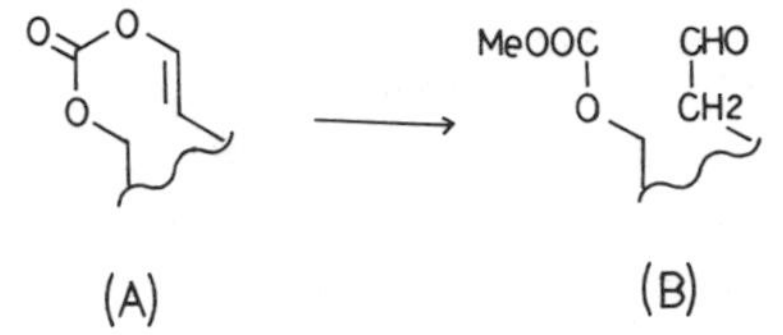

FIG. 9

cytochalasin E with LiAlH$_4$ yields alcohol **136** and lactol **137**. The reactions are summarized in Scheme 24. The absolute configuration was determined by X-ray analysis of the Ag salt (*218*).

Cytochalasin K (**138**) was isolated by Steyn *et al.* (*219*) as a cometabolite of cytochalasin E from *Asp. clavatus* MRC 1181. The structure, C$_{28}$H$_{33}$NO$_7$, mp 246–248°C, was determined on the basis of its physicochemical properties. In comparison with cytochalasin E, it has one more tertiary methyl group (in place of one less secondary methyl group) and also has a secondary hydroxyl group, the presence of which is confirmed by acetylation in which a strong downfield shift in H-7 ($\Delta1.55$ ppm) is recognized. Oxidation of cytochalasin K with chromium trioxide in pyridine affords the α,β-unsaturated ketone. The assumed structure was confirmed by acid-catalyzed rearrangement (2 N H$_2$SO$_4$ in THF by refluxing for 30 min) of cytochalasin E. Two products are obtained, one of which is identified as cytochalasin K and the other as the $\Delta^{6,12}$ isomer of cytochalasin K.

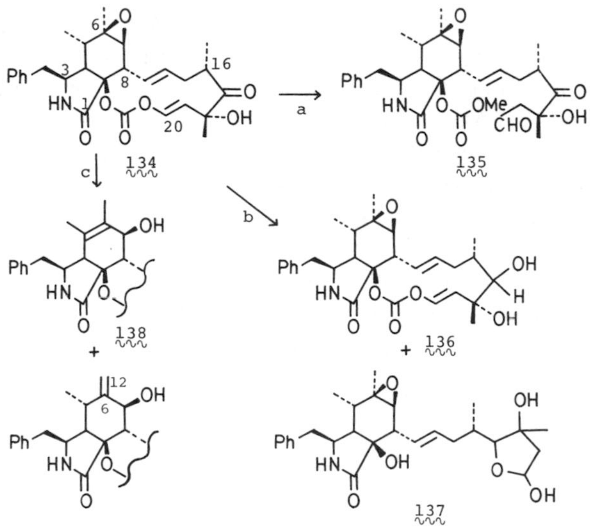

SCHEME 24. Chemical degradation of cytochalasin E (**134**). Reaction conditions: a, NaHCO$_3$–MeOH; b, LiAlH$_4$; c, 2 N H$_2$SO$_4$–THF.

Cytochalasin E kills rats within a few hours after dosing, and the LD_{50} values are 2.6 (ip) or 9.1 mg/kg (oral) in a single dose. Death is due to circulatory collapse caused by massive extravascular effusion of plasma.

b. Aspochalasins. Aspochalasin B (asposterol) was isolated as an antibiotic from culture filtrates of *Asp. microcysticus* in 1974 (*220*) by Heberle *et al.* Later, in 1979, Keller-Schierlein and Kupfer (*221*) isolated three inactive cometabolites (A, C, and D). Only aspochalasin B inhibits glucose uptake in the gram-positive bacterium, *Arthrobacter sialophilus*. Recently, Yahara *et al.* (*211*) reported that aspochalasin D strongly induces formation of actin-containing rodlets in the cytoplasm of treated fibroblasts.

The elucidation of chemical structure is performed on aspochalasin D, $C_{24}H_{35}NO_4$, mp 148°C (dec.), because it is a main component and the only one available in crystals. The chemical structures of other metabolites are induced by correlations. The spectroscopic data indicate four oxygen atoms in aspochalasin D (**143**) reside in an amidocarbonyl group ($\delta 9.01$), an α,β-unsaturated ketone

140 Aspochalasin A

141 Aspochalasin B

142 and 143
(diastereoisomers
at C-17 or C-18)

Aspochalasin C and D

(UV, 248 nm; ^{13}C, 196.7 ppm), and two secondary hydroxyl groups. The presence of the two secondary hydroxyl groups at adjacent carbon atoms is determined by $NaIO_4$ treatment, and the yielding of only one degradative product (dialdehyde) from tetrahydroaspochalasin D (**144**) shows that these are present in a macrocyclic ring (Scheme 25). The methylester derived from the above dialdehyde by oxidation and subsequent methylation affords a γ-lactone (**145**) by acid hydrolysis, which shows the existence of a 13,14 double bond. Oxidation of aspochalasin D with MnO_2 affords aspochalasin B, in which two protons at a trans double bond ($J = 15.5$ Hz) appear as very sharp doublet. The structure of a tetraol produced by ozonolysis of diacetylaspochalasin D and subsequent reduction with $NaBH_4$ clarifies the remaining moiety of macrocyclic ring.

These experimental results show similarities in many points to the known cytochalasin and chaetoglobosin groups, but the aromatic ring revealed in the other groups is not recognized in the UV (248 nm) and ^1H-NMR spectra. In-

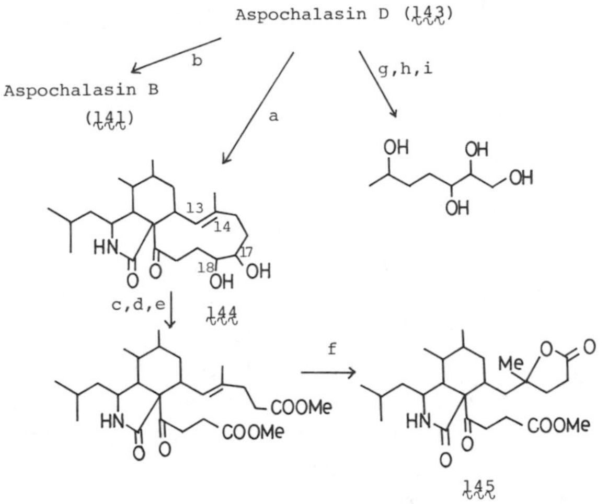

SCHEME 25. Chemical degradation of aspochalasin D (143). Reaction conditions: a, H_2–Pd/C; b, MnO_2; c, $NaIO_4$; d, Ag_2O; e, CH_2N_2; f, HCl; g, Ac_2O–Py; h, O_3; i, $NaBH_4$.

stead, the mass fragment at m/z 57 that often appears in members of the as-pochalasin group shows the presence of an isobutyl group in the side chain. X-ray crystal structure analysis of the 17,18-di-O-acetyl derivative of as-pochalasin C confirmed the structures of aspochalasin group and established the relative configurations (222). Aspochalasin C and D are diastereoisomers at C-18.

c. Biosynthesis.   A general biogenetic scheme for the cytochalasins was proposed in 1981 by Probst and Tamm (223) (Scheme 26). Acetyl-coenzyme A is condensed with seven or eight malonate units to form a C_{16} or C_{18} polyketide. The C-methylations may occur before stabilization of the polyketide. The methy-lated polyketide is combined with amino acid (L-Phe, L-Trp, or L-Leu, respec-tively), probably forming first the amide linkage and subsequently closing the lactam ring to yield the tetramic acid derivative (146). It is transformed to the substituted pyrrolinone (147) by reductions, dehydrations, and allylic oxidation. The latter compound undergoes an internal Diels–Alder-type cyclization to form the pentacyclic system, chaetoglobosin J (148), which possesses the required configurations at C-4, C-5, C-8, and C-9. Epoxidation of the cyclohexene ring yields chaetoglobosin A, which might be reduced to afford the unsaturated alcohol systems present in several chaetoglobosins.

Finally, cytochalasins possessing a macrocyclic lactone system, as represented by cytochalasin B, are obtained from carbocyclic precursors by a Baeyer–Vil-liger-type oxidation. In the case of a cyclic carbonate group, such as in cytochal-asin E and K, the carbonate moiety might result from a second Baeyer–Villiger-type insertion of an oxygen atom into a lactone ring.

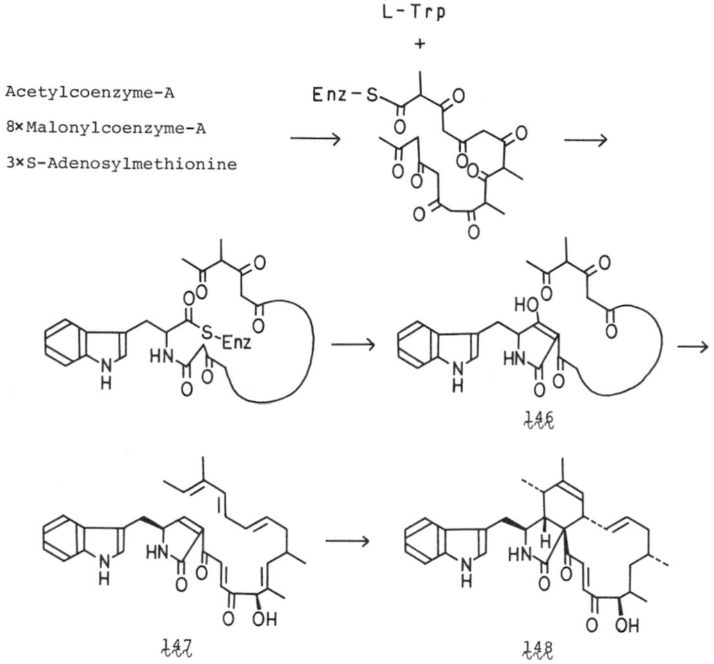

L-Trp

+

Acetylcoenzyme-A

8×Malonylcoenzyme-A

3×S-Adenosylmethionine

SCHEME 26. Biosynthesis of chaetoglobosin J (**148**).

3. Nigragillin and Nigerazine B

Nigragillin, oil, $[\alpha]_D^{24}$ $+107°$ (CHCl$_3$), was isolated together with L-alanyl-L-leucine anhydride, mp 259°C, from culture filtrates of *Asp. phoenicis* (*224*). The molecular formula was found to be C$_{13}$H$_{22}$N$_2$O, with high-resolution mass spectrometry. It has an UV absorption maximum at 262 nm, and the IR spectrum shows an amidocarbonyl group at 1650 cm^{-1} but no NH group. The signals of four olefinic protons and the absorption maximum in the ^{1}H-NMR and UV spectra, respectively, of nigragillin disappear in those of the tetrahydro derivative, which suggests the conjugation of two –CH=CH– and CO groups. The formation of all-trans sorbic acid by hydrolysis with 25% HCl confirms the presence of sorbic acid amide in the molecule. In the NMR spectrum two methyl protons (δ0.91 and 1.31) are both doublets, and another methyl group at δ2.28 (s) is characterized to *N*-CH$_3$. Two methylenes at δ2.51 (2*H*, dd) and δ3.67 (2*H*, dd) are both assigned to *N*-CH$_2$. One methine proton each (m, δ2.91 and 4.48) attaches to the above doublet CH$_3$ and *N*-CH$_2$ groups, respectively. On the basis of the above data and synthesis starting from *trans*-2,5-dimethylpiperazine, the structure of nigragillin was determined as (+)-*N*-methyl-*trans*-2,5-dimethyl-*N*-sorbylpiperazine (**149**). Later, nigragillin was also isolated as an active principle from *Asp. niger* (*225*), whose culture filtrates reveal a strong insecticidal effect on the silkworm larvae. Even at a dose of 5 μg/g (oral), nigragillin causes immediate poisoning and knockdown, whereas the activity of tetrahydro derivative is one-tenth that of nigragillin.

In 1983, a structurally related metabolite, nigerazine B, oil, $[\alpha]_D^{14}$ $+103°$, $C_{16}H_{22}N_2O$, was isolated from the mycelium of *Asp. niger* (226). The IR spectrum shows the absorption of a phenyl group (1606 cm^{-1}), an amide carbonyl group (1650 cm^{-1}), and an *N*-methyl group (2800 cm^{-1}). The presence of an *N*-CH$_3$ group (δ2.26) and five aromatic protons are confirmed by the NMR spectrum. Two olefinic protons at δ6.80 and 7.64 are coupled and have a trans conformation ($J = 15.5$ Hz). These data and the UV absorption maximum at 280 nm suggest the presence of a *trans*-cinnamic acid amide moiety. By irradiation of other protons, it is recognized that nigerazine B contains two sets of H$_3$C–(C–)H–CH$_2$ moieties in the molecule, and the methine and the methylene groups of each set are adjacent to a nitrogen atom. Therefore, the structure of nigerazine B is designated as *N*-methyl-2,5-dimethyl-*N'*-cinnamoylpiperazine (**150**). In comparison with nigragillin in coupling constants and chemical shifts in the NMR spectrum, it is judged that two *C*-methyl groups on the piperazine ring of nigerazine B have the trans position as in nigragillin. The spectral properties of *dl*-nigerazine B synthesized from trans-2,5-dimethylpiperazine as the starting material are in agreement with those of metabolite expect for specific rotation. Nigerazine B exhibits some inhibitory activity toward root growth of lettuce seedings (226).

149   150

4. Flavipucine and Isoflavipucine

Flavipucine (**151**), mp 130–131°C, $[\alpha]_D^{21}$ $-88°$, $C_{12}H_{15}NO_4$, was isolated from culture filtrates of *Asp. flavipes* (227) that demonstrate activity against gram-positive and -negative bacteria. The CHCl$_3$ extracts were chromatographed on alumina, and the initial elute (benzene–ether, 3:1) was a mixture of flavipucine and a probably isomeric compound. Purification was complicated by this cometabolite, but repeated recrystallization from MeOH–H$_2$O gives flavipucine in a homogeneous state. The later elution is a mixture of flavipucine and isoflavipucine, which is separated by rechromatography on silica gel column (228).

Flavipucine (229) is optically active, and the IR spectrum shows the presence of an NH (3200 cm^{-1}) and a saturated (1725 cm^{-1}) and two unsaturated carbonyl (1645 and 1629 cm^{-1}) groups. The presence of (CH$_3$)$_2$C=CH–CH$_2$–CO– moiety

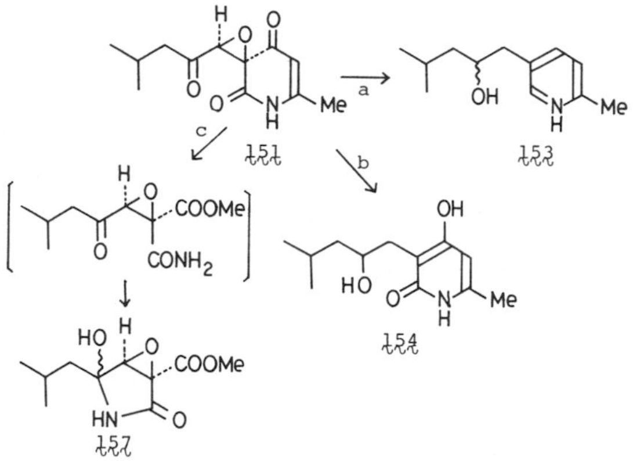

SCHEME 27. Chemical degradation of flavipucine (**151**). Reaction conditions: a, LiAlH$_4$; b, H$_2$–PtO$_2$; c, O$_3$.

is suggested by the NMR spectrum and also by a mass fragment at m/z 86. Reduction of flavipucine with LiAlH$_4$ affords 2-methyl-5-(2′-hydroxy-4′-methyl-pentyl)pyridine (**153**) in very low yield (Scheme 27). Catalytic hydrogenation of flavipucine yields a tetrahydrodeoxy derivative (**154**), C$_{12}$H$_{19}$NO$_3$, mp 189–190°C. It possesses the same side chain as **153** and contains conjugated CO–NH (1640, 1600, and 3250 cm^{-1}) and OH (3380 cm^{-1}) groups in the ring. Its UV spectrum shows a hypsochromatic shift from λ_{max} 287 to 260 nm in alkaline solution, as does that of 3-hexyl-4-hydroxy-6-methyl-2(1H)-pyridone (from λ_{max} 288 to 269 nm). To explain the optical activity and the remaining oxygen and nonexchangeable proton at $\delta 3.80$ in flavipucine, an epoxide bridge located between C-3 of ring and C-1′ of the side chain was assumed, and this spirobicyclic epoxide structure (**151**) was confirmed by synthesis (*230*).

Condensation of 4-hydroxy-6-methyl-2-pyridone with isobutylglyoxal in MeOH containing NaOMe yields quantitatively the Na salt of **155**. Its diacetate,

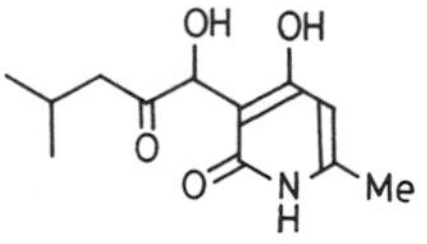

(**155**)

mp 129–130°C, is treated with KO-*t*-Bu–*t*-BuOH in *tert*-butylhydroperoxide at 0–25°C. The two isomers of the product (±)-flavipucine are separated by fractional crystallization, and the isomer crystallized initially as plates, mp 154–155°C, is identical to natural (−)-flavipiucine.

The still unsolved relative configuration of the oxirane bridge of flavipucine

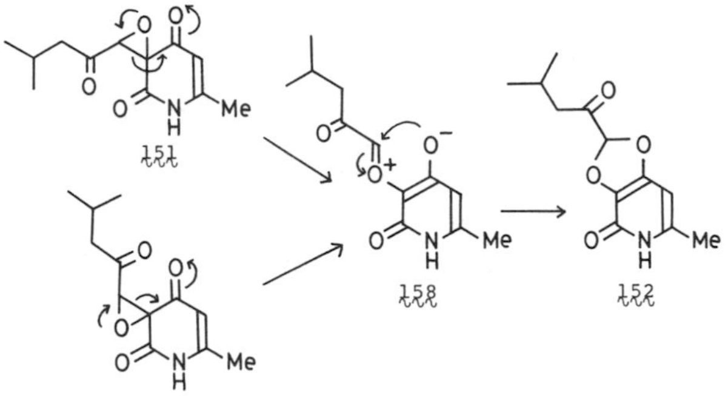

Epiflavipucine

SCHEME 28. Rearrangement of flavipucine (**151**).

was determined by X-ray analysis to be Z (*231*). The cis relationship of the isovaleroyl and amide functions of a product yielded by ozonolysis of (±)-flavipucine in MeOH was established by ring closure to the carbinol amide (**157**) (*232*).

Isoflavipucine (*233*), mp 166–171°C, optically inactive, $C_{12}H_{15}NO_4$, is obtained from flavipucine by melting, by treating on alumina column (35%), or best by refluxing in xylene for 10 hr (80%) (*234*). Thus, it seems to be an artifact produced during purification. The synthetic epiflavipucine (*E* configuration of the oxirane ring) also undergoes this facile rearrangement. A mechanism involving a common achiral intermediate (**158**) was proposed (*230*) (Scheme 28). The UV spectrum of isoflavipucine (λ_{max} 221 and 307 nm) is substantially different from that of flavipucine, indicating the presence of a new chromophore. The mass spectrum is practically indistinguishable from that of flavipucine, implying the same fragmentation pathway. The base peak m/z 152 [$M^+ - (CH_3)_2CHCH_2CO$] confirms the presence of the isobutylketone moiety. The NMR spectrum is remarkably similar to that of flavipucine, but the proton of the C-7 related oxirane bridge ($\delta 3.80$) in flavipucine is moved to lower field ($\delta 6.04$). Further, in the ^{13}C-NMR spectrum, the signal at 68.8 ppm (C-7) in flavipucine is found at 107.6 ppm in isoflavipucine. These displacements are consistent with the change at C-7 from mono- to dioxygen substitution. Isoflavipucine is acid-hydrolyzed to afford 3,4-dihydroxy-6-methyl-2-pyrazine and isobutylglyoxal. Based on the above chemical and spectroscopic evidence the structure **152** is determined.

In 1983, flavipucine, $[\alpha]_D^{18} -96°$, was also isolated from *Macrophoma* species (*235*) as a phytotoxin causing apple rot. A more toxic metabolite "Toxin A" (**159**), mp 134.5–135.5°C, $C_{12}H_{15}NO_4$, $[\alpha]_D^{18} +13°$, was also found. Isomer **159** seems to be identical to a cometabolite described in flavipucine purification. Separation was accomplished by using HPLC with an analytical Finepak SIL C18 column eluted with acetonitrile–water (1:3). The spectral data of flavipucine and Toxin A are quite similar, and Toxin A was determined to be a side chain

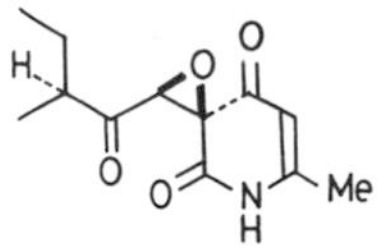

159

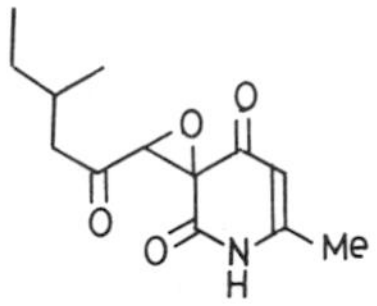

Omoflavipucine

isomer of flavipucine. Light brown necrotic spots are induced within a few hours at 23–24°C by application of a solution containing 1 μg/liter of "Toxin A" on noninjured apples.

Omoflavipucine has reported as a metabolite of *Asp. flavus*. It has a structure similar to that of flavipucine (*236*).

5. Nucleosides

The adenosine-related metabolites cordycepin (3'-deoxyadenosine, **160**), mp 230–231°C, $[\alpha]_D^{25}$ −35°, and 2'-deoxycoformycin (**161**), mp 220–225°C, $[\alpha]_D^{25}$ +76.4°, have been isolated from the culture broth of *Asp. nidulans* (*237,238*). 2'-Deoxycoformycin (**161**) had been isolated previously from *Streptomyces anti-*

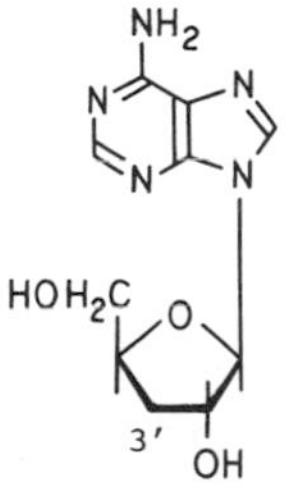

160 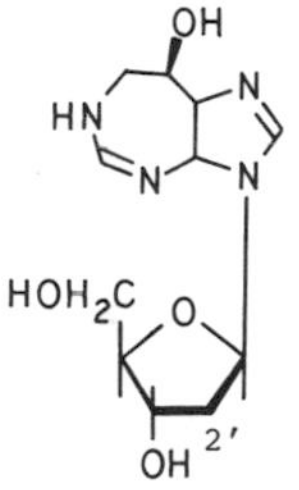   161

bioticus (*239*), and the structure was elucidated by spectroscopic and single-crystal X-ray diffraction techniques (*238*). It is an inhibitor of adenosine deaminase. Cordycepin (**160**) is cytotoxic against mouse leukemia L-5178Y cells. The structure is suggested from the ¹H- and ¹³C-NMR data and determined by comparison to the authentic sample.

6. Pteridine Glycosides

Two blue, fluorescent substances, asperopterin-A, mp 193°C (dec.), and asperopterin-B, mp >310°C, have been isolated from *Asp. oryzae* T-17 (*240*), the former being D-riboside of the latter (*241*). Asperopterin B has a UV spectrum similar to that of 2-amino-4,7-dihydroxypteridine (isoxanthopterin) and bears a CH$_2$OH group (δ4.52) and a N-CH$_3$ group (δ3.22) in the molecule. The carboxyl group at C-6 of hydroxypteridine ring is known to be eliminated by treatment with Na–Hg in alkaline solution. Asperopterin B is oxidized to carboxylic acid with $KMnO_4$, and subsequent Na–Hg reduction confirms the location of CH_2OH group at C-6. Comparison of the 6-carboxylic acid with three synthetic N-methylisoxanthopterin-6-carboxylic acids determines the position of methyl group at N-8.

The structure of asperopterin-B (163) was eventually confirmed by the following synthesis (242). 6,8-Dimethylisoxanthopterin is brominated, and the bromo derivative is converted with 0.5 N NaOH to the 6-hydroxymethyl compound, which is identical to asperopterin B in R_f and pK_a values and also in its UV and IR spectra. The hydroxymethyl group in the side chain together with the ring nitrogen atoms are methylated with $(CH_3)_2SO_4$ in asperopterin B, whereas, those of asperopterin A are not methylated under the same conditions, showing that asperopterin A is protected by a ribosyl group. The anomeric configuration of asperopterin A (162) is deduced to be β by a comparison of its molecular rotation (−233°) with those of known D-ribofuranoside.

162   R=β-D-ribofuranosyl
163   R=H

7. Versimide

Versimide, $C_9H_{11}NO_4$, was isolated from *Asp. versicolor* as an optically active metabolite (243). It was purified by molecular distillation at 60°C (0.003 mmHg). The NMR spectrum shows the presence of an OCH_3 (δ3.77), a terminal methylene (δ5.78 and 6.47), and three protons (δ2.0–3.2) one of which is coupled to a methyl group (δ1.35, J = 7 Hz). Dihydroversimide, $C_9H_{13}NO_4$, is formed by catalytic hydrogenation of versimide in which a new CH—CH_3 moiety occurs in place of the above terminal methylene. Hydrolysis of dihydroversimide with concentrated HCl at 100°C for 16 hr gives alanine and (R)-methylsuccinic acid. The latter acid is also obtained by hydrolysis of versimide itself. This evidence suggests that versimide is methyl (+)-(R)-α-(methylsuccinimido)acrylate (164). Examination of the IR spectra of versimide (1780 and 1720 cm⁻¹) and the dihydro-derivative (1790 and 1720 cm⁻¹) confirms the presence of a five-membered imide. Condensation of methylsuccinic acid anhydride with alanine at 180–200°C and subsequent methylation with CH_2N_2 gives DL-dihydroversimide (244). Versimide possesses insecticidal activity aganist *Drosophila melanogaster* (245). A related metabolite, pencolide (*cis*-α-citraconimidecrotonic acid, 165) was isolated from *Pen. multicolor* (246).

$$\underset{164}{\text{Me}\overset{O}{\underset{O}{\bigg|}}N-\overset{CH_2}{\underset{COOMe}{\diagdown}}}\qquad\underset{165}{\text{Me}\overset{O}{\underset{O}{\bigg|}}N-\overset{Me}{\underset{COOH}{\diagdown}}}$$

8. Silvaticamide

Silvaticamide, $C_{25}H_{29}NO_5$, mp 191–192°C (dec.), was isolated as a colorless amorphous compound by Yamazaki *et al.* from *Asp. silvaticus* cultured on rice (*247*). It is toxic, and intraperitoneal administration (208 mg/kg) in mice caused peritonitis and death within 4 days. Silvaticamide is unstable to light and gradually decomposes to a mixture of purple–red substances. It is optically inactive and gives a positive color reaction with $FeCl_3$, but negative ones with $Mg(OAc)_2$, Gibb's, and Dragendorff's reagents.

The absorption maxima at 210, 290, and 310 nm and those at 3390 (OH), 3305 (NH), 1660 (CO), 1598, and 1490 cm^{-1} (benzene ring) are observed in the UV and IR spectra, respectively. The molecular formula together with ^{13}C- and ^{1}H-NMR data suggest that silvaticamide (**166**) is a tricyclic compound containing two benzene rings and substituted with a methyl group and two 3,3-dimethylallyl groups. The methylene proton in the one isopentyl group is suggested to be attached to an oxygen atom from the chemical shift (δ4.69). This is confirmed by the formation of a new phenolic hydroxyl group with elimination of an isopentenyl group by catalytic hydrogenation. Other signals of one methine (H$_a$; δ6.21, br s), three aromatic protons, and four protons (which disappear upon addition of D$_2$O) are also observed in the NMR spectrum of silvaticamide. Two of three aromatic protons are indicated to be located at ortho positions by the coupling constant ($J = 8$ Hz). The third aromatic proton couples at long range ($J < 1$ Hz) with an aromatic methyl group (δ2.18), and a methine proton (H$_a$) couples ($J < 2$ Hz) with a hydroxyl proton (δ7.44). The shift of this peak (H$_a$, δ6.21 in silvaticamide and δ5.41 in triacetate) to δ7.07 in tetraacetate suggests that H$_a$ attaches to the carbon bearing a secondary hydroxyl group. From this evidence it is expected that the amide group indicated in the IR spectrum links two benzene rings together with the carbon atom bearing the secondary hydroxyl group to form a tricyclic ring system.

The assumed structure of **166** is closely related to the metabolites called secoanthraquinones, especially to arugosin B (**167**), a metabolite from *Asp. rugulosus* (*248*). Physicochemical properties of silvaticamide and arugosin B are very close, except the former contains a nitrogen atom in the molecule. Comparison of ^{1}H-NMR spectra indicates beyond doubt that the location of each functional group in both compounds is similar. Pyrolysis of the triacetate of **166** affords **168**, mp 131–132°C, which has the same molecular formula,

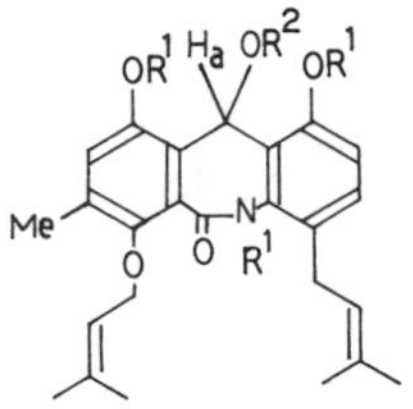

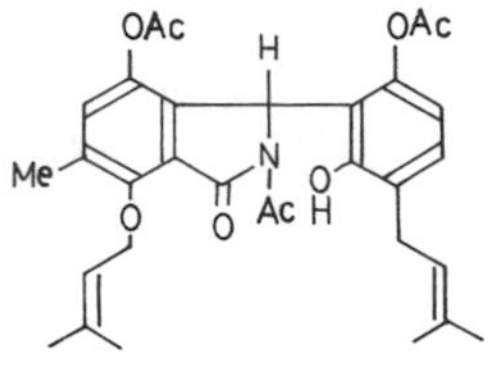

166   R¹=R²=H
      R¹=Ac R²=H (Triacetate)
      R¹=R²=Ac (Tetraacetate)

168

demonstrating that the two compounds are isomers. The structure of **168** was determined by the direct X-ray diffraction analysis, which confirmed that the location of the functional group in silvaticamide is correct. Silvaticamide is shown to be optically inactive. In the NMR spectra of the tri- and tetraacetates, the signals of H_a and acetyl methyl protons are shown in a doubled form, and the integral ratio of these signals corresponds to half of the protons. These findings suggest the assumption that silvaticamide is isolated as a mixture of C-8 hydroxyl stereoisomers.

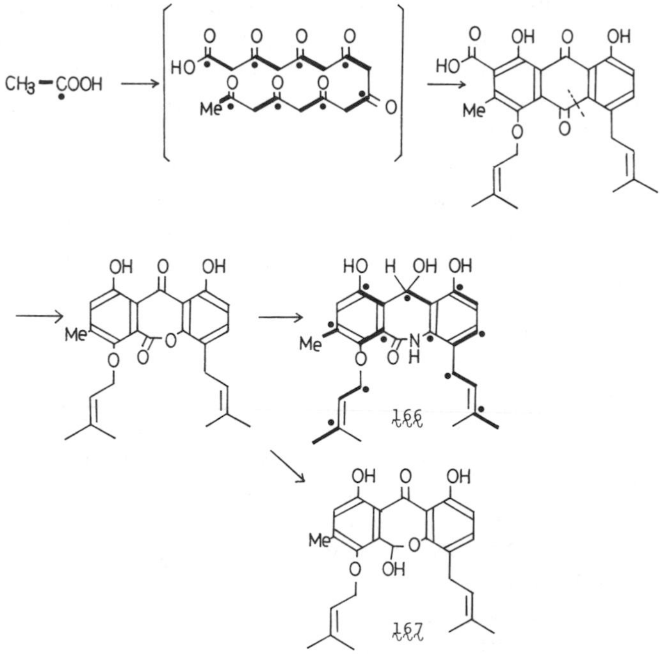

SCHEME 29. Biosynthesis of silvaticamide (**166**).

Experiments using [1-^{13}C]- and [1,2-^{13}C]-acetate indicate that oxidative cleavage of the anthraquinone ring between C-2 and C-15 occurs. The pathway depicted in Scheme 29 via the formation of "seco-anthraquinone" is proposed (*249*).

9. Nitro Compounds

A plant growth regulatory metabolite, 1-amino-2-nitrocyclopentanecarboxylic acid (**169**), $C_6H_{10}N_2O_4$, $[\alpha]_D$ 0°, was extracted from culture filtrates of *Asp. wentii* (*250,251*) and purified by using several kinds of ion-exchange resins (*252*). It decomposes at ~150°C without melting. The IR spectrum shows the presence of NH_2 (3639 and 3350 cm^{-1}), COOH (2570 and 1675 cm^{-1}), and NO_2 (1570 cm^{-1}) groups. In the NMR spectrum a one-proton triplet at δ5.54 and a six-proton multiplet at δ1.8–3.2 are observed.

Hydrolysis of **169** with boiling water yields ammonia and a mixture of carboxylic acids. The major product (**170**), $C_6H_9NO_5$, mp 101–103°C, contains a hydroxyl group (3500 cm^{-1}) which must be tertiary, since there is no change in the 4-6 region of the NMR spectrum on acetylation. The chemical shift of one proton (t, δ5.3) indicates that the secondary carbon bears a nitro group. The minor product (**171**), $C_6H_7NO_4$, mp 102–103°C, possesses no hydroxyl group,

but has a carboxyl group, a nitro group, and a double bond (1668 cm^{-1}) that is conjugated (λ_{max} 220 and 265 nm) and tetrasubstituted, as no olefinic proton is present in the NMR spectrum. This unsaturated acid (**171**) is obviously a dehydration product of **170**. This relationship is confirmed by formation of **171** upon heating **170** with dimethylsulfoxide and by its conversion to **170** with boiling water. Upon treating with ammonia, both compounds regenerate **169**. Compound **170** is hydrogenated to afford a hydroxyamino acid, $C_6H_{11}NO_3$, which reacts with periodic acid, so that the hydroxyl and a newly occurring amino group are vicinal. Further, **170** contains no double bond, so it must be monocyclic. The above evidence is consistent only with the structure **169**, which was confirmed by synthesis. Reaction of cyclopent-1-ene-carboxylic acid with dinitrogen tetroxide in the presence of iodine gives 1-iodo-2-nitrocyclopentanecarboxylic acid as the major product. The iodo acid can be converted to **169** with ammonia, and to the unsaturated acid (**171**) with pyridine. Metabolite **169** obtained from this fungus is optically inactive, which may be a result of racemization during the isolation procedure.

In feeding experiments, lysine was specifically incorporated into **169**, and diaminopimelic acid was not a precursor *(251)*. The fungus *Asp. wentii* also produces β-nitropropionic acid *(253)*. This compound is known to be biosynthesized from aspartic acid through decarboxylation, however, and does not derive from β-alanine in higher plants and *Pen. atrovenetum (254)*.

10. Xanthocillin Group

a. Xanthocillin-X and Xanthoascin. Several metabolites containing an isocyano or cyano group have been isolated from *Aspergillus* species. These show a characteristic absorption at 2150 (isocyanide) or 2250 cm⁻¹ (cyanide) in the IR spectrum.

Xanthocillin-X, a typical metabolite of this group, was first isolated from *Pen. notatum* as an antimicrobial principle *(255)*. The structure was elucidated by Hagedron and Tonjes *(256)* to be 1,4-di(*p*-hydroxyphenyl)-2,3-diisocyano-1,3-butadiene (**172**). Acid hydrolysis of **172** affords 1,4-di(*p*-hydroxyphenyl)-

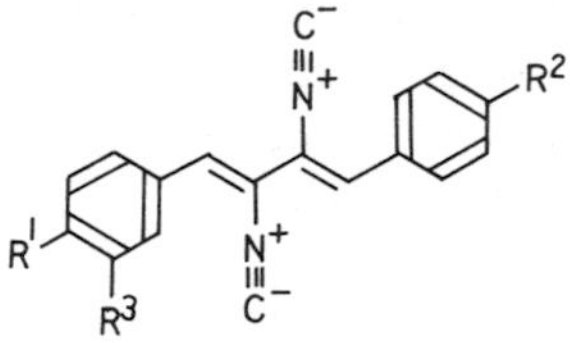

172   R¹=R²=OH   R³=H
XME   R¹=OH   R²=OMe   R³=H
XDE   R¹=R²=OMe   R³=H
XTE   R¹=R²=R³=OMe

butan-2,3-dione, mp 235°C, which affords the corresponding quinoxaline, mp 235°C, with *o*-phenylenediamine. The dimethylether (XDE), mp 181°C, is reduced with LiBH₄ to give the diol, mp 161°C. Diol was treated with Pb(OAc)₄ to give *p*-methoxyphenylacetaldehyde, and it afforded *p*-methoxyphenylacetic acid with H₂O₂ in alkaline solution. By hydrolysis of xanthocillin-X, the diformylamino derivative is obtained, which supports the presence of isocyano group. The structure of diketone derivative of XDE is determined by synthesis *(257)*.

Xanthocillin-X monomethylether (XME), dimethylether (XDE), and methoxy xanthocillin-X dimethylether (XTE) have been isolated from *Dendotomyces albus* and from two strains of *Asp. chevalieri* by Suzuki and co-workers *(258)*. All these compounds decompose near 183°C and are effective against Newcastle disease and inhibit the growth of *Bacillus subtilis*. XME specifically inhibits the conversion of arachidonic acid to prostaglandin H₂ *(259)* and shows an inhibitory activity against platelet aggregation.

Xanthoascin (**173**), C₂₃H₂₀N₂O₃, yellow needles, mp 165–170°C (dec.), was isolated from *Asp. candidus* by Natori and co-workers *(260)*. It has a characteristic

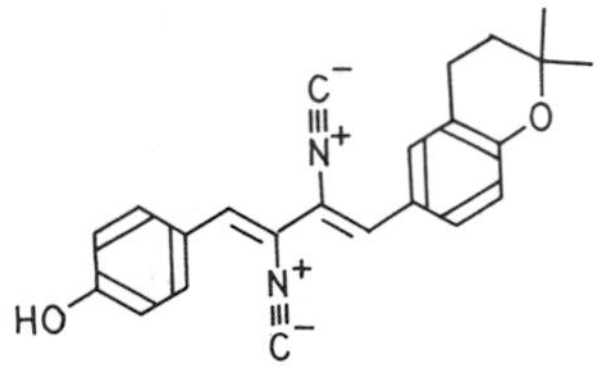

173

absorption at 2150 cm^{-1} in the IR spectrum, and the UV spectrum [λ_{max} 244, 302, 365, and 385(sh) nm] is nearly superimposable on those of the 1,4-diphenyl-1,3-butadienes. Oxidation of xanthoascin with chromic trioxide in AcOH affords 4-hydroxybenzoic acid, mp 215–218°C, and 2,2-dimethylchroman-6-carboxylic acid, mp 171–174°C. Formation of the diformylamino derivative by hydrolysis of xanthoascin confirms the presence of the isocyano group in the molecule. Xanthoascin exhibits cytotoxicity in cultured HeLa cells. It also exhibits not only hepto- and cardiotoxicity but also teratogenecity in experimental animals (*261*).

b. WF-5239. WF-5239, $C_9H_9NO_2$, mp 142–145°C, was isolated from the culture broth of *Asp. fumigatus* Fresenius (*262*). The AcOEt extracts are loaded onto a silica gel column, and the active fractions eluted with AcOEt are concentrated. The obtained crude solid is crystallized from EtOH, yielding WF-5239 as colorless needles. From the mass, IR, and ^{1}H-NMR spectra the structure **174** is

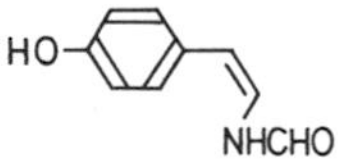

174 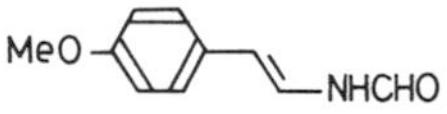   Tuberin

determined. While the double bond is assigned as *cis* on the basis of the coupling constant ($J = 10$ Hz) of the olefinic protons ($\delta 5.60$ and 6.65) together with the absence of the characteristic band at 965 cm^{-1} due to a trans double bond. The structurally related metabolite tuberin was isolated from *Streptomyces* species as an antibiotic; its double bond is trans (*263*). WF-5239 shows no antimicrobial activity, but it is a potent inhibitor of platelet aggregation induced by arachidonic acid and collagen.

c. Emerin. From the mycelium of *Asp. nidulans* Winter, a new metabolite containing cyano groups was isolated and named emerin (**175**) by Hatsuda and

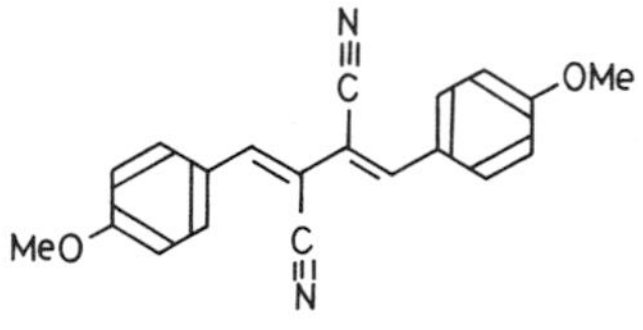

175

co-workers (264). It is extracted with benzene from the fungus and isolated by silica gel chromatography. It is crystallized from acetone as greenish yellow columnar crystals, mp 225–226°C. The UV spectrum closely resembles that of the dimethylether of xanthcillin-X (XDE). These metabolites also share similar IR spectra, except for the absorption band of triple bonds. Emerin shows a band at 2225 cm^{-1}, indicating the presence of a cyano group. The NMR spectrum suggests that it has a symmetrical structure, which is also supported by the presence of an intense fragmentation at m/z 158 in the mass spectrum arising from due to the loss of half the molecule. Emerin is synthesized by condensation of p-anisaldehyde and succinonitrile in potassium methoxide and identified.

d. Biogenesis of Xanthocillins. DL-[2-^{14}C]tyrosine has proved to be an excellent precursor of xanthocillin-X (18–25% incorporation). Greater than 90% of the radioactivity is found at C-2 and C-3 of the butadiene moiety (265), whereas [1-^{14}C]tyrosine is not incorporated into xanthocillin-X, which suggests that the carboxyl group of tyrosine is lost during incorporation. The nitrogen of the isocyanide does not come from the amino group of tyrosine either, since ^{14}C/^{15}N ratio increases from 10.2 to 61.6 when DL-[2-^{14}C,^{15}N]tyrosine is incorporated into xanthocillin-X. [2-^{14}C]Acetate, [^{14}C]formate and [^{14}CH$_3$]methionine were also tested (266) as possible precursors for the C of the isocyano group, without success. 1,4-Di-(4-hydroxyphenyl)-2,3-diformamido-1,3-butadiene (176), mp 240°C (dec.), was also found in *Pen. notatum* (267), but this formamide metabolite could not be proved to be a precursor of xanthocillin-X. Hagedron reported that the formamide group of a sesquiterpene isolated from a marine sponge is not transformed to the isocyanide, but it is biosynthesized from isocyanide (268).

REFERENCES

1. A. Stoll and A. Hofmann, in "The Alkaloids" (R. H. F. Manske, ed.), Vol. 8, p. 726. Academic Press, New York, 1965.
2. P. A. Stadler and P. Stutz, in "The Alkaloids" (R. H. F. Manske, ed.), Vol. 15, p. 1. Academic Press, New York, 1975.
3. W. B. Turner, "Fungal Metabolites." Academic Press, New York, 1971; W. B. Turner and D. C. Aldridge, "Fungal Metabolites II." Academic Press, New York, 1983.
4. E. C. White, *Science* **92**, 127 (1940); E. C. White and J. H. Hill, *J. Bacteriol.* **45**, 433 (1943).
5. J. C. MacDonald, in "Antibiotics II, Biosynthesis" (D. Gottlieb and P. D. Shaw, eds.), p. 43. Springer-Verlag, Berlin, New York, 1967.
6. B. J. Wilson, in "Microbial Toxins" (A. Ciegler, S. Kadis, and S. J. Ajl, eds.), Vol. 6, Chap. 3, Academic Press, New York, 1971.
7. M. Yamazaki, in "Biosynthesis of Mycotoxin-A Study in Secondary Metabolites", (P. S. Steyn, ed.), p. 199. Academic Press, New York, 1980.
8. S. Nakamura, *Bull. Agric. Chem. Soc., Jpn.* **24**, 629 (1960); S. Nakamura and T. Shiro, *Agric. Biol. Chem.* **25**, 573 (1961); S. Nakamura, *Agric. Biol. Chem.* **25**, 658 (1961); S. Nakamura, *Agric. Biol. Chem.* **25**, 665 (1961).
9. G. Dunn, G. T. Newbold, and F. S. Spring, *J. Chem. Soc.* 2586 (1949); T. Yokotsuka,

Kagaku to Seibutsu **5,** 150 (1967); M. Sasaki, T. Kikuchi, Y. Asao, and T. Yokotsuka, *Nippon Nogei Kagaku Kaishi* **41,** 154 (1967).

9a. T. Yokotsuka, Y. Asao, and M. Sasaki, *Nippon Nogei Kagaku Kaishi* **42,** 346 (1968).

9b. T. Yokotsuka, M. Sasaki, T. Kikuchi, Y. Asao, and A. Nobuhara, *Nippon Nogei Kagaku Kaishi* **41,** 32 (1967); M. Sasaki, Y. Asao, and T. Yokotsuka, *Nippon Nogei Kagaku Kaishi* **42,** 351 (1968).

9c. J. C. MacDonald, *Can. J. Biochem.* **48,** 1165 (1970).

10. J. C. MacDonald, *Can. J. Chem.* **50,** 543 (1972).

11. Y. Maebayashi, M. Sumita, K. Fukushima, and M. Yamazaki, *Chem. Pharm. Bull.* **26,** 1320 (1978).

12. G. Assante, L. Camarda, R. Locci, L. Merlini, G. Nasini, and E. Papadopoulos, *J. Agric. Food Chem.* **29,** 785 (1981).

13. J. D. Dutcher and O. Wintersteiner, *J. Biol. Chem.* **155,** 359 (1944).

14. J. D. Dutcher, *J. Biol. Chem.* **171,** 321 (1947); J. D. Dutcher, *J. Biol. Chem.* **171,** 341 (1947).

15. G. Dunn, G. T. Newbold, and F. S. Spring, *J. Chem. Soc.,* S131 (1949).

16. G. Dunn, J. J. Gallagher, G. T. Newbold, and F. S. Spring, *J. Chem. Soc.,* S126 (1949).

17. G. T. Newbold, W. Sharp, and F. S. Spring, *J. Chem. Soc.,* 2679 (1951).

18. U. Weiss, F. Strelitz, H. Flon, and I. N. Ashehov, *Arch. Biochem. Biophys.* **74,** 150 (1958).

19. J. D. Dutcher, *J. Biol. Chem.* **232,** 785 (1958).

20. S. Nakamura and T. Shiro, *Bull. Agric. Chem. Soc., Jpn.* **23,** 65 (1959); S. Nakamura and T. Shiro, *Bull. Agric. Chem. Soc., Jpn.* **23,** 418 (1959); S. Nakamura, Y. Kurimura, and T. Shiro, *Nippon Nogei Kagaku Kaishi* **35,** 1058 (1961).

21. M. Yamazaki, Y. Maebayashi, and K. Miyaki, *Chem. Pharm. Bull.* **20,** 2274 (1972).

22. T. Ueno, A. Nishimura, and F. Yoshizako, *Agric. Biol. Chem.* **41,** 901 (1977).

23. R. G. Micetich and J. C. MacDonald, *J. Chem. Soc.,* 1507 (1964).

24. J. C. MacDonald, *J. Biol. Chem.* **236,** 512 (1961).

25. J. C. MacDonald, *J. Biol. Chem.* **237,** 1977 (1962).

26. J. C. MacDonald, G. D. Micetich, and R. H. Haskin, *Can. J. Microbiol.* **10,** 90 (1964).

27. R. G. Micetich and J. C. MacDonald, *J. Biol. Chem.* **240,** 1692 (1965).

28. J. C. MacDonald, *Biochem. J.* **96,** 533 (1965).

29. A. Quilico and L. Panizzi, *Chem. Ber.* **76,** 348 (1943).

30. A. Quilico and C. Cardani, *Atti Accad. Nazl. Lincei, Rend. Classe Sci. Fis. Mat. Nat.* **9,** 220 (1950); *Chem. Abstr.* **45,** 3909 (1951).

31. Z. Kitamura, U. Kurimoto, and M. Yokoyama, *Yakugaku Zasshi* **76,** 972 (1956).

32. A. Quilico, *Gazz. Chim. Ital.* **78,** 111 (1948); *idem, ibid.* **83,** 155 (1953); A. Quilico, C. Cardani, and F. Piozzi, *Gazz. Chim. Ital.* **85,** 179 (1955); *idem, ibid.* **85,** 3 (1955); *idem, ibid.* **86,** 211 (1956); *idem, ibid.* **88,** 125 (1958); C. Cardani, G. Casnati, B. Carvalleri, and A. Quilico, *Atti Accad. Nazl. Lincei, Rend. Classe Sci. Fis. Mat. Nat.* **24,** 488 (1958); A. Quilico, F. Piozzi, and M. Dubini, *Gazz. Chim. Ital.* **88,** 1308 (1958); C. Casnati, G. Casnati, F. Piozzi, and A. Quilico, *Tetrahedron Lett.,* 1 (1959); F. Piozzi, G. Casnati, A. Quilico, and C. Casnati, *Gazz. Chim. Ital.* **90,** 451 (1960); *idem, ibid.* **90,** 476 (1960); C. Casnati, R. Cavalleri, F. Piozzi, and A. Quilico, *Gazz. Chim. Ital.* **92,** 105 (1962); G. Casnati, A. Quilico, and A. Risca, *Gazz. Chim. Ital.* **92,** 129 (1962); A. Quilico, *Res. Prog. Org. Biol. Med. Chem.* **1,** 225 (1964) and references therein.

33. A. J. Birch, G. E. Blance, S. David, and H. Smith, *J. Chem. Soc.,* 3128 (1961).

34. A. J. Birch and K. R. Farrar, *J. Chem. Soc.,* 4277 (1963).

35. R. Nakashima and G. P. Slater, *Tetrahedron Lett.,* 2649 (1971); cf. R. Nakashima and G. P. Slater, *Can. J. Chem.* **47,** 2069 (1969).

36. N. Takamatsu, S. Inoue, and Y. Kishi, *Tetrahedron Lett.,* 4661 (1971); N. Takamatsu, S. Inoue, and Y. Kishi, *Tetrahedron Lett.,* 4665 (1971); S. Inoue, N. Takamatsu, and Y. Kishi, *Yakugaku Zasshi* **97,** 558 (1977).

37. H. Nagasawa, A. Isogai, A. Suzuki, and S. Tamura, *Agric. Biol. Chem.* **43**, 1759 (1979).
38. G. Gatti and C. Fuganti, *J. Chem. Research (S)*, 366 (1979).
39. R. R. Fraser, S. Passannanti, and F. Piozzi, *Can. J. Chem.* **54**, 2915 (1976).
40. T. Hamasaki, K. Nagayama, and Y. Hatsuda, *Agric. Biol. Chem.* **40**, 203 (1976); T. Hamasaki, K. Nagayama, and Y. Hatsuda, *Agric. Biol. Chem.* **40**, 2487 (1976).
41. R. D. Stipanovic and H. W. Schroeder, *Trans. Br. Mycol. Soc.* **66**, 178 (1976).
42. C. M. Allen, Jr., *Biochemistry* **11**, 2154 (1972).
43. C. M. Allen, Jr., *J. Am. Chem. Soc.* **95**, 2386 (1973).
44. M. Barbetta, G. Casnati, A. Pochini, and A. Selva, *Tetrahedron Lett.*, 4457 (1969); G. Casnati, A. Pochini, and O. Ungaro, *Gazz. Chim. Ital.* **103**, 141 (1973).
45. A. Dossena, R. Marchelli, and A. Pochini, *J. Chem. Soc., Chem. Commun.*, 771 (1974).
46. A. Dossena, R. Marchelli, and A. Pochini, *Experientia* **31**, 1249 (1975).
47. H. Itokawa, Y. Akita, and M. Yamazaki, *Yakugaku Zasshi* **93**, 1251 (1973).
48. H. Nagasawa, A. Isogai, K. Ikeda, S. Sato, S. Murakoshi, A. Suzuki, and S. Tamura, *Agric. Biol. Chem.* **39**, 1901 (1975).
49. R. Cardillo, C. Fuganti, G. Gatti, D. Ghiringhelli, and P. Grasselli, *Tetrahedron Lett.*, 3163 (1974).
50. R. Marchelli, A. Dossena, A. Pochini, and E. Dradi, *J. Chem. Soc., Perkin Trans. 1*, 713 (1977).
51. S. Inoue, N. Takamatsu, and Y. Kishi, *Yakugaku Zasshi* **97**, 564 (1977).
52. S. Nakatsuka, H. Miyazaki, and T. Goto, *Tetrahedron Lett.*, **18**, 2817 (1980).
53. H. Nagasawa, A. Isogai, A. Suzuki, and S. Tamura, *Tetrahedron Lett.*, 1601 (1976).
54. G. Gatti, R. Cardillo, and C. Fuganti, *Tetrahedron Lett.*, 2605 (1978).
55. R. Cardillo, C. Fuganti, D. Ghiringhelli, and P. Grasselli, *Chim. Ind.* (Milan) **57**, 678 (1975); *Chem. Abstr.* **84**, 40511h (1976); G. Gatti, R. Cardillo, C. Fuganti, and D. Ghiringhelli, *J. Chem. Soc., Chem. Commun.* 435 (1976).
56. S. Inoue, S. Murata, N. Takamatsu, H. Nagano, and Y. Kishi, *Yakugaku Zasshi* **97**, 576 (1977); S. Inoue, N. Takamatsu, K. Hashizume, and Y. Kishi, *Yakugaku Zasshi* **97**, 582 (1977).
57. G. P. Slater, J. C. MacDonald, and R. Nakashima, *Biochemistry* **9**, 2886 (1970).
58. R. Marchelli, A. Dossena, and G. Casnati, *J. Chem. Soc., Chem. Commun.*, 779 (1975).
59. R. Cardillo, C. Fuganti, D. Ghiringhelli, P. Grasselli, and G. Gatti, *J. Chem. Soc., Chem. Commun.*, 778 (1975).
60. G. Casnati, M. F. Franciori, A. Guareschi, and A. Pochini, *Tetrahedron Lett.*, 2485 (1969); G. Casnati and A. Pochini, *J. Chem. Soc., Chem. Commun.*, 1328 (1970); G. Casnati, R. Marchelli, and A. Pochini, *J. Chem. Soc., Perkin Trans. 1*, 754 (1974); G. Casnati, G. P. Gardini, G. Palla, and C. Fuganti, *J. Chem. Soc., Perkin Trans. 1*, 2397 (1974); V. Bocchi, G. Casnati, and R. Marchelli, *Tetrahedron* **34**, 929 (1978).
61. M. F. Grundon, M. R. Hamblin, D. M. Harrison, J. N. Derry Logue, M. Maguire, and J. A. McGrath, *J. Chem. Soc., Perkin Trans. 1*, 1294 (1980); K. J. Baird, M. F. Gtundon, D. M. Harrison, and M. G. Magee, *Heterocycles* **15**, 713 (1981); D. M. Harrison and P. Quinn, *J. Chem. Soc., Chem. Commun.*, 879 (1983).
62. A. H. Jackson and A. E. Smith, *Tetrahedron* **21**, 989 (1965).
63. B. W. Bycroft and W. Landon, *Chem. Commun.*, 967 (1970).
64. S. Inada, K. Nagai, Y. Takayanagi, and M. Okazaki, *Bull. Chem. Soc., Jpn.* **49**, 833 (1976).
65. P. G. Sammes and A. C. Weedon, *J. Chem. Soc., Perkin Trans. 1*, 3053 (1979).
66. J. K. Allen, K. D. Barrow, and A. J. Jones, *J. Chem. Soc., Chem. Commun.*, 280 (1979).
67. C. Fuganti, P. Grasselli, and G. P.-Fantoni, *Tetrahedron Lett.*, 2453 (1979).
68. K. Arai, S. Sato, S. Shimizu, K. Nitta, and Y. Yamamoto, *Chem. Pharm. Bull.* **29**, 1510 (1981).

69. A. J. Birch and R. A. Russell, *Tetrahedron* **28**, 2999 (1972).
70. P. S. Steyn, *Tetrahedron Lett.*, 3331 (1971); P. S. Steyn, *Tetrahedron* **29**, 107 (1973).
71. P. S. Steyn and R. Vleggaar, *Phytochemistry* **15**, 355 (1976).
72. A. J. Birch and J. J. Wright, *Tetrahedron* **26**, 2329 (1970).
73. R. Ritchie and J. E. Saxton, *J. Chem. Soc., Chem. Commun.*, 611 (1975); R. Ritchie and J. E. Saxton, *Tetrahedron* **37**, 4295 (1981).
74. T. Kametani, N. Kanaya, and M. Ihara, *J. Am. Chem. Soc.* **102**, 3974 (1980).
75. C. L. Deyrup and C. M. Allen, Jr., *Phytochemistry* **14**, 971 (1975).
76. A. W. Sangster and K. L. Stuart, *Chem. Rev.*, 69 (1965).
77. J. Coetzer, *Acta Crystallog., Sect. B* **B29**, 685 (1973).
78. A. J. Hutchison and Y. Kishi, *J. Am. Chem. Soc.* **101**, 6786 (1979).
79. B. J. Wilson and C. H. Wilson, *Science* **144**, 177 (1964).
80. M. Yamazaki, S. Suzuki, and K. Miyaki, *Chem. Pharm. Bull.* **19**, 1739 (1971); M. Yamazaki, H. Fujimoto, T. Kawasaki, E. Okuyama, and T. Kuga, Proc. 19th Symp. on Chemistry of Natural Products, Hiroshima, Japan (1975).
81. M. Yamazaki, H. Fujimoto, and E. Okuyama, *Tetrahedron Lett.*, 2861 (1976).
82. M. Yamazaki, H. Fujimoto, and E. Okuyama, *Chem. Pharm. Bull.* **25**, 2554 (1977).
83. M. Yamazaki, S. Suzuki, and K. Kukita, *J. Pharm. Dyn.* **2**, 119 (1979); M. Yamazaki, S. Suzuki, and N. Ozaki, *J. Pharm. Dyn.* **6**, 748 (1983); S. Suzuki, K. Kikkawa, and M. Yamazaki, *J. Pharm. Dyn.* **7**, 935 (1984).
84. M. Yamazaki, K. Sasago, and K. Miyaki, *J. Chem. Soc., Chem. Commun.*, 408 (1974).
85. M. Yamazaki, H. Fujimoto, T. Akiyama, U. Sankawa, and Y. Iitaka, *Tetrahedron Lett.*, 27 (1975).
86. D. T. Dix, J. Martin, and C. E. Moppett, *J. Chem. Soc., Chem. Commun.*, 1168 (1972).
87. M. Yamazaki, H. Fujimoto, and T. Kawasaki, *Tetrahedron Lett.*, 1241 (1975).
88. N. Eickman, J. Clardy, R. J. Cole, and J. W. Kirksey, *Tetrahedron Lett.*, 1051 (1975).
89. R. J. Cole, J. W. Kirksey, J. H. Moore, B. R. Blankenship, U. L. Diener, and N. D. Davis, *Appl. Microbiol.* **24**, 248 (1972).
90. H. W. Schroeder, R. J. Cole, H. Hein, Jr., and J. W. Kirksey, *Appl. Microbiol.* **29**, 857 (1975).
91. J. Fayos, D. Lokensgard, J. Clardy, P. J. Cole, and J. W. Kirksey, *J. Am. Chem. Soc.* **96**, 6785 (1974).
92. R. T. Gallagher and G. C. M. Latch, *Appl. Environ. Microbiol.* **117**, 405 (1977).
93. J. B. Day and P. G. Mantle, *Appl. Environ. Microbiol.* **43**, 514 (1982).
94. M. Yamazaki, H. Fujimoto, and T. Kawasaki, *Chem. Pharm. Bull.* **28**, 245 (1980).
95. R. J. Cole and J. W. Kirksey, *J. Agric. Food Chem.* **21**, 927 (1973).
96. P. S. Steyn, R. Vleggaar, and C. J. Rabie, *Phytochemistry* **20**, 538 (1981).
97. R. J. Cole, J. W. Kirksey, J. W. Dorner, D. M. Wilson, J. C. Johnson, Jr., A. N. Johnson, D. M. Bedell, J. P. Springer, K. K. Chexal, J. C. Clardy, and R. H. Cox, *J. Agric. Food Chem.* **25**, 826 (1977); D. S. P. Patterson, B. J. Shreeve, B. A. Roberts, and S. M. MacDonald, *Appl. Environ. Microbiol.* **42**, 916 (1981).
98. J. Willingale, K. P. W. C. Perera, and P. G. Mantle, *Biochem. J.* **214**, 991 (1983).
99. K. P. W. C. Perera, J. B. Day, P. G. Mantle, and L. Rodrigues, *Appl. Environ. Microbiol.* **43**, 503 (1982).
100. M. Uramoto, M. Tanabe, K. Hirotsu, and J. Clardy, *Heterocycles* **17**, 349 (1982).
101. R.J. Cole, Proceeding US-Jpn Conf. on Mycotoxin in Human and Animal Health, University of Maryland, College Park.
102. B. J. Wilson, C. H. Wilson, and A. W. Hayes, *Nature (London)* **220**, 77 (1968).
103. R. F. Bond, J. C. Boeyens, C. W. Holzapel, and P. S. Steyn, *J. Chem. Soc., Perkin Trans. 1*, 1751 (1979).

104. J. P. Springer, G. Büchi, B. Kobbe, A. L. Demain, and J. Clardy, *Tetrahedron Lett.*, 2403 (1977).
105. M. Nakagawa, H. Sugumi, S. Kodato, and T. Hino, *Tetrahedron Lett.* **22**, 5323 (1981).
106. R. D. Stipanovic, H. W. Schroeder, and H. Hein, Jr., *Lloydia* **39**, 158 (1976).
107. A. Hirano, Y. Iwai, R. Masuma, K. Tei, and S. Omura, *J. Antibiotics* **32**, 781 (1979); Y. Konda, M. Onda, A. Hirano, and S. Omura, *Chem. Pharm. Bull.* **28**, 2987 (1980).
108. L. Maat and H. C. Beyerman, *in* "The Alkaloids" (A. Brossi, ed.), Vol. 22, p. 281. Academic Press, New York, 1984.
109. D. W. Nagel, K. G. R. Pachler, P. S. Steyn, R. Vleggaar, and P. L. Wessels, *Tetrahedron* **32**, 2625 (1976).
110. P. S. Steyn and R. Vleggaar, *J. Chem. Soc., Chem. Commun.*, 560 (1983).
111. P. M. Scott, M. A. Merrien, and J. Polonsky, *Experientia* **32**, 140 (1976); P. M. Scott and B. P. C. Kennedy, *J. Agric. Food Chem.* **24**, 865 (1976).
112. R. Weindling and O. H. Emerson, *Phytopathology* **26**, 1068 (1936).
113. G. A. Glister and T. I. Williams, *Nature (London)* **153**, 651 (1944); J. G. Kidd, *Science* **105**, 511 (1947).
114. S. Wilkinson and J. F. Spilsbury, *Nature (London)* **206**, 619 (1965).
115. J. R. Johnson, W. F. Bruce, and J. D. Dutcher, *J. Am. Chem. Soc.* **65**, 2005 (1943); W. A. Rightsel, H. G. Schneider, B. J. Sloan, P. R. Graf, F. A. Miller, Q. R. Bartz, and J. Ehrlich, *Nature (London)* **204**, 1333 (1964).
116. R. P. Mull, R. W. Townley, C. R. Scholtz, *J. Am. Chem. Soc.* **67**, 1626 (1945); A. Brcken and H. Raistrick, *Biochem. J.* **41**, 569 (1947); J. R. Johnson, A. R. Kidwai, and J. S. Warner, *J. Am. Chem. Soc.* **75**, 2110 (1953).
117. M. R. Bell, J. R. Johnson, B. S. Wildi, and R. B. Woodward, *J. Am. Chem. Soc.* **80**, 1001 (1958).
118. A. F. Beecham and J. M. Mathieson, *Tetrahedron Lett.*, 3131 (1966).
119. P. M. Hardy and B. Ridge, *in* "Progress in Organic Chemistry" (W. Carruthers and J. K. Sutherland, eds.), Vol. 8, p. 129. Butterworths, London, 1973.
120. P. G. Sammes, *in* "Progress in the Chemistry of Organic Natural Products" (W. Herz, H. Grisebach, and G. W. Kirby, eds.), Vol. 32, p. 51. Springer-Verlag, Vienna and New York, 1975.
121. S. Johne and D. Groger, *Pharmazie* **32**, 1 (1977).
122. A. Tayler, *in* "Microbial Toxins" (S. Kadis and C. Ciegler, eds.), Vol. 7, p. 337. Academic Press, New York, 1971.
123. G. W. Kirby and D. J. Robins, *in* "The Biosynthesis of Mycotoxins-A Study in Secondary Metabolites" (P. S. Steyn, ed.), p. 301. Academic Press, New York, 1980.
124. T. Fukuyama and Y. Kishi, *J. Am. Chem. Soc.* **98**, 6723 (1976); T. Fukuyama, S. Nakatsuka, and Y. Kishi, *Tetrahedron,* **37**, 2045 (1981).
125. G. W. Kirby, D. J. Robins, M. A. Sefton, and R. R. Talekar, *J. Chem. Soc., Perkin Trans. 1*, 119 (1980).
126. R. J. Suhadolnik and R. G. Chenoweth, *J. Am. Chem. Soc.* **80**, 4391 (1958); A. K. Bose, K. S. Khanchandani, R. Tavares, and P. T. Funke, *J. Am. Chem. Soc.* **90**, 3593 (1968); J. A. Winstead and R. J. Suhadolnik, *J. Am. Chem. Soc.* **82**, 1644 (1960).
127. J. C. MacDonald and G. P. Slater, *Can. J. Biochem.* **53**, 475 (1975).
128. J. D. Bu'Lock and A. P. Ryles, *Chem. Commun.*, 1404 (1970).
129. J. D. Bu'Lock and C. Leigh, *J. Chem. Soc., Chem. Commun.*, 628 (1975).
130. G. W. Kirby, G. L. Patrick, and D. J. Robins, *J. Chem. Soc., Perkin Trans. 1*, 1336 (1978).
131. G. W. Kirby, W. Lösel, P. S. Rao, D. J. Robins, M. A. Sefton, and R. R. Talekar, *J. Chem. Soc., Chem. Commun.*, 810 (1983).
132. P. J. Curtis, D. Greatbanks, B. Hesp, A. F. Cameron, and A. A. Freer, *J. Chem. Soc., Perkin Trans. 1*, 180 (1977).

133. J. P. Ferezou, A. Quesneau-Thierry, M. Barbier, A. Kollmann, and J. F. Bousquet, *J. Chem. Soc., Perkin Trans. 1,* 113 (1980).
133a. P. Waring, R. D. Eichner, U. T. Palni, and A. Mullbacher, *Tetrahedron Lett.* 735 (1986).
134. R. Nagarajan, L. L. Huckstep, D. H. Lively, D. C. DeLong, M. M. Marsh, and N. Neuss, *J. Am. Chem. Soc.* **90,** 2980 (1968).
135. N. Neuss, R. Nagarajan, B. B. Molloy, and L. L. Huckstep, *Tetrahedron Lett.,* 4467 (1968).
136. P. A. Miller, P. W. Trown, W. Fulmor, G. O. Morton, and J. Karliner, *Biochem. Biophys. Res. Commun.* **33,** 219 (1968).
137. S. Kamata, H. Sakai, and A. Hirota, *Agric. Biol. Chem.* **47,** 2637 (1983).
138. R. Nagarajan, N. Neuss, and M. M. Marsh, *J. Am. Chem. Soc.* **90,** 6518 (1968).
139. J. W. Moncrief, *J. Am. Chem. Soc.* **90,** 6517 (1968).
140. D. B. Cosulich, N. R. Nelson, and J. H. van den Hende, *J. Am. Chem. Soc.* **90,** 6519 (1968).
141. M. I. P. Boente, G. W. Kirby, and D. J. Robins, *J. Chem. Soc., Chem. Commun.,* 619 (1981).
142. G. W. Kirby, D. J. Robins, and W. M. Stark, *J. Chem. Soc., Chem. Commun.,* 812 (1983).
143. R. Baute, G. Deffieux, M. A. Baute, M. J. Filleau, and A. Neveu, *Tetrahedron Lett.,* 3943 (1976).
144. C. M. Mass, P. S. Steyn, P. H. van Rooyen, and C. J. Robie, *J. Chem. Soc., Chem. Commun.,* 350 (1982).
145. H. Seya, K. Nozawa, S. Nakajima, K. Kawai, and S. Udagawa, *J. Chem. Soc., Perkin Trans. 1.,* 109 (1986).
146. Y. Maebayashi, Y. Horie, and M. Yamazaki, *Proc. Jpn. Assoc. Mycotoxicol.* **20,** 28 (1984).
147. K. Sakata, H. Masago, A. Sakurai, and N. Takahashi, *Tetrahedron Lett.* **23,** 2095 (1982).
148. D. H. Berg, R. P. Massing, M. M. Hoehn, L. D. Boeck, and R. L. Hamill, *J. Antibiotics* **29,** 394 (1976).
149. D. H. Berg, R. L. Hamill, and M. M. Hoehn, *Chem. Abstr.* **86,** 41802g (1977); D. H. Berg, R. L. Hamill, and M. M. Hoehn, *Chem. Abstr.* **86,** 87671s (1977).
150. A. Kato, T. Saeki, S. Suzuki, K. Ando, G. Tamura, and K. Arima, *J. Antibiotics* **22,** 322 (1969).
151. K. Sakata, T. Kuwatsuka, A. Sakurai, N. Takahashi, and G. Tamura, *Agric. Biol. Chem.* **47,** 2673 (1983).
152. K. Sakata, 89th Meeting of Chubu Branch of Japan, Agricultural Chemical Society, (1984).
153. J. Clardy, J. P. Springer, G. Büchi, K. Matsuo, and R. Wightman, *J. Am. Chem. Soc.* **97,** 663 (1975).
154. J. Büchi, K. C. Luk, B. Kobbe, and J. M. Townsend, *J. Org. Chem.* **42,** 244 (1977).
155. M. Yamazaki, H. Fujimoto, and E. Okuyama, *Chem. Pharm. Bull.* **26,** 111 (1978).
156. M. Yamazaki, E. Okuyama, and Y. Maebayashi, *Chem. Pharm. Bull.* **27,** 1611 (1979).
157. J. P. Springer, *Tetrahedron Lett.,* 339 (1979).
158. G. Büchi, P. R. DeShong, S. Katsumura, and Y. Sugimura, *J. Am. Chem. Soc.* **101,** 5084 (1979).
159. T. Ohnuma, Y. Kimura, and Y. Ban, *Tetrahedron Lett.* **22,** 4969 (1981).
160. M. Nakagawa, M. Ito, Y. Hasegawa, S. Akashi, and T. Hino, *Tetrahedron Lett.* **25,** 3865 (1984).
161. M. Nakagawa, M. Ito, Y. Hasegawa, S. Akashi, M. Taniguchi, and T. Hino, *Heterocycles* **23,** 224 (1985).
162. M. Nakagawa, M. Taniguchi, M. Sodeoka, M. Ito, K. Yamaguchi, and T. Hino, *J. Am. Chem. Soc.* **105,** 3709 (1983).
163. G. A. Ellestad, P. Mirando, and M. P. Kunstmann, *J. Org. Chem.* **38,** 4204 (1973).
164. Y. Kimura, T. Hamasaki, and H. Nakajima, *Tetrahedron Lett.* **23,** 225 (1982).
164a. R. S. L. Chang, V. J. Lotti, R. L. Monaghan, J. Birnbaum, E. O. Stapley, M. A. Goetz, G. Albers-Schönberg, A. A. Patchett, J. M. Liesch, O. D. Hensens, and J. M. Springer, *Science* **230,** 177 (1985); M. A. Goetz, M. Lopez, R. L. Monaghan, R. S. L. Chang, V. J. Lotti, and

T. B. Chen, *J. Antibiotics* **38**, 1634 (1985); J. M. Leisch, O. D. Hensens, J. P. Springer, R. S. L. Chang, and V. J. Lotti, *J. Antibiotics* **38**, 1638 (1985).

165. T. Fehr and W. Acklin, *Helv. Chim. Acta* **49**, 1907 (1966).
166. R. T. Gallagher, J. Clardy, and B. J. Wilson, *Tetrahedron Lett.* **21**, 239 (1980); R. T. Gallagher and B. J. Wilson, *Mycopathologia* **66**, 183 (1978).
167. R. J. Cole, J. W. Dorner, J. P. Springer, and R. H. Cox, *J. Agric. Food Chem.* **29**, 293 (1981).
168. R. J. Cole, J. W. Kirksey, and J. M. Wells, *Can. J. Chem.* **20**, 1159 (1974); J. P. Springer, J. Clardy, J. M. Wells, R. J. Cole, and J. W. Kirksey, *Tetrahedron Lett.*, 2531 (1975).
169. R. J. Cole, J. W. Dorner, J. A. Lansden, R. H. Cox, C. Pape, B. Cunfer, S. S. Nickolson, and D. M. Bedell, *J. Agric. Food Chem.* **25**, 1197 (1977).
170. R. T. Gallagher, J. Finer, J. Clardy, A. Leutwiler, F. Weibel, W. Acklin, and D. Arigoni, *Tetrahedron Lett.* **21**, 235 (1980).
171. A. B. Smith and R. Mewshaw, *J. Am. Chem. Soc.* **107**, 1769 (1985).
172. M. Tanabe, 26th IUPAC Symposium, Tokyo (1977).
173. W. Acklin, F. Weibel, and D. Arigoni, *Chimia* **31**, 63 (1977).
174. A. E. de Jesus, P. S. Steyn, F. R. van Heerden, and R. Vleggaar, *J. Chem. Soc., Perkin Trans. 1*, 697 (1984).
175. R. F. Vesonder, L. Tjarks, W. Rohwedder, and D. O. Kieswetter, *Experientia* **36**, 1308 (1908).
176. A. E. de Jesus, P. S. Steyn, F. R. van Heerden, R. Vleggaar, P. L. Wessels, and W. E. Hull, *J. Chem. Soc., Chem. Commun.*, 289 (1981); idem, *J. Chem. Soc., Perkin Trans. 1*, 1847 (1983).
177. R. T. Gallagher, A. D. Hawkes, P. S. Steyn, and R. Vleggaar, *J. Chem. Soc., Chem. Commun.*, 614 (1984).
178. A. E. de Jesus, C. P. Gorst-Allman, P. S. Steyn, F. R. van Heerden, R. Vleggaar, P. L. Wessels, and W. E. Hull, *J. Chem. Soc., Perkin Trans. 1*, 1863 (1983).
179. R. T. Gallagher, T. McCabe, K. Hirotsu, J. Clardy, J. Nicholson, and B. J. Wilson, *Tetrahedron Lett.* **21**, 243 (1980).
180. S. Danishefsky, S. Chackalmannil, P. Harrison, M. Silvestri, and P. Cole, *J. Am. Chem. Soc.* **107**, 2474 (1985).
181. D. Brewer, W. A. Jerram, D. Meiler, and A. Taylor, *Can. J. Microbiol.* **16**, 433 (1970); D. Brewer, W. A. Jerram, and A. Taylor, *Can. J. Microbiol.* **14**, 861 (1968); D. Meiler and A. Taylor, *Can. J. Microbiol.* **17**, 83 (1971).
182. S. Sekita, K. Yoshihira, S. Natori, S. Udagawa, T. Muroi, Y. Sugiyama, H. Kurata, and M. Umeda, *Can. J. Microbiol.* **27**, 766 (1981); S. Udagawa, T. Muroi, H. Kurata, S. Sekita, K. Yoshihira, S. Natori, and M. Umeda, *Can. J. Microbiol.* **25**, 170 (1979); S. Sekita, *Chem. Pharm. Bull.* **31**, 2998 (1983).
183. M. A. O'Leary and J. R. Hanson, *Tetrahedron Lett.* **23**, 1855 (1982); M. A. O'Leary, J. R. Hanson, and B. L. Yeoh, *J. Chem. Soc., Perkin Trans. 1*, 567 (1984).
184. Y. Yamamoto, K. Nishimura, and N. Kiriyama, *Chem. Pharm. Bull.* **24**, 1853 (1976).
185. W. A. Jerram, A. G. McInnes, W. S. G. Maass, D. G. Smith, A. Taylor, and J. A. Walter, *Can. J. Chem.* **53**, 727 (1975).
186. Y. Yamamoto, N. Kiriyama, S. Shimizu, and S. Koshimura, *Gann* **67**, 623 (1976); S. Shimizu, Y. Yamamoto, and S. Koshimura, *Chem. Pharm. Bull.* **30**, 1896 (1982); S. Shimizu, Y. Yamamoto, J. Inagaki, and S. Koshimura, *Gann* **73**, 642 (1982).
187. K. Arai, K. Masuda, K. Kiriyama, N. Nitta, Y. Yamamoto, and S. Shimizu, *Chem. Pharm. Bull.* **29**, 961 (1981).
188. K. Arai, S. Shimizu, Y. Taguchi, and Y. Yamamoto, *Chem. Pharm. Bull.* **29**, 991 (1981).
189. E. Schwenner, Ph.D. Thesis, Westfälischen Wilhelms University, Münster (1982).
190. A. W. Hason, *Acta Cryst., Sect. B* **B33**, 293 (1977).

191. K. Arai, S. Shimizu, and Y. Yamamoto, *Chem. Pharm. Bull.* **29,** 1005 (1981).
192. K. Arai, Ph.D. Thesis, Kyoto University, Kyoto (1984).
193. S. Ohmomo, M. Sugita, and M. Abe, *Nippon Nogei Kagaku Kaishi* **47,** 57 (1973).
194. R. T. Gallagher, J. L. Richard, H. M. Stahr, and R. J. Cole, *Mycopathologia* **66,** 31 (1978); K. C. Luk, B. Kobbe, and J. M. Townsend, *Appl. Environ. Microbiol.* **33,** 211 (1977).
195. T. Yokota, A. Sakurai, S. Iriuchijima, and N. Takahashi, *Agric. Biol. Chem.* **45,** 53 (1981).
196. C. W. Holzapfel, *Tetrahedron* **24,** 2101 (1968); C. W. Holzapfel, R. D. Hutchison, and D. C. Wilkins, *Tetrahedron* **26,** 5239 (1970).
197. I. F. H. Purchase, *Toxicol. Appl. Pharmacol.* **18,** 114 (1971).
198. C. H. Holzapfel and P. J. Gildenhuys, *S. Afr. J. Chem.* **30,** 125 (1977); *Chem. Abstr.* **87,** 151941e (1977).
199. M. Somei, S. Tokutake, and C. Kaneko, *Chem. Pharm. Bull.* **31,** 2153 (1983); A. P. Kozikowski, M. N. Greco, and J. P. Springer, *J. Am. Chem. Soc.* **106,** 6873 (1984); H. Muratake and M. Natsume, *Heterocycles* **23,** 1111 (1985).
200. C. W. Holzapfel and D. C. Wilkins, *Phytochemistry* **10,** 351 (1971); R. M. McGrath, P. S. Steyn, and N. P. Ferreira, *J. Chem. Soc., Chem. Commun.,* 812 (1973); A. E. de Jesus, P. S. Steyn, R. Vleggaar, G. W. Kirby, M. J. Varley, and N. P. Ferreira, *J. Chem. Soc., Perkin Trans. 1,* 3292 (1981); A. A. Chalmers, C. P. Gorst-Allman, and P. S. Steyn, *J. Chem. Soc., Chem. Commun.,* 1367 (1982).
201. J. C. Schabort, D. C. Wilkins, C. W. Holzapfel, D. J. J. Potgieter, and A. W. Neitz, *Biochim. Biophys. Acta* **250,** 311 (1971); J. C. Schabort and D. J. J. Potgieter, *Biochim. Biophys. Acta* **250,** 329 (1971); D. J. Steenkamp, J. C. Schabort, and N. P. Ferreira, *Biochim. Biophys. Acta* **309,** 440 (1973); R. M. McGrath, P. S. Steyn, N. P. Ferreira, and D. C. Neethling, *Bioorg. Chem.* **4,** 11 (1976).
202. C. H. Holzapfel, *in* "The Biosynthesis of Mycotoxins-A Study in Secondary Metabolites" (P. S. Steyn, ed.), Academic Press, New York, 1980.
203. P. S. Steyn and P. L. Wessels, *Tetrahedron Lett.,* 4707 (1978); M. J. Nolte, P. S. Steyn, and P. L. Wessels, *J. Chem. Soc., Perkin Trans. 1,* 1057 (1980).
204. P. S. Steyn and C. J. Rabie, *Phytochemistry* **15,** 1 (1976).
205. M. Binder and Ch. Tamm, *Angew. Chem. Int. Ed. Engl.* **12,** 370 (1973).
206. D. C. Aldridge, J. J. Armstrong, R. N. Speake, and W. B. Turner, *Chem. Commun.,* 26 (1967); D. C. Aldridge, J. J. Armstrong, R. N. Speake, and W. B. Turner, *J. Chem. Soc., C,* 1667 (1967).
207. S. Masamune, Y. Hayase, W. Schilling, K. W. Chan, and G. S. Bates, *J. Am. Chem. Soc.* **99,** 6756 (1977); D. Sherling, I. Csendes, and Ch. Tamm, *Helv. Chim. Acta* **59,** 914 (1976); T. Schmidlin, W. Zurchner, and Ch. Tamm, *Helv. Chim. Acta* **64,** 235 (1981); J. Auerbach and S. M. Weinreb, *J. Org. Chem.* **40,** 3311 (1975); M. Kim and S. M. Weinreb, *Tetrahedron Lett.,* 579 (1979); M. Kim, J. E. Starrett, and S. M. Weinreb, *J. Org. Chem.* **46,** 5383 (1981).
208. W. Rothweiler and Ch. Tamm, *Experientia* **22,** 750 (1966); W. Rothweiler and Ch. Tamm, *Helv. Chim. Acta* **53,** 696 (1970).
209. G. Stork, Y. Nakahara, Y. Nakahara, and W. J. Greenlee, *J. Am. Chem. Soc.* **100,** 7775 (1978); G. Stork and E. Nakamura, *J. Am. Chem. Soc.* **105,** 5510 (1983).
210. M. Flashner, J. Rasmussen, B. H. Patwardhan, and S. W. Tanenbaum, *J. Antibiotics* **35,** 1345 (1982).
211. I. Yahara, F. Harada, S. Sekita, K. Yoshihira, and S. Natori, *J. Cell Biol.* **92,** 62 (1982).
212. Y. Sato, Y. Saito, T. Tezuka, S. Sekita, K. Yoshihira, and S. Natori, *J. Pharm. Dyn.* **5,** 418 (1982).
213. S. B. Carter, *Endeavour* **113,** 77 (1972).
214. A. L. Demain, N. A. Hunt, V. Malik, B. Kobbe, H. Hawkins, K. Matsuo, and G. N. Wogan, *Appl. Environ. Microbiol.* **31,** 138 (1976).

215. D. C. Aldridge, B. F. Burrows, and W. B. Turner, *J. Chem. Soc., Chem. Commun.*, 148 (1972).

216. T. Fujishima, M. Ishikawa, H. Ishige, H. Yoshino, J. Ohishi, and S. Ikegami, *Hakko Kogaku Kaishi* **57,** 15 (1979).

217. D. C. Aldridge, D. Greatbanks, and W. B. Turner, *J. Chem. Soc., Chem. Commun.*, 551 (1973).

218. G. Büchi, Y. Kitaura, S.-S. Yuan, H. E. Wright, J. Clardy, A. L. Demain, T. Glinsukon, N. Hunt, and G. N. Wogan, *J. Am. Chem. Soc.* **95,** 5423 (1973).

219. P. S. Steyn, F. R. van Heerden, and C. J. Rabie, *J. Chem. Soc., Perkin Trans. 1*, 541 (1982).

220. W. Heberel, W. Loeffler, and W. A. Koenig, *Arch. Microbiol.* **100,** 73 (1974).

221. W. K.-Schierlein and E. Kupfer, *Helv. Chim. Acta* **62,** 1501 (1979).

222. K. N.-Laves and M. Dobler, *Helv. Chim. Acta* **65,** 1426 (1982).

223. A. Probst and Ch. Tamm, *Helv. Chim. Acta* **64,** 2056 (1981).

224. F. Caesar, K. Jansson, and E. Mutschler, *Pharm. Acta Helv.* **44,** 676 (1969); *Chem. Abstr.* **72,** 12937k (1970).

225. A. Isogai, T. Horii, A. Suzuki, S. Murakoshi, K. Ikeda, S. Sato, and S. Tamura, *Agric. Biol. Chem.* **39,** 739 (1975).

226. T. Iwamoto, A. Hirota, S. Shima, H. Sakai, and A. Isogai, *Agric. Biol. Chem.* **49,** 3323 (1985).

227. C. G. Casinovi, G. Grandolini, R. Mercantini, N. Oddo, R. Olivieri, and A. Tonolo, *Tetrahedron Lett.*, 3175 (1968).

228. J. A. Findlay and L. Radics, *J. Chem. Soc., Perkin Trans. 1*, 2071 (1972).

229. J. A. Findlay and D. Kwan, *J. Chem. Soc., Perkin Trans. 1*, 2962 (1972).

230. N. N. Girotra, Z. S. Zelawski, and N. L. Wendler, *J. Chem. Soc., Chem. Commun*, 566 (1976).

231. P. S. White, J. A. Findlay, and W. H. J. Tam, *Can. J. Chem.* **56,** 1904 (1978).

232. N. N. Girotra and N. L. Wendler, *Tetrahedron Lett.*, 4793 (1979).

233. J. A. Findlay, J. Krepinsky, A. Shum, C. G. Casinovi, and L. Radics, *Can. J. Chem.* **55,** 600 (1977).

234. N. N. Girotra, A. A. Patchett, and N. L. Wendler, *Heterocycles* **6,** 1299 (1977).

235. T. Sassa and Y. Onuma, *Agric. Biol. Chem.* **47,** 1155 (1983); T. Sassa, *Agric. Biol. Chem.* **47,** 1417 (1983).

236. C. G. Casinovi, G. Grandolini, and L. Radics, 12th IUPAC Int. Symp. Chem. Nat. Prod., Tenerife, Abstr. A7 (1980).

237. E. A. Kaczka, E. L. Dulaney, C. O. Gitterman, H. B. Woodruff, and K. Folkers, *Biochem. Biophys. Res. Commun.* **14,** 452 (1964).

238. K. Kodama, H. Kusakabe, H. Machida, Y. Midorikawa, S. Shibuya, A. Kuninaka, and H. Yoshino, *Agric. Biol. Chem.* **43,** 2375 (1979).

239. P. W. K. Woo, H. W. Dion, S. M. Lange, L. F. Dahl, and L. J. Durham, *J. Heterocyclic. Chem.* **11,** 641 (1974).

240. Y. Kaneko, *Agric. Biol. Chem.* **29,** 965 (1965); Y. Kaneko, *Nippon Nogei Kagaku Kaishi* **40,** 227 (1966).

241. Y. Kaneko and M. Sanada, *Hakko Kougaku Zasshi* **47,** 8 (1969).

242. S. Matsuura, M. Yamamoto, and Y. Kaneko, *Bull. Chem. Soc., Jpn.* **45,** 492 (1972).

243. A. G. Brown, *J. Chem. Soc., C,* 2572 (1970).

244. A. G. Brown and T. C. Smale, *J. Chem. Soc., Perkin Trans. 1*, 65 (1972).

245. M. Cole and G. N. Rolinson, *Appl. Microbiol.* **24,** 660 (1972).

246. J. H. Birkinshaw, M. G. Kalyanpur, and C. E. Stickings, *Biochem. J.* **86,** 237 (1963); G. M. Strunz and W. -Y. Ren, *Can. J. Chem.* **54,** 2862 (1976); A. Srinivasan, K. D. Richards, and R. K. Olsen, *Tetrahedron Lett.*, 891 (1976).

247. M. Yamazaki, H. Fujimoto, Y. Ohta, Y. Iitaka, and A. Itai, *Heterocycles* **15,** 889 (1981).
248. J. A. Ballantine, D. J. Francis, C. H. Hassall, and J. L. C. Wright, *J. Chem. Soc., C,* 1175 (1970).
249. Y. Maebayashi and M. Yamazaki, *Chem. Pharm. Bull.* **33,** 4296 (1985).
250. B. F. Burrows, S. D. Mills, and W. B. Turner, *Chem. Commun.,* 75 (1965).
251. P. W. Brian, G. W. Elson, H. G. Hemming, and M. Radley, *Nature (London)* **207,** 998 (1965).
252. B. F. Burrows and W. B. Turner, *J. Chem. Soc., C,* 255 (1966).
253. A. J. Birch, B. J. McLoughlin, H. Smith, and J. Winter, *Chem. Ind.,* 840 (1960).
254. E. Candlish, L. D. La Croix, and A. M. Unroce, *Biochemistry* **8,** 182 (1969); J. M. Hylin and H. Matsumoto, *Arch. Biochem. Biophys.* **93,** 542 (1961).
255. W. Rothe, *Pharmazie* **5,** 190 (1950); W. Rothe, *Deut. Med. Wochschr.* **79,** 1080 (1954).
256. I. Hagedorn and H. Tonjes, *Pharmazie* **12,** 567 (1957); *idem, ibid.* **11,** 409 (1956).
257. I. Hagedorn, U. Eholzer, and A. Luttringhaus, *Chem. Ber.* **93,** 1584 (1960).
258. A. Takatsuki, S. Suzuki, K. Ando, G. Tamura, and K. Arima, *J. Antibiotics* **21,** 671 (1968); K. Ando, S. Suzuki, A. Takatsuki, K. Arima, and G. Tamura, *J. Antibiotics* **21,** 582 (1968).
259. N. Kitahara and A. Endo, *J. Antibiotics* **34,** 1556 (1981).
260. C. Takahashi, S. Sekita, K. Yoshihira, and S. Natori, *Chem. Pharm. Bull.* **24,** 2317 (1976).
261. Y. Ito, K. Ohtsubo, K. Yoshihira, S. Sekita, S. Natori, and H. Tsunoda, *Jpn. J. Exp. Med.* **48,** 187 (1978).
262. K. Umezawa, K. Yoshida, M. Okamoto, M. Iwami, H. Tanaka, M. Kohsaka, and H. Imanaka, *J. Antibiotics* **37,** 469 (1984).
263. K. Ohkuma, K. Anzai, and S. Suzuki, *J. Antibiotics* **15,** 115 (1962).
264. M. Ishida, T. Hamasaki, and Y. Hatsuda, *Agric. Biol. Chem.* **36,** 1847 (1972); *idem, ibid.* **39,** 2181 (1975).
265. H. Achenbach and H. Grisebach, *Z. Naturforsch.* **20,** 137 (1965); H. Achenbach and F. Konig, *Experientia* **27,** 1250 (1971); H. Achenbach and F. Konig, *Chem. Ber.* **105,** 784 (1972).
266. R. B. Herbert and J. Mann, *J. Chem. Soc., Chem. Commun.,* 1008 (1983); R. B. Herbert and J. Mann, *J. Chem. Soc., Chem. Commun.,* 1474 (1984).
267. S. Pfeifer, H. Bar, and J. Zarnacck, *Pharmazie* **27,** 536 (1972).
268. M. R. Hagadone and P. J. Scheuer, *J. Am. Chem. Soc.,* **106,** 2447 (1984).

—— Chapter 5 ——

DAPHNIPHYLLUM ALKALOIDS

Shosuke Yamamura

Department of Chemistry
Faculty of Science and Technology
Keio University
Hiyoshi, Yokohama, Japan

I. Introduction

In 1975, the first review on daphniphyllum alkaloids appeared in this treatise (Vol. 15, p. 41) (*1*). As described therein, a number of alkaloids with a novel structure have been isolated from the family Daphniphyllaceae, particularly from the species *Daphniphyllum macropodum* Miq. and *D. teijsmanni* Zollinger. From a structural point of view, these alkaloids are derived from six molecules of mevalonic acid through a squalene-like intermediate and have been mainly divided into five types of nitrogen heterocyclic skeletons. However, further investigation of the alkaloidal component of the plant *Daphniphyllum gracile* Gage (*2*) growing in the Trust Territory of New Guinea resulted in the isolation of nine alkaloids with a new nitrogen heterocyclic skeleton (*3,4*), strongly suggesting that the daphniphyllum alkaloids so far obtained must be classified into six skeletal types. From a chemotaxonomical point of view; the species *Daphniphyllum glaucescens* Bl. (*5*) growing in Taiwan is more similar to the plant *D. gracile* than to the species in Japan, because it contains mainly the sixth type of alkaloid, represented by daphnigracine and already structurally determined (*3,4*).

Many grazing cattle have long been known to die from eating poisonous plants such as bracken fern and others. In 1978, a disease with jaundice, colic, and photophobia as the main symptoms broke out among cattle in Hokkaido, the northern island of Japan. It was demonstrated by Sonoda *et al.* to be plant poisoning caused by *Daphniphyllum humile* Maxim. (Ezo-Yuzuriha) (*6*), from which some new alkaloids were isolated in addition to several known ones. Pharmacological properties are also described briefly in Section III.

From a viewpoint of organic synthesis, these alkaloids with their challenging, complex structure seem to be quite attractive to organic chemists. However, none of them has not yet been synthesized, although some attempts have been made. In this chapter, all of the newly isolated alkaloids from *D. gracile*

THE ALKALOIDS, VOL. 29

TABLE I

Alkaloids of *Daphniphyllum gracile, D. humile,* and *D. teijsmanni*

Name	Empirical formula	Melting point (°C)	Ref.
Yuzurine	$C_{24}H_{37}O_4N$	229–230 (B MeI)	3,8
Daphnigracine	$C_{24}H_{37}O_4N$	198–199 (B MeI)	3
Daphnigraciline	$C_{23}H_{35}O_4N$	76–78	3
Oxodaphnigracine	$C_{24}H_{35}O_5N$	116–117	3
Oxodaphnigraciline	$C_{23}H_{33}O_5N$	107–109	3
Epioxodaphnigraciline	$C_{23}H_{33}O_5N$	102–104	3
Daphgracine	$C_{24}H_{35}O_4N$	—	4
Daphgraciline	$C_{23}H_{33}O_4N$	—	4
Hydroxydaphgraciline	$C_{23}H_{33}O_5N$	—	4
Daphniteijsmanine	$C_{30}H_{49}O_3N$	228–232	9
Isodaphnilactone-B	$C_{22}H_{31}O_2N$	196–198 (B MeI)	7
Zwitterionic alkaloid	$C_{22}H_{33}O_3N$	247–248	10
Deoxyyuzurimine	$C_{27}H_{37}O_6N$	132–134	7
Yuzurimine-C	$C_{23}H_{29}O_5N$	186–187	11

and *D. humile* will be described, including their spectral and chemical properties (Table I).

II. Structures and Properties

From a structural point of view, the 38 alkaloids so far obtained from the Daphniphyllaceae are classified into six types of nitrogen heterocyclic skeleton represented by daphniphylline (**1**), secodaphniphylline (**2**), daphnilactone-A (**3**), daphnilactone-B (**4**), yuzurimine (**5**), and daphnigracine (**6**), as shown in Fig. 1. Of the alkaloids, the 9 daphnigracine-type alkaloids are distinct from the other types in the following point: they have no 2-azabicyclo[3.3.1]nonane ring system in their structure.

A. DAPHNIPHYLLINE-TYPE ALKALOIDS

Of the many alkaloids that have been isolated from *Daphniphyllum macropodum* Miq. growing in Japan, seven bases belong to the daphniphylline-type group of alkaloids, whose structures have been already elucidated on the basis of spectral and chemical evidence together with an X-ray crystallographic analysis (*1*). Chemotaxonomically, it is interesting that daphniphylline (**1**) has also been found in *D. gracile* growing in New Guinea, although it occurs only as a minor component (*4*).

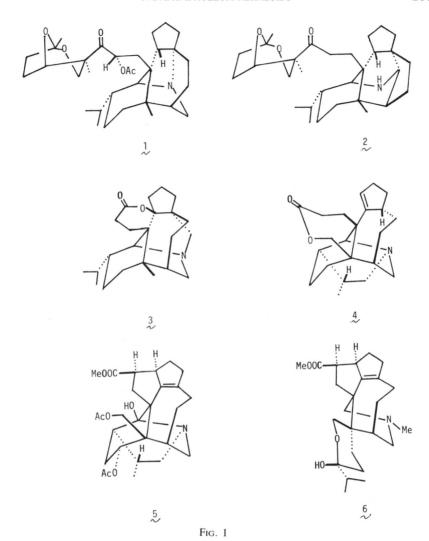

FIG. 1

Structurally, remarkable variations in the remaining partial structure aside from the main nitrogen heterocyclic skeleton are found in this group. Codaphniphylline (7) and daphmacrine (8) (Fig. 2) have a 2,8-dioxa- and a 6-oxabicyclo[3.2.1]octane ring system, respectively. Biogenetically, these two ring systems are derived from a squalene-like intermediate, as demonstrated in Scheme 1, and accordingly they have been synthesized from geraniol (12).

The benzoate (9) derived from geraniol (13) was readily converted to the desired cyclization product (10) in four steps (Scheme 2) [(1) Hg(OCOCF$_3$)$_2$ in CH$_3$NO$_2$ ($-15°$C, 3 hr); (2) aqueous NaCl (room temp., 20 hr); (3) NaBH$_4$–O$_2$

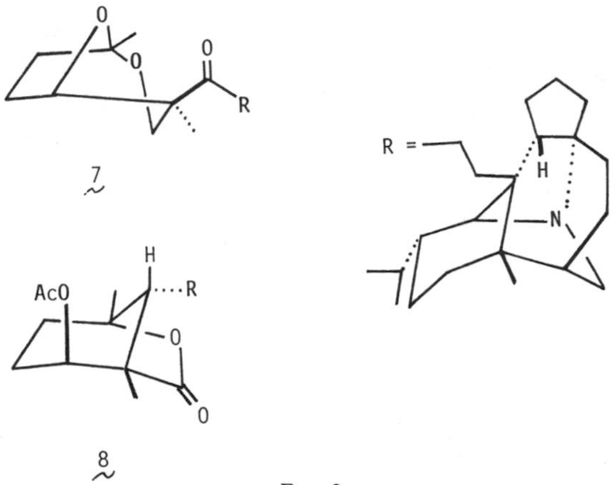

FIG. 2

in DMF (0°C, 30 min); (4) PCC in CH_2Cl_2 (room temp., 4 hr) (60% overall yield)]. Compound **10** was further converted to a thioacetate (**11**) in four steps [(1) sat. NH_3 in MeOH (0°C, 5 hr); (2) PDC in DMF (0°C, 1 hr); (3) $(CH_2SH)_2$–BF_3 Et_2O) in CH_2Cl_2 (0°C, 5.5 hr); (4) K_2CO_3 in MeOH (0°C, 2 hr) (64% overall yield)]. On oxidation with PDC in DMF (room temp., 1 hr), this thio-acetal was spontaneously cyclized to the desired lactone [mp 162–164°C; ν_{max} (film) 1780 and 1715 cm^{-1}] in 54% yield, which was then subjected to $NaBH_4$

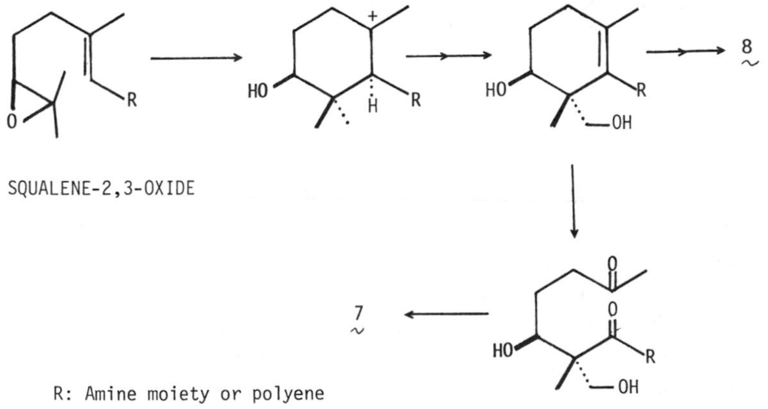

SQUALENE-2,3-OXIDE

R: Amine moiety or polyene

SCHEME 1. Biogenesis of the oxygen containing parts of codaphniphylline (**7**) and daphma-crine (**8**).

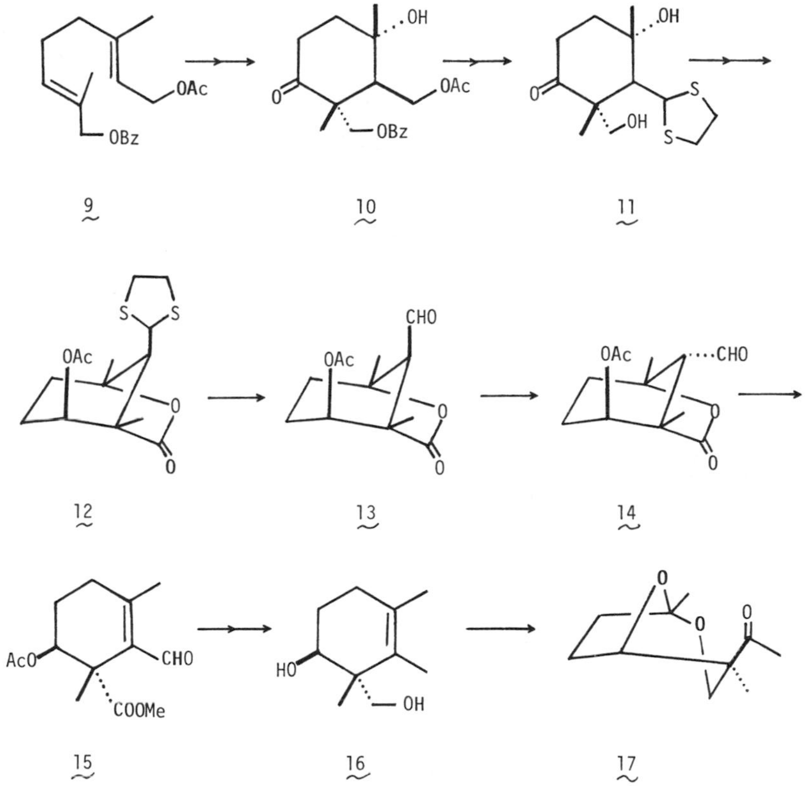

SCHEME 2. Syntheses of some oxygenated fragments.

reduction followed by acetylation to give rise to the corresponding acetate [**12**; mp 146.5–147.5°C; ν_{max} (film) 1770 and 1735 cm^{-1}] in 78% yield. This lactone was treated with TTN in CHCl$_3$–MeOH (room temp., 1 hr) and then with CF$_3$COOH in CHCl$_3$ containing a small amount of water (0°C, 1 hr) to afford an aldehyde (**13**), which was subjected to isomerization with SiO$_2$ (room temp., overnight) followed by methylation to afford an α,β-unsaturated aldehyde (**15**) in 48% overall yield. In this case, the desired aldehyde (**14**) is too unstable to be isolated and seems to be simultaneously converted to **15**. Finally, this ester (**15**) was readily converted to a diol (**16**) in two steps [(1) excess NaBH$_4$ in EtOH (room temp., overnight); (2) H$_2$–Pt/C in MeOH (room temp., 1.5 hr)], from which the desired acetal (**17**) was synthesized, in 72% yield, on ozonization of **16** in MeOH (−60°C, 30 min).

Many research groups have attempted synthesis of the nitrogen heterocyclic skeletons. No successful results have been reported yet.

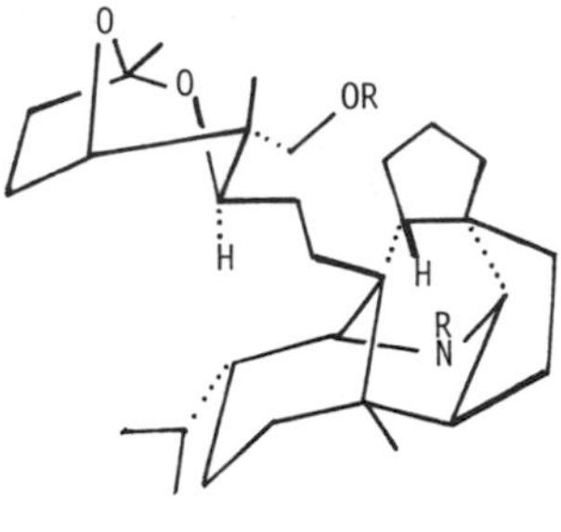

18   R = H
19   R = Ac

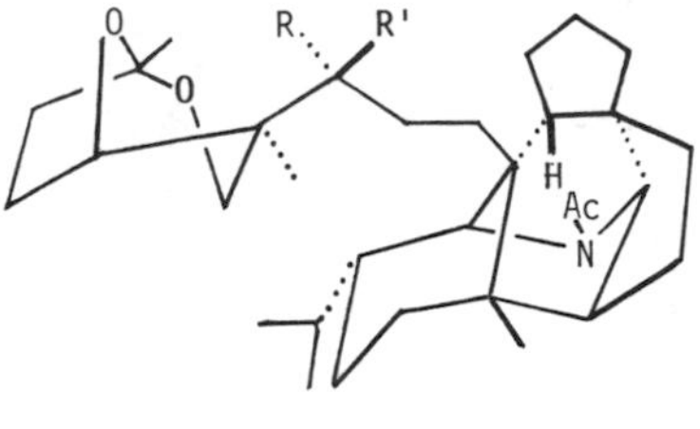

21   R·R' = O
22   R = OH, R'= H
23   R = OMs, R'= H

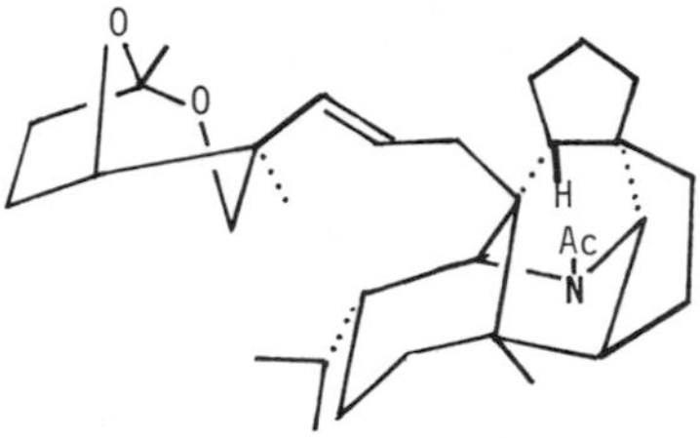

24

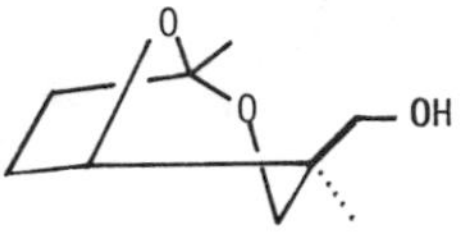

20

FIG. 3

B. SECODAPHNIPHYLLINE-TYPE ALKALOIDS

Four alkaloids belong to a group of secodaphniphylline-type ones represented by secodaphniphylline (**2**), which is considered to be the central biosynthesis precursor to the triterpenoid alkaloids that constitute the daphniphyllum alkaloids. Of them, daphniteijsmanine (**18**) (Fig. 3) has been newly isolated from the species *Daphniphyllum teijsmanni*, and its structure also been elucidated as follows (*9*).

Daphniteijsmanine (**18**; mp 228–232°C) has the molecular formula $C_{30}H_{49}O_3N$ [m/z 471 (M$^+$) and 286]. Judging from the ^{1}H-NMR and mass spectra ($\delta 0.76, 0.92, 0.94, 2.54$, and 3.06; m/z 286], this alkaloid must have the same nitrogen heterocyclic skeleton as that of secodaphniphylline (**2**), as seen in Table II. However, a remarkable difference is seen in the nonnitrogen-containing moiety of the structure: although both **2** and **18** display two methyl singlets ($\delta 0.89$ and 1.42 in **2**; $\delta 1.12$ and 1.48 in **18**), the latter compound has an NMR signal at $\delta 3.42$ (2*H*, s) arising from a hydroxymethyl group, which is shifted to $\delta 3.80$ on acetylation with Ac_2O–pyridine giving rise to *N*-acetyl-daphniteijsmanine acetate [**19**; $C_{34}H_{53}O_5N$; m/z 555 (M$^+$) and 328]. In addition, two singlets at $\delta 3.37$ and 4.22 are assignable to the two protons of the

TABLE II
^{1}H-NMR Spectra of Secodaphniphylline (2), Daphniteijsmanine (18), and Acetal 20^a

2	18	20
0.77 (3H, s)	0.76 (3H, s)	0.74 (3H, s)
0.89 (3H, s)	1.12 (3H, s)	
0.89 (3H, d, J = 6 Hz)	0.92 (3H, d, J = 6.2 Hz)	
0.90 (3H, d, J = 6 Hz)	0.94 (3H, d, J = 6.2 Hz)	
1.42 (3H, s)	1.48 (3H, s)	1.46 (3H, s)
2.51 (1H, d, J = 4.2 Hz)	2.54 (1H, d, J = 4.4 Hz)	
2.6–2.9 (2H, m)		
3.01 (1H, br s)	3.06 (1H, br s)	
	3.37 (1H, dd, J = 8, 5 Hz)	
3.49 (1H, d, J = 12 Hz)		3.40 (1H, d, J = 12 Hz)
4.23 (1H, dd, J = 12, 2 Hz)		3.65 (1H, br d, J = 12 Hz)
	3.42 (2H, s)	3.78 (2H, br s)
4.63 (1H, m)	4.22 (1H, m)	4.18 (1H, m)

^{a} In CDCl$_3$.

(—CH—O—) groups. From these data, daphniteijsmanine seems to have an acetal grouping similar to that of the degradation product (20) of daphniphylline (1) except for its stereochemistry (14). Particularly, the stereochemistry at the carbon atom bearing both methyl and hydroxymethyl groups is different [δ1.12 and 3.42 in 18; δ0.74 and 3.78 in 20]. In addition, the proton corresponding to the signal at δ3.37 is in an axial configuration, as depicted in the structure 18, which has been unambiguously determined by the following chemical evidence.

N-Acetylsecodaphniphylline (21) was reduced with NaBH$_4$ in THF (room temp., overnight) to give a hydroxy compound [22; mp 216–217°C; C$_{32}$H$_{51}$O$_4$N m/z 513 (M$^+$); δ (CDCl$_3$) 4.06 (1H, m)], in 72% yield. When heated in AcOH at 110°C for 4.5 hr, compound 22 was recovered completely in spite of conditions vigorous enough to cleave an acetal ring, suggesting that the newly formed hydroxyl group in 22 adopts an unfavorable configuration. Thus, 22 was further treated with mesyl chloride–pyridine (room temp., 2.5 hr) to give the corresponding mesylate [23; δ3.10 (3H, s) and 5.47 (1H, m)] which was also heated in AcOH under the same condition as described above to afford N-acetyldaphniteijsmanine (19) and an alkene (24) in 28 and 27% yields, respectively. The formation process of 19 from 23 is shown in Scheme 3.

In connection with Scheme 1, the nonnitrogen-containing moiety of daphniteijsmanine (18) may be formed in vivo, as seen in Scheme 4. The direct precursor of 18 must take the stereochemistry as depicted in 25, because compound 22 has been recovered on acid treatment.

The daphniphyllum alkaloids with complex structures seem to be quite attractive to synthetic organic chemists. However, no successful synthesis of these

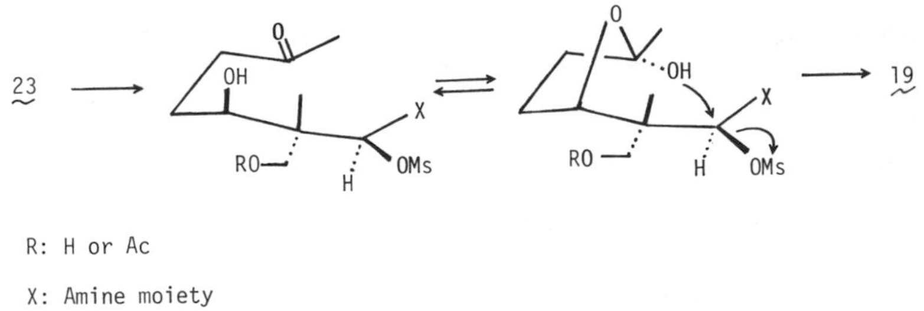

R: H or Ac

X: Amine moiety

SCHEME 3. Reaction mechanism of the acetal formation of N-acetyldaphniteijsmanine acetate (19).

alkaloids has yet been carried out. In 1983, Orban and Turner (*15*) attempted a synthesis of secodaphniphylline (**2**), on the basis of a retrosynthesis analysis of **2** (**2** → **25** → **26** → **27** + **28**), which included a Diels–Alder reaction between the siloxy diene (**27**) and 2-carbomethoxy-3-methylcyclohexanone (**28**), as shown in Scheme 5. The Diels–Alder reaction between **27** and **28** was carried out by heating them (equimolar amounts) at 140°C for 66 hr in a dry, base-washed glass tube (which had been purged of oxygen, evacuated, and then sealed) to afford a desired endo adduct (**26**) in 20% yield. Furthermore, on fluoride-mediated alkylation (*n*-Bu$_4$NF, −15°C, 0.5 hr), the adduct (**26**) was stereo- and regioselectively converted to the corresponding alkene (**29**), whose predicted stereostructure was unambiguously confirmed by an X-ray cyrstallographic analysis. Finally, the intermediate aldehyde (**30**) derived from **29** [(1) O$_3$ in EtOH–CH$_2$Cl$_2$; (2) Zn in AcOH] would readily undergo the desired

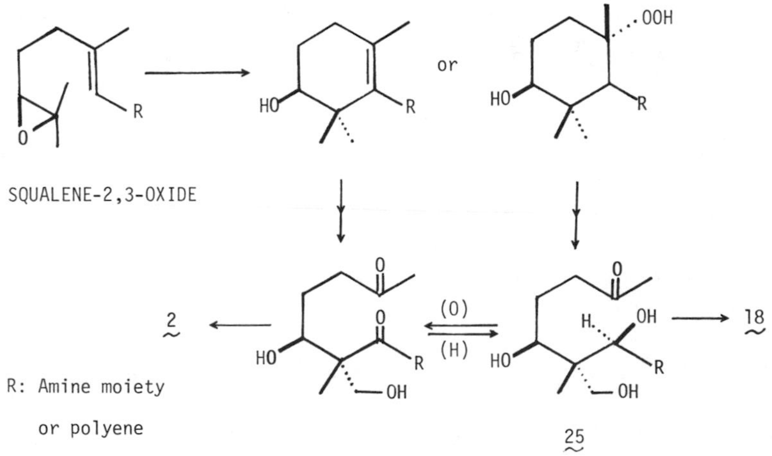

SQUALENE-2,3-OXIDE

R: Amine moiety

or polyene

SCHEME 4. Biogenesis of the oxygen heterocyclic skeletons of secodaphniphylline (**2**) and daphniteijsmanine (**18**).

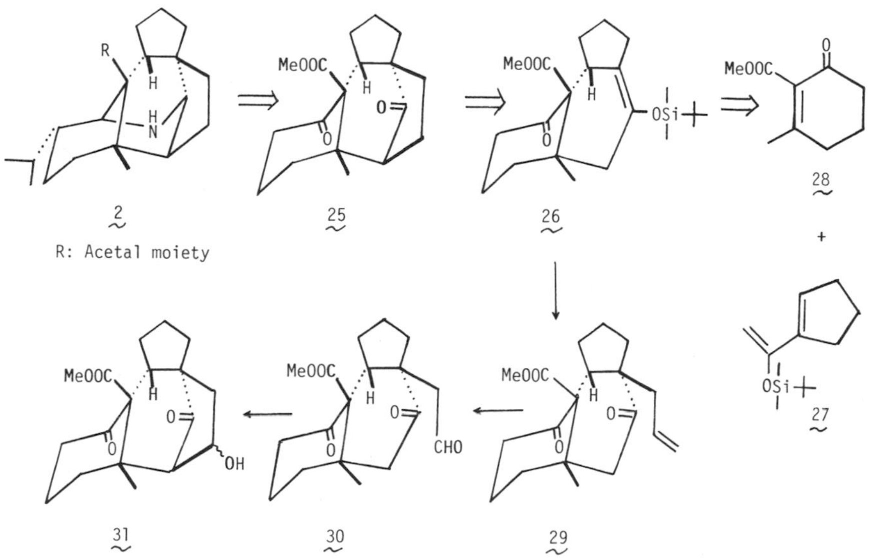

SCHEME 5. Retrosynthesis of secodaphniphylline (**2**).

aldol cyclization (n-Bu$_4$NF in THF, 25°C) to complete the main carbocyclic framework (**31**) of secodaphiphylline (**2**). Further synthetic study of **2** starting from **31** is desirable.

C. DAPHNILACTONE-A

As described in the previous review (*1*), the stereostructure of daphnilactone-A (**3**) was determined by X-ray crystallographic analysis, and its biogenesis was also demonstrated. Since then, no other new alkaloids with the same nitrogen heterocyclic skeleton have yet been found in nature.

D. DAPHNILACTONE-B-TYPE ALKALOIDS

Daphilactone-B (**4**) is a major component of several alkaloids isolated from the fruits of three kinds of plants growing in Japan (*1*). Further examination of the alkaloidal components in the fruits of the species *Daphniphyllum teijsmanni* Zollinger (*10*) led to the isolation of a zwitterionic alkaloid (**32**) (Fig. 4), which was recently proved to be only one major component of the water-soluble fraction of the methanol extract (*16*). This alkaloid [**32**; mp 247–248°C; C$_{22}$H$_{33}$O$_3$N; m/z 359 (M$^+$)] produces the following spectral data: ν_{max} (Nujol) 3300, 1595 and 1575sh cm^{-1}; δ (CD$_3$OD–D$_2$O) 1.09 (3H, d, J = 6 Hz), 3.4–4.05 (5H, complex), 3.70 (1H, d, J = 10.5 Hz), 4.27 (1H, d, J = 10.5 Hz), and

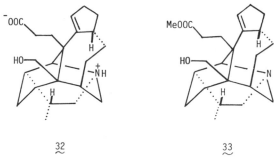

FIG. 4

5.74 (1*H*, br s, Wh 5 Hz). From these spectral data, indicating the presence of a
hydroxymethyl group ($\delta 3.70$ and 4.29) and a carboxylate (ν_{max} 1595 and 1575sh
cm^{-1}), this zwitterionic alkaloid is regarded as the hydration product of
daphnilactone-B (**4**). In fact, **32** was readily converted to a mixture of daphnilactone-B (**4**) and a methyl ester (**33**) on treatment with anhydrous MeOH containing HCl gas (room temp., 5 hr) followed by NaHCO$_3$.

   In connection with a cattle poisoning event caused by the plant *Daphniphyllum
humile*, further examination of its toxic substances led to the isolation of isodaphnilactone-B [**34**; C$_{22}$H$_{31}$O$_2$N; *m/z* 341 (M$^+$); ν_{max} (film) 1735 cm^{-1} and
no hydroxyl group] which was characterized as the corresponding methiodide
(mp 196–198°C; C$_{23}$H$_{34}$O$_2$NI) (*7*). The spectral data of this new alkaloid are
quite similar to those of daphnilactone-B (**4**). Particularly, the ^{1}H-NMR spectra
of both compounds are quite similar to each other except for slight differences in
their chemical shifts, as seen in Table III. These data show clearly that the newly
isolated compound is a double bond isomer of **4,** and its structure must be
depicted as **34** on the basis of the NMR signal at $\delta 5.78$, which is assignable to the
olefinic proton of a trisubstituted double bond accommodated in a five-membered ring. On the other hand, the possibility that the double bond is included in
the seven-membered ring can be ruled out, because its NMR spectrum would

TABLE III
^{1}H-NMR Spectra^a of Daphnilactone-B (**4**)
and Isodaphnilactone-B (**34**)

4	34
1.00 (3*H*, d, *J* = 6 Hz)	1.08 (3*H*, d, *J* = 7 Hz)
3.63 (1*H*, d, *J* = 13 Hz)	3.75 (1*H*, d, *J* = 13 Hz)
4.73 (1*H*, d, *J* = 13 Hz)	4.50 (1*H*, d, *J* = 13 Hz)
5.67 (1*H*, br s, Wh = 5.5 Hz)	5.78 (1*H*, br s, Wh = 5 Hz)

^{a} In CDCl$_3$.

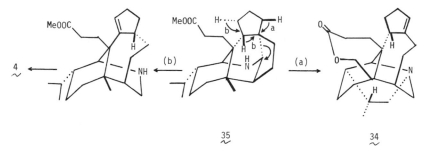

SCHEME 6. Biogenesis of daphnilactone-B (**4**) and isodaphnilactone-B (**34**).

be expected to show a pretty broad signal due to the olefinic proton (Wh $\simeq$ 14 Hz) (*17*).

Presumably, daphnilactone-B (**4**) and isodaphnilactone-B (**34**) are both produced from a common intermediate such as methyl homosecodaphniphyllate (**35**), as shown in Scheme 6.

E. YUZURIMINE-TYPE ALKALOIDS

Of 10 structurally known alkaloids belonging to the yuzurimine group, deoxyyuzurimine [**36**; mp 132–124°C; $C_{27}H_{37}O_6N$; m/z 471 (M^+)] (Fig. 5) was isolated as a minor component from the species *Daphniphyllum humile* (*7*). The spectral data of **26** were completely identical to those of the reduction product obtained from yuzurimine (**5**) on zinc reduction in acetic acid (90–95°C, 2 hr).

On the basis of spectral data (IR, ^{1}H NMR, and mass) coupled with co-occurrence of yuzurimine (**5**), a tentative structure was proposed for yuzurimine-C (mp 186–187°C; $C_{23}H_{29}O_5N$), as cited in the previous review (*1*). Further chemical and spectral evidence was obtained, supporting the structure **37** (Fig. 6) for yuzurimine-C, as follows (*11*). Yuzurimine-C methiodide (**38**) was readily

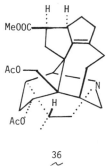

36

FIG. 5

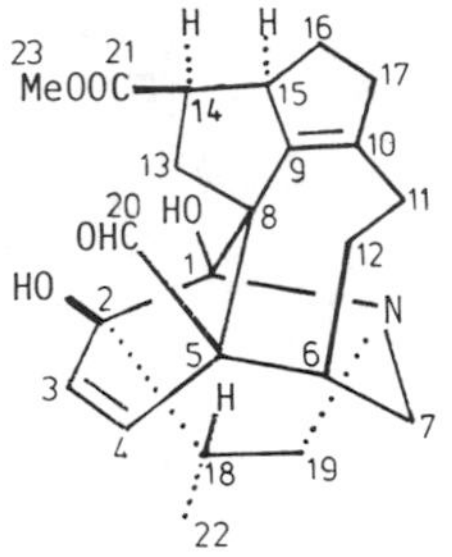

FIG. 6

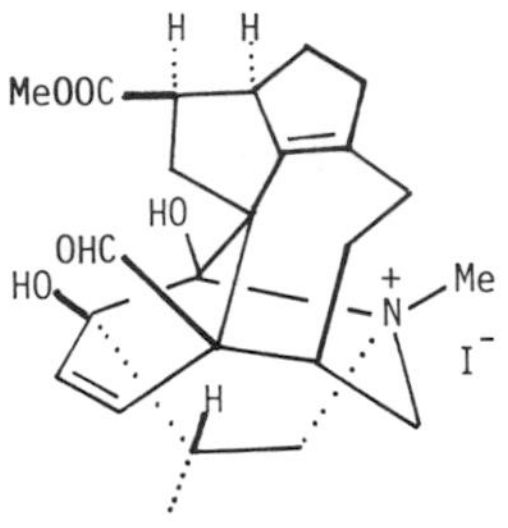

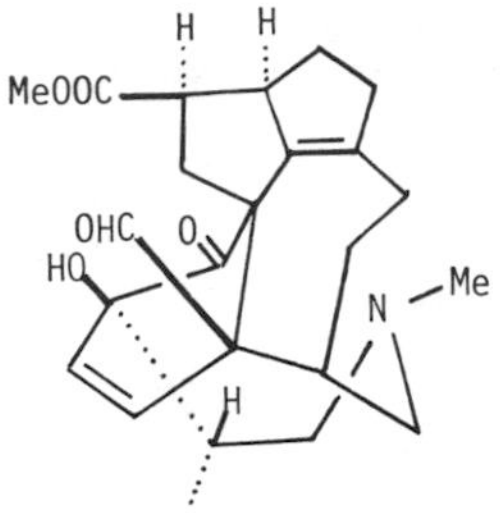

converted to a keto amine (**39;** mp 199–200°C; $C_{24}H_{31}O_5N$) only by contact
with aqueous alkaline solution. In the IR spectrum of **39,** the extraordinary low
C=O frequency (ν_{max} 1640 cm^{-1}) arising from the newly formed keto group
must result from proximity of the nitrogen atom. On oxidation with NaIO$_4$ in
aqueous MeOH–THF (room temp., 16 hr), yuzurimine-C (**37**) was converted to
a keto lactam (**40;** mp 179–181°C; $C_{23}H_{27}O_5N$) in good yield. Particularly,
important information on the disubstituted double bond can be obtained from the
^{1}H-NMR spectrum of **40,** in which there are two doublets due to the
—(H)C=C(H)— (cis) grouping [δ5.61 (1*H*, d, *J* = 14 Hz) and 6.01 (1*H*, d, *J* =
14 Hz)]. Clearly, the double bond in **40** (*J* = 14 Hz) must be included in a
medium ring and not in the original six-membered one (*J* = 10 Hz in **37, 38,**
and **39**).

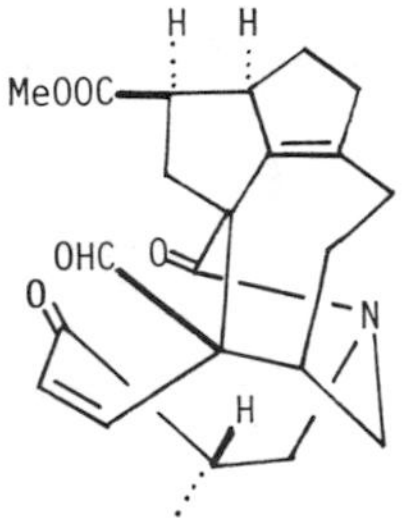

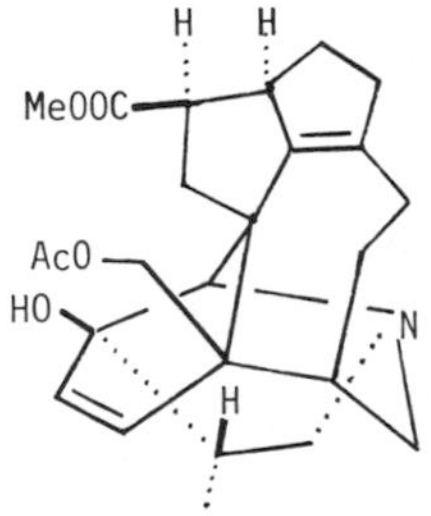

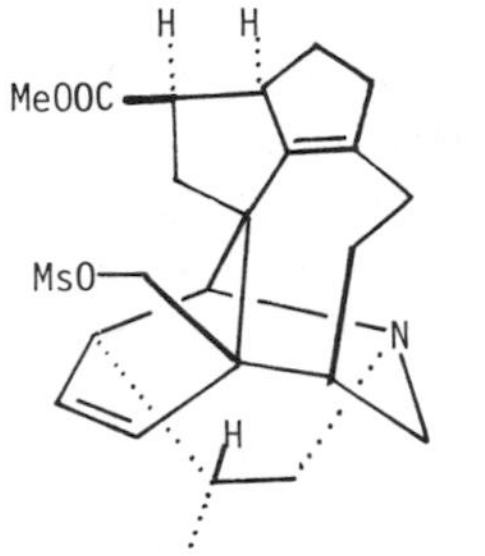

Yuzurimine-C has an aldehyde group [ν_{max} 1723 cm^{-1}; δ9.99 (1*H*, s)] instead of the acetoxymethyl group as in yuzurimine (**5**). On reduction with limited
amounts of LiAlH$_4$ in THF (room temp., 5 hr) followed by acetylation, **37** was
converted to the corresponding acetate (**41;** mp 171–173°C; $C_{25}H_{33}O_5N$) having
^{1}H-NMR signals due to the newly formed acetoxymethyl group [δ2.09 (3*H*, s)
and 4.48 (2*H*, br s)]. Furthermore, the structure **37** for yuzurimine-C is supported by comparison of its ^{13}C-NMR spectrum with those of yuzurimine (**5**),
deoxyyuzurimine (**36**), and its derivative (**42**).

TABLE IV

^{13}C-NMR Spectra[a] of Yuzurimine (**5**), Deoxyyuzurimine (**36**),
Yuzurimine-C (**37**), and Mesylate **42**

Functional group	5	36	37	42	Assignment
CH$_3$ —	14.7	15.5	11.4	18.4	C-22
	51.0	51.1	51.2	51.3	C-23
				37.2	MeSO$_3$ —
— CH$_2$ —	25.2	25.4	24.8	25.1	C-3, C-11, C-12, C-13,
	27.1	26.9	25.6	25.6	C-16, C-17
	27.3	27.2	29.6	27.7	
	28.7	28.2	35.5	39.9	
	37.4	39.1	42.4	43.3	
	43.1	42.8			
	58.6	58.1	49.1	51.7	C-7, C-19, C-20
	64.4	64.8	56.8	63.7	
	67.0	67.1		71.6	
C = O			207.0		C-20
$\begin{array}{c}\|\\ - CH -\\ \|\end{array}$	34.3	35.1[b]	36.0	32.9	C-6
	34.3	37.6[b]	40.6	35.1[b]	C-18
	42.2[b]	38.3[b]		39.2[b]	C-2
	43.1[b]	42.8	43.0	42.9	C-15
	57.5	54.0	55.9	54.8	C-14
	72.9	73.4			C-4
		67.0		66.8	C-1
$\begin{array}{c}\|\\ - C -\\ \|\end{array}$	45.0	41.1	48.7	39.9[b]	C-5
	52.1	46.4	55.2	42.9[b]	C-8
	96.8		91.6		C-1
			80.1		C-2
COOMe	175.3	175.0	175.6	175.2	C-21
C = C	136.7	133.4	125.0	138.6	C-9, C-10
	144.0	144.5	136.9	140.2	
			125.0	127.8	C-3, C-4
			132.5	131.5	
AcO —	21.0	21.1			MeCOO —
	21.0	21.1			MeCOO —
	170.1	170.1			MeCOO —
	170.8	170.8			MeCOO —

[a] In CDCl$_3$.
[b] Assignment of chemical shifts for close-lying peaks may be reversed.

Signals of a quaternary carbon at C-1 connected to the hydroxyl and the nitrogen atom are found in the spectra of **5** (δ96.8) and **37** (δ91.6), whereas the spectra of **36** and **42** lack the corresponding signal (Table IV). Nine methylene and six methine signals are observed in the spectrum of **5**, while seven methylene, four methine, and one carbonyl signals are found in the case of yuzurimine-C (**37**). Furthermore, four olefinic signals are observed in both **37** and **42**. Clearly, the presence of the hydroxyl group at C-2 in yuzurimine-C (**37**) rationalizes the upfield shift of the C-19 and C-22 signals and the downfield shift of the C-18 signal. As seen in Table IV, the other signals show a one to one correspondence for each carbon atom. Of the 10 known yuzurimine-type alkaloids, yuzurimine-C (**37**) is the most highly oxygenated one.

F. DAPHNIGRACINE-TYPE ALKALOIDS

As cited in the previous review (*1*), a number of daphniphyllum alkaloids are structurally divided into five types of nitrogen heterocyclic skeleton represented by daphniphylline (**1**), secodaphniphylline (**2**), daphnilactone-A (**3**), daphnilactone-B (**4**), and yuzurimine (**5**). All of these compounds, however, possess in common the 2-azabicyclo[3.3.1]nonane system, part of which constitutes a portion of the bicyclo[5.3.0]decane system [A] (*3*). On the other hand, the nine novel alkaloids described in this section constitute the group of daphnigracine-type alkaloids, which have no 2-azabicyclo[3.3.1]nonane system and differ from the other structurally known bases. Among them, yuzurine (**43**) was first isolated as one of the minor components from the species *Daphniphyllum macropodum*, and the remaining ones were obtained from *D. gracile*. An extensive survey was made of the occurrence of alkaloids in New Guinea plants (*2*). According to general procedure, each alkaloidal component was extracted from the leaves and the bark of *D. gracile*, independently. Interestingly, analytical TLC of the leaf extracts did not show any spot corresponding to daphniphylline (**1**) or yuzurimine (**5**), both of which had been already isolated from the Japanese species as major products. Careful separation of this crude material afforded five alkaloids [daphnigracine (**6**), daphnigraciline (**44**), oxodaphnigracine (**45**), oxodaphnigraciline (**46**), and epioxodaphnigraciline (**47**)], whereas daphgracine (**48**), daphgraciline (**49**), and hydroxydaphgraciline (**50**) were obtained from the bark of the plant along with daphniphylline (**1**) and daphgraciline (**44**). From a chemotaxonomical viewpoint, it is noted that **1** has been found in the bark of *D. gracile*, although it has not been detected in the leaves of the same plant.

1. Yuzurine

Yuzurine [**43**; $C_{24}H_{37}O_4N$; m/z 403 (M$^+$)] (Fig. 7) is a colorless viscous liquid [ν_{max} 1740 cm^{-1}; δ (CDCl$_3$) 0.85 (3*H*, t, *J* = 7.4 Hz), 2.17 (3*H*, s), 3.21

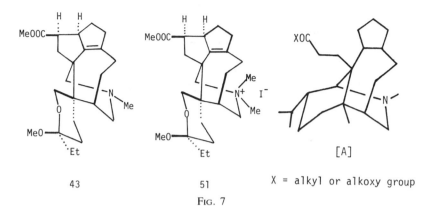

FIG. 7

43   51   X = alkyl or alkoxy group

(3H, s), 3.64 (3H, s), and 3.93 (2H, s)] which was characterized as the corresponding methiodide (51; mp 229–230°C; $C_{25}H_{40}O_4NI$). On the basis of its spectral data, indicating the presence of one each of ethyl, N-methyl, and methoxyl groups, the carbon skeleton of yuzurine is considerably different from those of the other daphniphyllum alkaloids. Thus, the complex structure of yuzurine (43) was elucidated by means of an X-ray crystallographic analysis of the methiodide (51).

2. Daphnigracine and Daphnigraciline

Daphnigracine [6; $C_{24}H_{37}O_4N$; m/z 403 (M$^+$) and 385 (M$^+$ − 18)] (Fig. 8) is a colorless viscous liquid, which has one each of hydroxyl and methoxycarbonyl

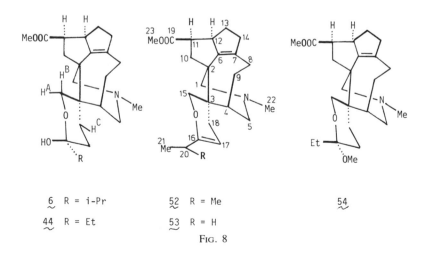

6   R = i-Pr   52   R = Me   54

44 R = Et   53   R = H

FIG. 8

TABLE V

^{1}H-NMR Spectra[a] of Daphnigracine (**6**) and Daphnigraciline (**44**)

6	44
0.93 (6*H*, d, *J* = 7 Hz)	0.94 (3*H*, t, *J* = 7 Hz)
2.15 (3*H*, s)	2.19 (3*H*, s)
3.62 (3*H*, s)	3.63 (3*H*, s)
3.89 (1*H*, dd, *J* = 12.5, 2 Hz)	3.90 (1*H*, dd, *J* = 12.5, 2 Hz)
4.32 (1*H*, d, *J* = 12.5 Hz)	4.35 (1*H*, d, *J* = 12.5 Hz)

[a] In CDCl$_3$.

TABLE VI

^{13}C-NMR Spectra[a] of the Dehydro Compounds **52**, **53**, and **55**

Functional group	52	53	55	Assignment
CH$_3$ —	20.4 (2 Me)	11.6	11.4	C-21
	32.1	26.6	26.5	C-22
	51.0	51.0	51.5	C-23
— CH$_2$ —	27.2	27.2	21.1	C-9, C-13, C-14, C-20
	27.2	27.2	26.3	
	28.1	27.4	27.7	
		28.0	38.2[b]	
	39.8[b]	39.9[b]	38.6[b]	C-18
	42.6[b]	42.6[b]		C-8
	46.6[b]	46.6[b]	46.1	C-10
	56.5	56.4	56.1	C-5
	62.1	62.2	62.9	C-1
	69.4	69.6	69.1	C-15
— CH —	34.2	34.3	34.4	C-4
	42.6	42.6	42.6	C-12
	54.9	54.9	45.0	C-11
	27.4			C-20
— C —	36.6	36.6	37.1	C-3
	46.5	46.4	50.3	C-2
C = C	90.3	92.0	91.8	C-17
	133.6	133.5	138.6	C-7
	159.5	156.0	156.1	C-16
	145.8	145.9	183.3	C-6
C = O	175.4	175.5	173.7	C-19
			208.9	C-8

[a] In CDCl$_3$.

[b] Assignment of chemical shifts for close-lying peaks may be reversed.

groups [ν_{max} 3450 cm^{-1}; ν_{max} 1730 cm^{-1} and $\delta 3.62$ (3H, s)] and has been characterized as the corresponding methiodide (mp 198–199°C; $C_{25}H_{40}O_4NI$). Daphnigraciline [44; mp 76–78°C; $C_{23}H_{35}O_4N$; m/z 389 (M^+) and 371 (M^+ − 18)] also has both hydroxyl and methoxycarbonyl groups. As shown in Table V, the ^{1}H-NMR spectra of both alkaloids are quite similar except for the following points: an NMR signal due to an isopropyl group is observed at $\delta 0.93$ in 6, while daphnigraciline (44) has a methyl triplet at $\delta 0.94$. Furthermore, when treated with Ac_2O–AcOH (1:1), both 6 and 44 are readily converted to the corresponding dehydration products (52 and 53, respectively) in high yields.

In the ^{13}C-NMR spectra of the two dehydro compounds (52 and 53), the methyl, methylene, methine, and quaternary carbon signals are easily differentiated by measurements of their partially relaxed FT ^{13}C-NMR spectra coupled with off-resonance experiments, and the results are summarized in Table VI. In the spectrum of 52, nine methylene and four methine signals are observed, while ten methylene and three methine signals are found in that of 53.

From these data, daphnigracine and daphnigraciline must have the same carbon skeleton except for the different alkyl groups. Their structures were finally elucidated by the following chemical evidence: when treated with MeOH containing one drop of AcOH (room temp., overnight), daphnigraciline (44) was converted to yuzurine (43) in high yield, in addition to a small amount of its epimer (54), which was characterized as a methiodide (mp 144–146°C; $C_{25}H_{40}O_4NI$). The configuration of the hydroxyl group in 6 and 44 was determined by their ^{1}H-NMR spectra: a sharp doublet ($\delta 4.32$ in 6, $\delta 4.35$ in 44) arising from the geminal proton (H^A), which is in a 1,3-diaxial relationship to the hydroxyl group, is observed at lower field than the signal ($\delta 3.89$ in 6, $\delta 3.90$ in 44) of the equatorial germinal one (H^B), which can be coupled with a proton (H^C) along with "W" path.

3. Oxodaphnigracine, Oxodaphnigraciline, and Epioxodaphnigraciline

Oxodaphnigracine (45), oxodaphnigraciline (46), and epioxodaphnigraciline (47) (Fig. 9) display the following physical and spectral data: 45: mp 116–117°C, $C_{24}H_{35}O_5N$, ν_{max} 3400 br, 1730, 1685, and 1650 cm^{-1}, λ_{max} (MeOH) 253 nm (ϵ, 7500); 46: mp 107–109°C, $C_{23}H_{33}O_5N$, ν_{max} 3380 br, 1730, 1685, and 1650 cm^{-1}, λ_{max} (MeOH) 253 nm (ϵ, 8060); 47: mp 102–104°C, $C_{23}H_{33}O_5N$, ν_{max} 3400 br, 1735, 1685, and 1650 cm^{-1}, λ_{max} (MeOH) 252 nm (ϵ, 7000). The spectral data of both 45 and 46 indicate that these two alkaloids are quite similar to each other except for the alkyl groups [$\delta 0.88$ (6H, d, $J = 7$ Hz) in 45; $\delta 0.94$ (3H, t, $J = 7$ Hz) in 46] (see Table VII), similar to the case of daphnigracine (6) and daphnigraciline (44). In their UV spectra, the absorption maximum is observed at 253 nm, indicating the presence of an α,β-unsaturated

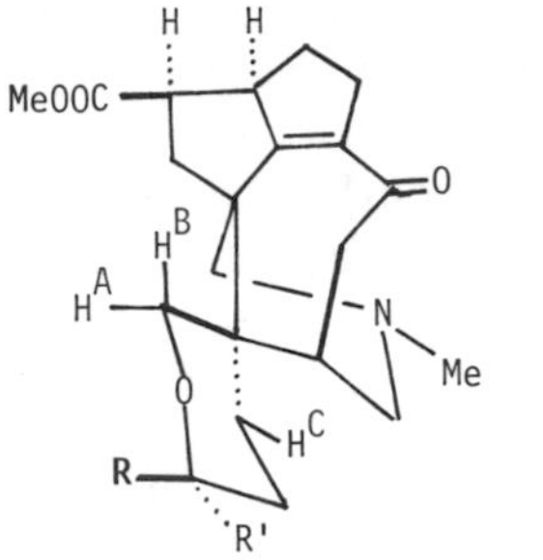

45 R = OH, R' = i-Pr

46 R = OH, R' = Et

47 R = Et, R' = OH

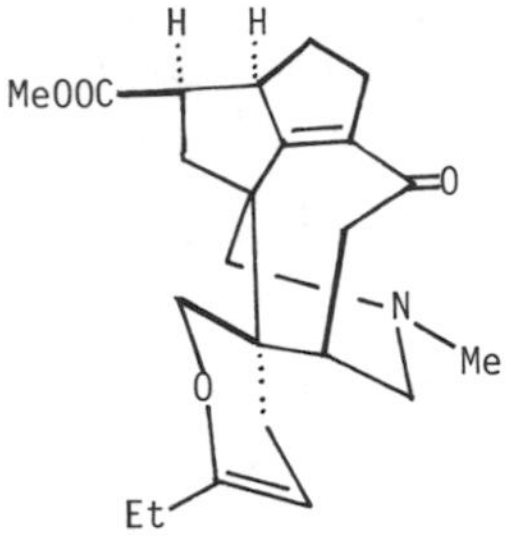

55

FIG. 9

keto grouping in which the CO group must be included in the seven-membered ring.

Epioxodaphnigraciline (**47**) is regarded as an epimer of oxodaphnigraciline (**46**) on the basis of their ¹H-NMR spectra: **47** has a sharp doublet ($\delta 3.88$) arising from the geminal proton (H^A) at slightly higher magnetic field than that of oxodaphnigraciline, as seen in Table VII, indicating that epioxodaphnigraciline must be represented by **47**. In fact, on treatment with $Ac_2O-AcOH$ (1:1), both **46** and **47** were converted in high yields to the same dehydro compound (**55**), the structure of which was confirmed by its ¹³C-NMR spectrum (see Table VI).

4. Daphgracine, Daphgraciline, and Hydroxydaphgraciline

Daphgracine (**48**) and daphgraciline (**49**) (Fig. 10) both are quite similar to each other in their structures, as judged from their spectral data [**48**: $C_{24}H_{35}O_4N$,

TABLE VII

¹H-NMR Spectra[a] of Oxodaphnigracine (**45**),
Oxodaphnigraciline (**46**), and Epioxodaphnigraciline (**47**)

45	46	47
0.88 (6H, d, J = 7 Hz)	0.94 (3H, t, J = 7 Hz)	0.92 (3H, t, J = 7.5 Hz)
2.10 (3H, s)	2.12 (3H, s)	2.15 (3H, s)
3.60 (1H, dd, J = 11, 2 Hz)	3.58 (1H, dd, J = 11.5, 3 Hz)	
3.69 (3H, s)	3.70 (3H, s)	3.61 (4H, br s)[b]
4.03 (1H, d, J = 11 Hz)	4.03 (1H, d, J = 11.5 Hz)	3.88 (1H, d, J = 12 Hz)

[a] In CDCl₃.
[b] One of the geminal protons (H^B) is included.

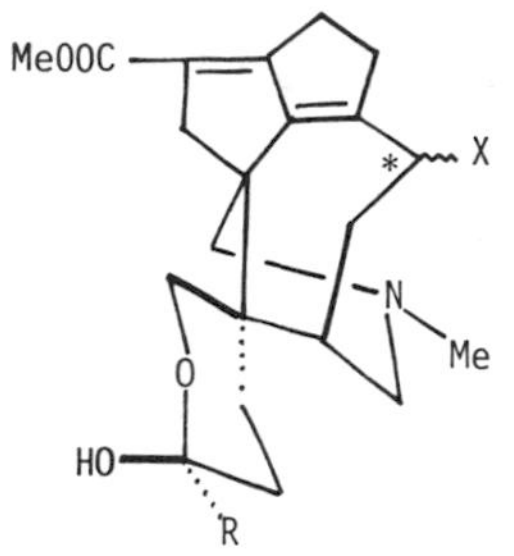

48 R = i-Pr, X = H

49 R = Et, X = H

50 R = Et, X = OH

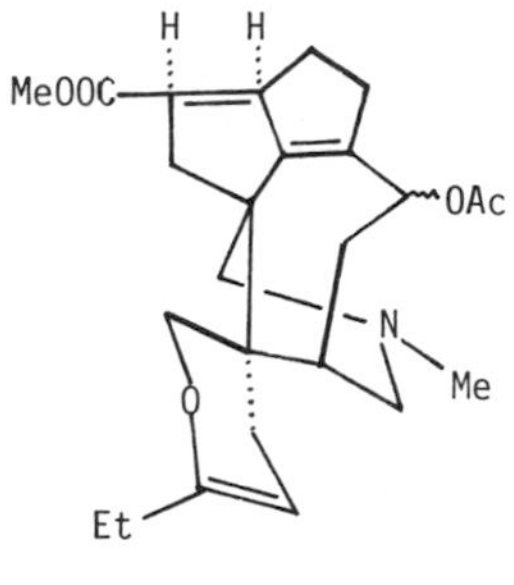

56

FIG. 10

m/z 401 (M$^+$), ν_{max} 3450, 1690, 1655, and 1620 cm^{-1}, λ_{max} (MeOH) 298 nm (ϵ, 7000); **49**: $C_{23}H_{33}O_4N$, m/z 387 (M$^+$), ν_{max} 3450, 1690, 1655, and 1620 cm^{-1}, λ_{max} (MeOH) 298 nm (ϵ, 9000)]. (see Table VIII). Remarkable differences are observed in their ^{1}H NMR spectra, however. The former has two doublets at $\delta 1.11$ and 1.17 due to an isopropyl group, while one methyl triplet is observed in **49**. Their IR and UV spectra indicate the presence of an $\alpha\beta,\gamma\delta$-unsaturated ester group, which is confirmed by the ^{13}C-NMR spectrum of daphgraciline (**49**): δ (CDCl$_3$) 7.68 (q), 21.6 (t), 25.8 (t), 26.1 (t), 27.4 (t), 27.8 (t), 32.8 (t), 35.6 (t), 35.9 (t), 42.0 (t), 42.9 (t), 46.0 (q), 46.3 (s), 50.7 (q), 55.8 (t), 61.6 (t), 62.5 (t), 96.0 (s), 115.7 (s), 149.4 (s), 153.5 (s), 166.6 (s), and 168.7 (s). At lower magnetic field, five singlets are observed at $\delta 115.7$, 149.4, 153.5, 166.6, and 168.7, four of which are assignable to fully substituted olefinic carbons. The remaining one is due to the ester carbonyl group. From these data, the stereostructure of daphgraciline must be that represented by **49**, in which the configuration of the hydroxyl group has been confirmed by the two signals at $\delta 3.90$ and 4.32. Daphgracine (**48**) also has the same stereochemistry as that of daphgraciline (**49**) based on the same reasoning.

Hydroxydaphgraciline [**50**; $C_{23}H_{33}O_5N$; m/z 403 (M$^+$)] has the same carbon skeleton as that of daphgraciline (**49**), as evidenced from its IR and UV spectra [ν_{max} 3400 br, 1695, 1660, and 1625 cm^{-1}; λ_{max} (MeOH) 294 nm (ϵ, 7000)] together with the ^{1}H-NMR spectrum. As seen in Table VIII, hydroxydaphgraciline has one secondary hydroxyl group: the signal at $\delta 4.23$ is shifted to $\delta 5.52$ (1H, br d, $J = 5$ Hz) on treatment with Ac$_2$O–pyridine, giving rise to a monoacetate (**56**). Both the chemical shift and the coupling constant of this signal suggest that the secondary hydroxyl group is located at the asterisk position in **50**.

From a biogenetic point of view, the daphnigracine-type alkaloids represented

TABLE VIII

^{1}H-NMR Spectra^a of Daphgracine (**48**), Daphgraciline (**49**), and Hydroxydaphgraciline (**50**)

48	49	50
1.11 (3H, d, J = 6 Hz)	1.12 (3H, t, J = 7 Hz)	1.12 (3H, t, J = 7 Hz)
1.17 (3H, d, J = 6 Hz)		
2.13 (3H, s)	2.16 (3H, s)	2.12 (3H, s)
3.67 (3H, s)	3.70 (3H, s)	3.68 (3H, s)
3.82 (1H, br d, J = 12 Hz)	3.90 (1H, br d, J = 12 Hz)	3.67 (1H, br d, J = 12 Hz)
4.21 (1H, d, J = 12 Hz)	4.32 (1H, d, J = 12 Hz)	4.29 (1H, d, J = 12 Hz)
		4.23 (1H, br d, J = 5 Hz)

^{a} In pyridine-d_5.

by daphnigracine (**6**) and daphnigraciline (**44**) may be produced from an intermediate such as **52** according to pathway a or b, as shown in Scheme 7 (*3*). Both daphgracine (**48**) and daphgraciline (**49**) are also formed from **53** in the same manner (*4*).

III. Pharmacological Properties

Decoctions of the bark and leaves of *Daphniphyllum macropodum* have long been used as a remedy for vermicide and asthma. This species is known to contain a number of alkaloids, general pharmacological tests of which have been cited briefly in the previous review (*1*).

Recently, a disease characterized by jaundice, colic, and photophobia broke out among range cattle in Hokkaido. It was proved by Sonoda *et al.* to be a plant poisoning caused by *D. humile,* which contained toxic substances affecting the liver specifically (*6*). In laboratory tests, a cow and a goat died suddenly following administration of at least 3.0 and 2.7 g/kg, respectively, of this plant (as doses of crude drug). For clarification of the clinical and pathological profiles of *D. humile,* acute intoxication was induced in rabbits, with compulsory oral administration of powdered plant extracts, and the results so far obtained are summarized here (*18*).

1. Clinical Findings

Decrease in body temperature just before the death, decrease in pH of urine, and change to positive in protein, bilirubin, and occult blood tests were observed. Shift to the left of the nuclei of the pseudoeosinophiles, rise in icteric index, marked increase in GPT, γ-GTP, and LDH activities were also recognized.

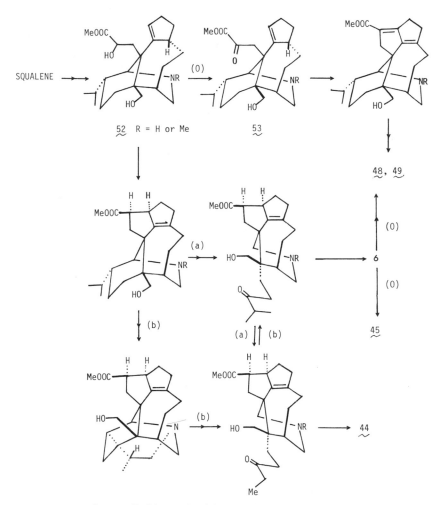

SCHEME 7. Biogenesis of the daphnigracine-type alkaloids.

2. Pathologic Findings

Cloudiness of the centrolobule and fragility of the liver were noted. Boiled fleshlike findings with rough-surface of myocardium and filling of incoagulable blood in ventricle were recorded, and increased cellularity in adrenal cortex and congestion of small intestine were observed.

3. Histopathological Findings

Centrolobule necrobiosis, necrosis, cell infiltration, and interstital hyperplasia of the liver were identified along with congestion of the spleen, degeneration and necrobiosis of the epithelial cells of the urinigerous tubules, degeneration and necrobiosis of the myocardial cells, pulmonary edema, and increased cellularity in the adrenal cortex. Falling off and necrobiosis of the epithelium of mucous membranes and cell infiltration in the submucosa of small intestine were also observed.

Further, the mutagenic activity of five daphniphyllum alkaloids (codaphniphylline, daphnilactone-B, yuzurimine, yuzurimine-C, and deoxyyuzurimine) was examined by Hashimoto and Ushijima (19), using *Salmonella typhimurium* variant TA 100 and TA 98. However, none of them were found to show any mutagenic activity.

REFERENCES

1. S. Yamamura and Y. Hirata, *in* "The Alkaloids," (R. H. F. Manske ed.), Vol. 15, p. 41. Academic Press, New York, 1975.
2. T. G. Hartley, E. A. Dunstone, J. S. Fitzgerald, S. R. Johns, and J. A. Lamberton, *Lloydia* **36**, 217 (1973).
3. S. Yamamura, J. A. Lamberton, H. Irikawa, Y. Okumura, and Y. Hirata, *Chem. Lett.*, 923 (1975); S. Yamamura, J. A. Lamberton, H. Irikawa, Y. Okumura, M. Toda, and Y. Hirata, *Bull. Chem. Soc. Jpn.* **50**, 1836 (1977).
4. S. Yamamura, J. A. Lamberton, M. Niwa, K. Endo, and Y. Hirata, *Chem. Lett.*, 393 (1980).
5. S. Yamamura, M. Niwa, and T. Wu, unpublished results.
6. M. Sonoda, M. Tasaka, K. Takahashi, M. Koiwa, S. Minami, M. Iwasa, and K. Yashiro, *J. Jpn. Vet. Med. Assoc.* **31**, 140 (1978).
7. S. Yamamura and Y. Terada, *Chem. Lett.*, 1381 (1976).
8. S. Yamamura, K. Sasaki, M. Toda, and Y. Hirata, *Tetrahedron Lett.*, 2023 (1974).
9. S. Yamamura and Y. Hirata, *Tetrahedron Lett.*, 3673 (1974).
10. S. Yamamura, M. Toda, and Y. Hirata, *Bull. Chem. Soc. Jpn.* **49**, 839 (1976).
11. S. Yamamura, H. Irikawa, Y. Okumura, and Y. Hirata, *Bull. Chem. Soc. Jpn.* **48**, 2120 (1975).
12. S. Nishiyama, Y. Ikeda, and S. Yamamura, *Bull. Chem. Soc. Jpn.* **59**, 875 (1986).
13. M. A. Umbreit and K. B. Sharpless, *J. Am. Chem. Soc.* **99**, 5526 (1977).
14. H. Irikawa, M. Toda, S. Yamamura, and Y. Hirata, *Tetrahedron* **24**, 5691 (1968).
15. J. Orban and J. V. Turner, *Tetrahedron Lett.* **24**, 2697 (1983).
16. M. Niwa and S. Yamamura, unpublished results.
17. M. Toda, H. Niwa, H. Irikawa, Y. Hirata, and S. Yamamura, *Tetrahedron* **30**, 2683 (1974); H. Niwa, M. Toda, S. Ishimaru, Y. Hirata, and S. Yamamura, *ibid.* **30**, 3031 (1974).
18. M. Mizuno, *J. Coll. Dairying* **10**, 257 (1984).
19. K. Hashimoto and J. Ushijima, *J. Coll. Dairying* **10**, 29 (1984).

Chapter 6

THE CULARINE ALKALOIDS

Luis Castedo

Departamento de Química Orgánica
Facultad de Química
Universidad de Santiago
Santiago de Compostela, Spain

AND

Rafael Suau

Departamento de Química Orgánica
Facultad de Ciencias
Universidad de Málaga
29071 Málaga, Spain

I. Introduction

Since the isolation of the first cularine alkaloid, more than 40 years ago, two reviews in Vols. 4 and 10 of this treatise (*1, 2*) and chapters in the books of Shamma (*3a, 3b*) and Kametani (*4a, 4b*) have dealt with four representative alkaloids of the cularine group. In the last decade, more than 25 other alkaloids have been isolated and characterized, including several with modified structures and different oxidation states. This group of alkaloids is regularly reviewed in the *Specialist Reports* of the Chemical Society (Alkaloid Series) (*5*). A complete listing of cularine alkaloids was published in 1984 (*6*). It is now accepted that the dibenzoxepine skeleton is the characteristic structure of cularine alkaloids, although these may have evolved from several well-differentiated biogenic routes. The cularine–morphine dimer cancentrine and its derivatives have been excluded from this chapter as there have been no reports since the last review in Vol. 14 of this treatise (*7*).

Although Chemical Abstracts favors the name 2,3,12,12a-tetrahydro-6,9,10-trimethoxy-1-methylbenzoxepine[2,3,4-*i,j*]isoquinoline for the parent alkaloid (cularine, **11**), trivial names and the following numbering system will be used in this chapter.

THE ALKALOIDS, VOL. 29

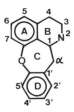

Early reports on simple cularines oxygenated at the 4' and 5' positions termed them isocularines, or cancentrine-type cularines, because of their analogous substitution pattern to that of the alkaloid cancentrine. To simplify nomenclature, we refer to them as 4',5'-substituted cularines.

II. Occurrence and Classification

The family Fumariaceae of higher plants is the source of most cularine alkaloids. A palynological study of 32 taxa of Fumariaceae present in the Iberian peninsula showed that pollen characters can be used to establish a hypothetical relationship between the genera of Fumariaceae (8). Apparently, only genera located at intermediate generic levels produce cularine alkaloids, namely, *Corydalis, Sarcocapnos,* and *Ceratocapnos.* In addition, outside Iberia the genus *Dicentra* (not mentioned in a previous study, Ref. 8) is also found to contain cularine alkaloids. In all cases, these alkaloids are related through a common biogenetic pathway originating from a tetraoxygenated tetrahydrobenzyliso-quinoline precursor. Other cularine bases, with aporphines as postulated biogenetic precursors, have been isolated from the family Annonaceae. More recently, cularines have been found among the Berberidaceae; they are thought to be oxidation products of protoberberinium salts. All known cularine alkaloids and their botanical sources are listed in Table I. They have been classified according to their structure and functional features.

III. Physical and Chemical Properties

Table II lists molecular formulas, melting points, specific rotations, and references related to physical and spectroscopic data for all naturally occurring cularine and related alkaloids isolated up to date. Most of the data included in previous reviews in this treatise (1, 2) have been excluded.

A. CULARINES

The cularine group is the largest representative class and presently consists of 14 alkaloids. Alkaloids of this group, with three oxygenated substitutents at positions 7, 3', and 4', have received the name normal cularines. For compounds with substituents at positions 7, 4', and 5', the generic name isocularines or cancentrine-type cularines has been used.

Other differences among this class of alkaloids resides in the type of substituent (hydroxy, methoxy, or methylenedioxy) and in the presence or absence of a methyl group at the nitrogen. They show an absorption maximum at about 283 nm in the UV spectrum, in common with many isoquinoline alkaloids. For those cularines with a phenolic group, under basic conditions, this band shifts to 294 nm.

^{1}H-NMR spectroscopy gives direct information about the structure of cularines. The presence (in deuterochloroform solution) of H-1 as a pair of doublets between 4.1 and 4.5 ppm with coupling constants of 11 and 4 Hz is quite characteristic; it is a signal not frequently observed in isoquinoline alkaloids. The substitution pattern can be deduced from the aromatic part of the spectrum. Thus for 7,3',4'-substituted cularines, H-2' and H-5' appear as two singlets which may overlap the AB spin system corresponding to H-5 and H-6. For 7,4',5'-substituted cularines, two AB quartets are expected, although H-2' and H-3' may be accidentally equivalent and appear as a singlet. To observe the expected coupling, solvent change or protonation is needed.

Mass spectroscopy (MS) is particularly useful to establish the nature and location of the oxygenated substituents. When a methoxy group is present at C-3', the base peak corresponds to the fragment M − 15, represented by the *para*-quinonium ion (31) indicated in Scheme 1. The presence of an intense molecular ion (usually with an intensity around 50%) and an important peak of

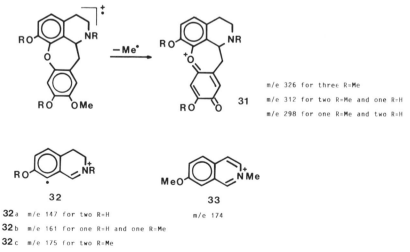

m/e 326 for three R=Me
m/e 312 for two R=Me and one R=H
m/e 298 for one R=Me and two R=H

32a   m/e 147 for two R=H
32b   m/e 161 for one R=H and one R=Me
32c   m/e 175 for two R=Me

m/e 174

SCHEME 1

TABLE I

Structures and Botanical Sources of Cularine Alkaloids

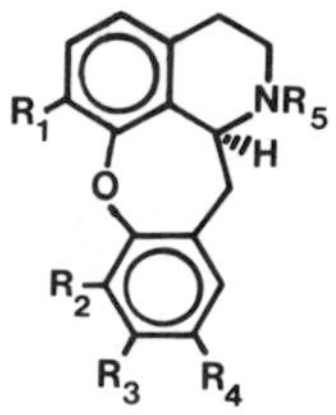

Alkaloid	Structure	Plant source	Reference
(+)-Norcularicine (1)	R_1 = OH, R_2 = R_5 = H, R_3 + R_4 = OCH_2O	*Corydalis claviculata*	9
(+)-Norcularidine (2)	R_1 = OH, R_2 = R_5 = H, R_3 = R_4 = OMe	*C. claviculata*	10
(+)-Cularimine (3)	R_1 = R_3 = R_4 = OMe, R_2 = R_5 = H	*Dicentra eximia*	11
(+)-Celtisine (4)	R_1 = R_3 = OH, R_2 = H, R_4 = OMe, R_5 = Me	*Sarcocapnos enneaphylla*	12
(+)-Breoganine (5) (culacorine)	R_1 = R_4 = OH, R_2 = H, R_3 = OMe, R_5 = Me	*C. claviculata, S. crassifolia*	9,12
(+)-Cularicine (6)	R_1 = OH, R_2 = H, R_3 + R_4 = OCH_2O, R_5 = Me	*C. claviculata*	2,13
(+)-Celtine (7)	R_1 = R_4 = OMe, R_2 = H, R_3 = OH, R_5 = Me	*S. enneaphylla*	12
(+)-Cularidine (8)	R_1 = OH, R_2 = H, R_3 = R_4 = OMe, R_5 = Me	*D. cucullaria, C. claviculata, Ceratocapnos heterocarpa*	1,11,13,14
(+)-Enneaphylline (9)	R_1 = R_3 = OMe, R_2 = H, R_4 = OH, R_5 = Me	*S. crassifolia, S. enneaphylla*	31
(+)-O-Methylcularicine (10)	R_1 = OMe, R_2 = H, R_3 + R_4 = OCH_2O, R_5 = Me	*C. claviculata*	15,16
(+)-Cularine (11)	R_1 = R_3 = R_4 = OMe, R_2 = H, R_5 = Me	*D. cucullaria, D. eximia D. formosa, D. oregana, C. claviculata*	1,2,11
(+)-Claviculine (12)	R_1 = R_2 = OH, R_3 = OMe, R_4 = H, R_5 = Me	*S. crassifolia, C. claviculata*	10,15–17
(+)-Sarcocapnidine (13)	R_1 = R_3 = OMe, R_2 = OH, R_4 = H, R_5 = Me	*S. crassifolia, C. claviculata*	15–17
(+)-Sarcocapnine (14)	R_1 = R_2 = R_3 = OMe, R_4 = H, R_5 = Me	*S. enneaphylla*	18

290

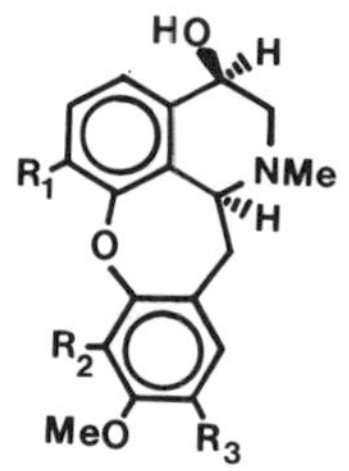

(+)-Limousamine (**15**) $R_1 = OH, R_2 = H, R_3 = OMe$ *C. claviculata*   19
(+)-4-Hydroxysarcocapnine (**16**) $R_1 = R_2 = OMe, R_3 = H$ *S. enneaphylla*   20

Gouregine (**17**)   *Guatteria ouregou*   21

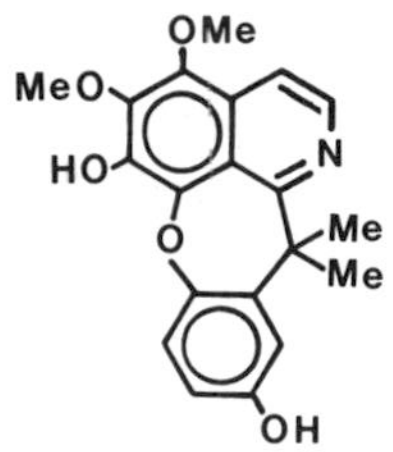

(±)-Linaresine (**18**)   *Berberis valdiviana*   22

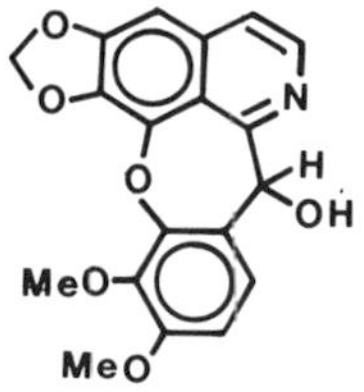

(continued)

TABLE I (*Continued*)

Alkaloid	Structure	Plant source	Reference
(±)-3,4-Dihydrolinaresine (**19**)	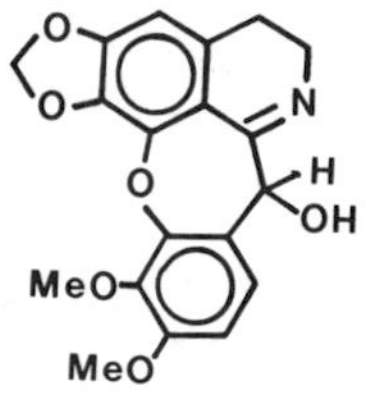	*Berberis valdiviana*	22
Oxocompostelline (**20**)	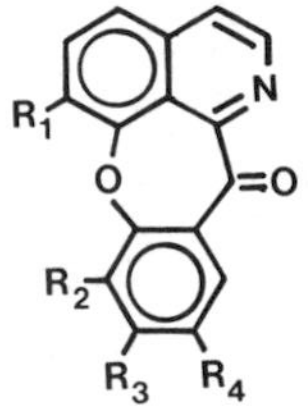 R$_1$ = OMe, R$_2$ = H, R$_3$ + R$_4$ = OCH$_2$O	*S. enneaphylla*	23
Oxocularine (**21**)	R$_1$ = R$_3$ = R$_4$ = OMe, R$_2$ = H	*C. claviculata*	23,9
Oxosarcocapnidine (**22**)	R$_1$ = R$_3$ = OMe, R$_2$ = OH, R$_4$ = H	*S. crassifolia*	17
Oxosarcophylline (**23**)	R$_1$ = OH, R$_2$ = R$_3$ = OMe, R$_4$ = H	*S. crassifolia*, *S. enneaphylla*	24
Oxosarcocapnine (**24**)	R$_1$ = R$_2$ = R$_3$ = OMe, R$_4$ = H	*S. enneaphylla*	18
Yagonine (**25**)	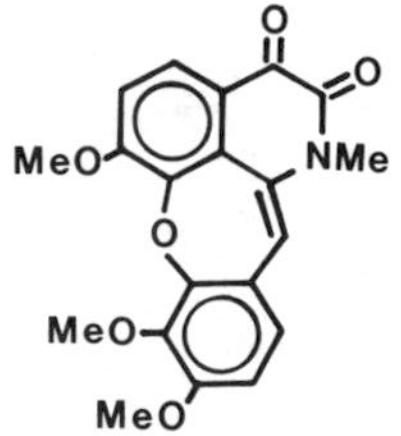	*S. enneaphylla*	24

Aristoyagonine (26)

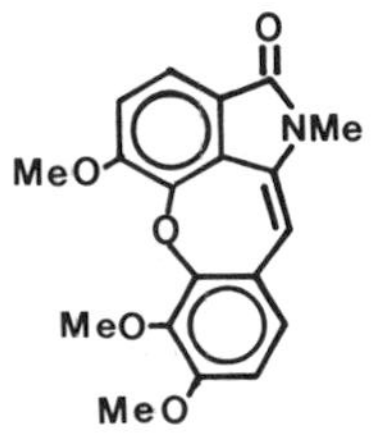

S. enneaphylla   24

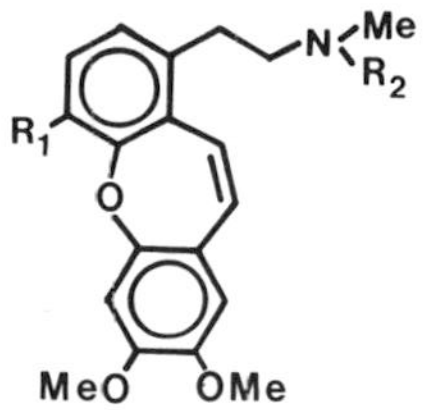

R_1 = OH, R_2 = Me
R_1 = OMe, R_2 = Me
R_1 = OMe, R_2 = H

Secocularidine (27)   *C. claviculata*   25
Secocularine (28)   *S. crassifolia*   25
Norsecocularine (29)   *C. claviculata*   36

Noyaine (30)   *C. claviculata*   36

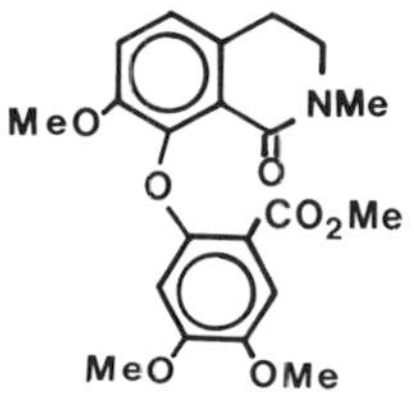

293

TABLE II

Physical Constants of Cularine and Related Alkaloids

Alkaloid name	Molecular formula (mol. wt.)	Melting point °C (solvent) (Ref.)	Optical rotation, ° (conc, solvent) (Ref.)	Additional data (Ref.)
Norcularicine (1)	$C_{17}H_{15}NO_4$ (297.10)	— 96–99 for the (±) isomer (26)	+216 (0.06, MeOH) (9)	UV (9), IR (26), ^{1}H NMR (9,26), MS (9,26)
Norcularidine (2)	$C_{18}H_{19}NO_4$ (313.13)	—	+216 (0.06, MeOH)) (10)	UV (10), ^{1}H NMR (10), MS (10)
Cularimine (3)	$C_{19}H_{21}NO_4$ (327.14)	100–101 (ethyl ether) (27)	+259.5 (0.94, MeOH) (27)	^{1}H NMR (28), MS (28)
Celtisine (4)	$C_{18}H_{19}NO_4$ (313.13)	158–160 (EtOH) (12)	+212 (0.025, MeOH) (12)	UV (12), IR (12), ^{1}H NMR (12), MS (12)
Breoganine (5)	$C_{18}H_{19}NO_4$ (313.13)	249–250 (EtOH) (12) 196–200 for the (±) isomer (28)	+278 (0.057, MeOH) (12) +188 (0.08, MeOH) (9)	UV (9,12), IR (12), ^{1}H NMR (9,12), MS (9,12)
Cularicine (6)	$C_{18}H_{17}NO_4$ (311.11)	185 (MeOH) (29) 155–156 for the (±) isomer (26)	+295 (0.96, CHCl₃) (29)	IR (26), ^{1}H NMR (26,28) MS (26,28)
Celtine (7)	$C_{19}H_{21}NO_4$ (327.14)	94–96 (EtOH) (12)	+181 (0.08, MeOH) (12)	UV (12), IR (12), ^{1}H NMR (12), MS (12)
Cularidine (8)	$C_{19}H_{21}NO_4$ (327.14)	156 (MeOH) (29) 201–203 for the (±) isomer (30)	+292 (0.99, CHCl₃) (29)	^{1}H NMR (28,30), MS (28)
Enneaphylline (9)	$C_{19}H_{21}NO_4$ (327.14)	205–207 (EtOH) (31)	+256 (0.8, EtOH) (31)	UV (31), ^{1}H NMR (31), MS (31)
O-Methylcularicine (10)	$C_{19}H_{19}NO_4$ (325.13)	Oil	+283 (0.3, EtOH) (15)	UV (16), ^{1}H NMR (16), MS (16)

294

Compound	Formula (mass)	mp (solvent) (ref)	$[\alpha]$	Spectral data
Cularine (11)	$C_{20}H_{23}NO_4$ (341.16)	113–114.5 (ethyl ether) (27)	+285 (0.8, MeOH) (32)	UV (32), ^{1}H NMR (28), MS (28,32), ^{13}C NMR (33)
Claviculine (12)	$C_{18}H_{19}NO_4$ (313.13)	112–113 (EtOH) (17) 126 (16)	+443 (0.41, MeOH) (17) +404 (0.1, EtOH) (16)	UV (16,17), ^{1}H NMR (16,17), MS (16,17), ^{13}C NMR (16,17)
Sarcocapnidine (13)	$C_{19}H_{21}NO_4$ (327.14)	126–127 (EtOH) (17) 118 (16)	+385 (0.07, MeOH) (17) +368 (0.1, EtOH)	UV (16,17), ^{1}H NMR (16,17), ^{13}C NMR (16,17), MS (16,17)
Sarcocapnine (14)	$C_{20}H_{23}NO_4$ (341.16)	213–215 (as HCl) (EtOH–Et$_2$O) (18)	+218 (0.3, EtOH) (18)	UV (18), ^{1}H NMR (18), MS (18)
Limousamine (15)	$C_{19}H_{21}NO_5$ (343.14)	—	+185 (0.074, MeOH) (19)	UV (19), ^{1}H NMR (19), MS (19)
4-Hydroxysarcocapnine (16)	$C_{20}H_{23}NO_5$ (357.14)	145–146 (EtOH) (20)	+314 (0.11, CHCl$_3$) (20)	UV (20), ^{1}H NMR (20), MS (20)
Gouregine (17)	$C_{20}H_{19}NO_5$ (353.12)	112–114 (MeOH) (21)	—	UV (21), ^{1}H NMR (21), ^{13}C NMR (21), MS (21)
Linaresine (18)	$C_{19}H_{15}NO_6$ (353.08)	215 (MeOH–benzene–Et$_2$O) (22)	—	UV (22), IR (22), ^{1}H NMR (22), MS (22)
Dihydrolinaresine (19)	$C_{19}H_{17}NO_6$ (355.10)	170 (MeOH–benzene–Et$_2$O) (22)	—	UV (22), IR(22), ^{1}H NMR (22), MS(22)
Oxocompostelline (20)	$C_{18}H_{11}NO_5$ (321.06)	295 (EtOH) (23)	—	UV (23), IR (23), ^{1}H NMR (23), MS (23)
Oxocularine (21)	$C_{19}H_{15}NO_5$ (337.09)	198–199 (EtOH) (23)	—	UV (23,26), IR (23), MS (23,9), ^{1}H NMR (23)

(continued)

295

TABLE II (Continued)

Alkaloid name	Molecular formula (mol. wt.)	Melting point °C (solvent) (Ref.)	Optical rotation, ° (conc, solvent) (Ref.)	Additional data (Ref.)
Oxosarcocapnidine (22)	$C_{18}H_{13}NO_5$ (323.07)	231–232 (MeOH) (17)	—	UV (17), IR (17), ^{1}H NMR (17), MS (17)
Oxosarcophylline (23)	$C_{18}H_{13}NO_5$ (323.07)	170–171 (EtOH) (24)	—	UV(24), IR (24), ^{1}H NMR (24), MS (24)
Oxosarcocapnine (24)	$C_{19}H_{15}NO_5$ (337.09)	202–203 (EtOH) (18)	—	UV (18), IR (18), ^{1}H NMR (18), MS (18)
Yagonine (25)	$C_{20}H_{17}NO_6$ (367.10)	226–227 (MeOH) (24)	—	UV (24), IR (24), ^{1}H NMR (24), ^{13}C NMR (24), MS (24)
Aristoyagonine (26)	$C_{19}H_{19}NO_6$ (325.13)	165–166 (MeOH) (24)	—	UV (24), IR (24), ^{1}H NMR (24), ^{13}C NMR (24), MS (24)
Secocularidine (27)	$C_{20}H_{23}NO_4$ (341.16)	189–190 (MeOH) (25)	—	UV (25), IR (25), ^{1}H NMR (25), MS (25)
Secocularine (28)	$C_{21}H_{25}NO_4$ (355.17)	194–196 (as $HClO_4$) (EtOH) (25)	—	UV (25), IR (25), ^{1}H NMR (25), MS (25)
Norsecocularine (29)	$C_{20}H_{23}NO_4$ (341.16)	203–205 (benzene) (36)	—	UV (36), ^{1}H NMR (36), MS (36)
Noyaine (30)	$C_{21}H_{23}NO_7$ (401.14)	—	—	UV (36), ^{1}H NMR (36), ^{13}C NMR (36), MS (36)

low intensity (arising from **32**), which corresponds to the upper part of the molecule, is essential to determine the type of substituent at C-7 (*28, 32, 34*). In those *N*-methyl cularines with a methoxy group at C-7, the peak at m/e 174 ($C_{11}H_{12}NO^+$) (*16*), which is presumably due to the fragment **33**, has more significance than the expected m/e 175 (**32c**). Similarly, the methylenedioxy cularines show an analogous fragmentation pattern. The 7,4',5'-substituted cularines invariably exhibit coincident molecular ion and base peaks.

1. Norcularines: (+)-Norcularicine and (+)-Norcularidine

The UV spectrum of the monophenolic base (+)-norcularicine (**1**) exhibits a bathochromic shift in alkaline solution (*9*). The presence of a methylenedioxy group (5.98 ppm) and two aromatic protons as singlets (6.63 and 6.68 ppm) in the ^{1}H-NMR spectrum readily establishes the 3',4'-substitution pattern of the D ring, leaving position 7 for the hydroxy function. The MS peak at m/e 147 (**32a**) confirms the substitution of ring A. Eschweiler–Clarke methylation of **1** gives (+)-cularicine (**6**). Synthesis of (±)-norcularicine has been achieved as an intermediate step in the total synthesis of **6** (*26*).

The amorphous base (+)-norcularidine (**2**) also exhibits the characteristic behavior of a phenolic cularine (*10*). The peak at m/e 147 (**32a**) in the mass spectrum locates the phenolic group at C-7, while ^{1}H-NMR signals of two methoxy groups and two aromatic singlets (6.62 and 6.70 ppm) established the substitution pattern for the D ring. N-Methylation furnishes (+)-cularidine (**8**).

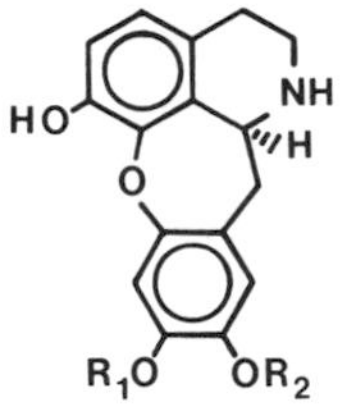

1 $R_1+R_2=CH_2$
2 $R_1=R_2=Me$
34 $R_1=Me, R_2=H$

Demethylation of (±)-cularimine (**3**) with hydrobromic acid does not give the monodemethylated (±)-norcularidine (**2**). Instead, the unnatural compound norbreoganine (**34**) was obtained as the result of the loss of methyl groups from positions 7 and 3' (*37*).

2. 7,3',4'-Substituted Cularines: (+)-Celtisine, (+)-Breoganine, (+)-Celtine, (+)-Enneaphylline, and (+)-O-Methylcularicine

The presence of an *N*-methyl group (2.5–2.6 ppm) in the ^{1}H-NMR spectrum immediately distinguishes these alkaloids from the nor compounds of the pre-

vious group. Conversion of the monophenolic compounds celtine (**7**) and enneaphylline (**9**) or the diphenolic celtisine (**4**) and breoganine (**5**) to (+)-cularine
(**11**) by diazomethane treatment establishes their skeletal and oxygenated substitution patterns (*12, 31*). Correlation of the ^{1}H-NMR methoxy signals of the
trideuteromethyl derivatives of celtisine, breoganine, and celtine (**38, 39,** and
37, respectively) with those of cularine (**11**) determines the positions of the
phenolic functional groups (Table III). The methoxy frequencies in cularine were
assigned by a correlation of the ^{1}H-NMR shifts of the *O*-methyl groups with
those of deuteromethylated cularidine (**35**) and a synthetic cularine derivative
(**36**), which corrected previous assignments (*27, 28*).

Demethylation of (+)-cularine (**11**) under acidic conditions (48% HBr/AcOH)
gives a mixture of demethylated products; breoganine (**5**) is the major component
(*12, 38*), in agreement with previous observations (*37*). Under conditions of
lower temperature and increased reaction time, selective demethylation affords **5**
and celtisine (**4**) (36 and 21%, respectively). Preferential demethylation at position 4' occurs under nucleophilic conditions (NaEtS–DMF), changing the proportions of **5** to **4** (18 and 43%, respectively). Under the same conditions, celtine
(**7**) yields **4** (42%) and enneaphylline (**9**) gives **5** (45%) (*38*).

TABLE III
^{1}H NMR Chemical Shifts for Methoxy Groups of Cularine Derivatives[a]

		C-7 OMe	C-3' OMe	C-4' OMe
11	$R_1 = R_2 = R_3 = Me$	3.85	3.79	3.87
35	$R_1 = CD_3, R_2 = R_3 = Me$	—	3.80	3.88
36	$R_1 = R_2 = Me, R_3 = CD_3$	3.86	—	3.90
37	$R_1 = R_3 = Me, R_2 = CD_3$	3.85	3.80	—
38	$R_1 = R_2 = CD_3, R_3 = Me$	—	3.81	—
39	$R_1 = R_3 = CD_3, R_2 = Me$	—	—	3.89
4	$R_1 = R_2 = H, R_3 = Me$			
5	$R_1 = R_3 = H, R_2 = Me$			
7	$R_1 = R_3 = Me, R_2 = H$			
9	$R_1 = R_2 = Me, R_3 = H$			
10	$R_1 = Me, R_2 + R_3 = CH_2$			

[a] In CDCl$_3$ solution. Data from Ref. *12.*

The nonphenolic O-methyl derivative cularicine (10) is obtained as an oil. It is characterized in the ^{1}H-NMR spectrum by the presence of a methylenedioxy group and two aromatic singlets (6.82 and 6.50 ppm) for the D ring. Compound 10 proved to be identical with the O-methylation product of cularicine (6) (15, 16).

3. 7,4′,5′-Substituted Cularines: (+)-Sarcocapnine, (+)-Sarcocapnidine, and (+)-Claviculine

(+)-Sarcocapnine (14) was the first natural 7,4′,5′-substituted cularine alkaloid isolated (18). Its UV spectrum is independent of the pH of the solution. In the ^{1}H-NMR spectrum, the presence of three methoxy groups (4.02, 3.82, and 3.82 ppm) reveals its nonphenolic nature. The two AB quartets account for the four aromatic protons and not only indicate the 4′,5′-substitution at the D ring but also readily differentiate the spectrum from that of the isomeric cularine (11). Total synthesis of 14 has already been described (39, 40).

The monophenolic base (+)-sarcocapnidine (13) yields 14 by O-methylation (17). The peak at m/e 174 (33) in the mass spectrum suggests a methoxy group at C-7, leaving the 4′ or 5′ position for the hydroxy group. The chemical shift displacement observed on comparing the ^{13}C-NMR spectrum of the phenol with that of the phenoxide (41, 42) shows a Δδ of −6.38 ppm (for a methine carbon) for C-2′, which shows that the phenolic function is para to it. Nuclear Overhauser effect (NOE) difference studies locate the methoxy groups at positions 7 and 4′ (16). Racemic 13 has been synthesized by two different approaches (39, 40, 43).

The mass spectrum of the diphenolic (+)-claviculine (12) shows a peak at m/e 161 (32b), locating one hydroxy group at C-7 (16, 17). The second phenolic group is located at C-5′, since a ^{13}C-NMR shift comparison of 12 with its diphenolate gives a Δδ of −6.8 ppm for C-4a (quaternary para carbon atom to C-7) and of −8 for C-2′ (methine para carbon atom to C-5′) (17). The location of the methoxy group at C-4′ is confirmed by NOE difference experiments.

Selective demethylation of (+)-sarcocapnine (14) is observed under nucleophilic conditions (NaEtS–DMF) and gives (+)-claviculine (12) in 77% yield. Less selectivity is observed under acidic conditions, and preferential demethylation at C-5′ [to give sarcocapnidine (13) in 58% yield] is followed by loss of the methyl group at C-4′. The catechol 40 was obtained in 10% yield (38).

12 $R_1 = R_2 = H$, $R_3 = Me$
13 $R_1 = R_3 = Me$, $R_2 = H$
14 $R_1 = R_2 = R_3 = Me$
40 $R_1 = Me$, $R_2 = R_3 = H$

B. HYDROXYCULARINES

The class of alkaloids represented by 4-hydroxycularines is readily distinguished by ¹H-NMR spectroscopy. Beside the characteristic quartet for H-1 (as the X part of an ABX spin system) between 4.0 and 4.6 ppm, a second low-field aliphatic proton (at ~4.6 ppm) appears as a quartet (X part of a second ABX spin system) with two small coupling constants to indicate the C-4 position of the hydroxy group.

1. (+)-4-Hydroxysarcocapnine

The presence of a band at 3511 cm⁻¹ in the IR spectrum of the nonphenolic compound (+)-4-hydroxysarcocapnine (16) suggests an intramolecularly hydrogen-bonded hydroxy group (20). ¹H-NMR spectroscopy shows the presence of three methoxy groups (4.05, 3.86, and 3.86 ppm), one N-methyl (2.62 ppm), and two pairs of ortho-coupled protons [6.6 and 6.74 (J = 8.6 Hz), H-3' and H-2'; 6.83 and 7.14 (J = 8.4 Hz), H-6 and H-5] to confirm a 7,4',5'-cularine substitution pattern. The X part of an ABX system at 4.58 ppm locates the hydroxy group at C-4.

Comparative ¹H-NMR studies have established that the C ring in 16 has the same twist–boat conformation as in cularine (11) and (+)-sarcocapnine (14), while the B ring is a half-chair, or a distorted form, easily interconverted by simple rotation of the N—C-3—C-4 bonds without involvement of the C ring.

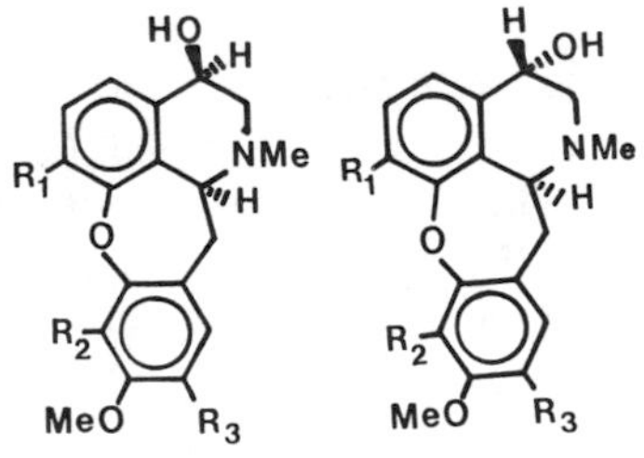

15 R_1=OH, R_2=H, R_3=OMe   15a
16 R_1=R_2=OMe, R_3=H   16a
41 R_1=R_3=OMe, R_2=H   41a
42 R_1=OAc, R_2=H, R_3=OMe   42a

The quasi-equatorial position for H-4, indicated by the small coupling constant with H-3α (J = 2.3 Hz) and H-3β (J = 3.9 Hz), establishes the quasi-axial position at C-4 for the hydroxy group with a syn stereochemical relationship between H-1 and H-4.

The synthesis of 16 and its C-4 epimer 16a, and of three other C-4 epimeric pairs of 4-hydroxycularines (15 and 15a, 41 and 41a, 42 and 42a), shows that the chemical shift of H-4 is hardly affected by the syn or anti geometry, and that the chemical shift of H-1 is sufficiently great to permit direct configurational assignments. Comparative study of both epimers by lanthanide-induced shift displacement has confirmed that in each series the hydroxy is quasi-axial, and the H-1 shift in the anti series is greater than that of naturally occurring (+)-hydroxysarcocapnine (16) (syn series) (20). In both epimers, acetylation of the hydroxy group gives the expected chemical shift differences for the geminal protons. This situation contrasts sharply with that found in 4-hydroxyaporphines such as (+)-cataline (43), where the hydroxy group is pseudoequatorial in the natural epimer (43) and pseudoaxial in its synthetic C-4 epimer (44) (44).

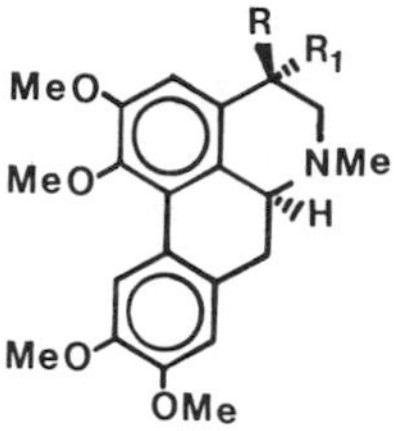

43 R=OH, R₁=H (cataline)
44 R=H, R₁=OH (epicataline)

2. (+)-Limousamine

The monophenolic base (+)-limousamine (15) has the phenolic group located at C-7 according to MS fragmentation (m/e 177 and 159), and the presence two aromatic protons as singlets (6.7 and 6.55 ppm) establishes the 7,3',4'-substitution pattern (19). A broad singlet (a quartet after addition of deuterium oxide) in the ¹H-NMR spectrum locates the hydroxy group at C-4. On the basis of simple comparison of the H-4 chemical shift with those of epimeric pairs in the aporphine (44) and berberine (45) series, syn stereochemistry between H-4 and H-1 has been postulated for compound 15: an argument which proved erroneous for the cularine series (20). The syn stereochemistry of 15 has been demonstrated by comparing the ¹H-NMR spectrum of synthetic limousamine with that of its C-4 epimer (15a), both having been obtained synthetically by lead tetraacetate oxidation of (+)-cularidine (8), followed by hydrolysis of the resulting diacetates (syn: anti ratio = 2:3) (20).

C. TETRADEHYDROCULARINES

Although structurally classified as cularine alkaloids tetradehydrocularines not only have anomalous substitution patterns and functional groups but they also originate from different plant sources. This suggests that they are products of biosynthetic pathways different from those of all other cularine alkaloids. They show characteristic bathochromic shifts under acidic conditions due to the presence of totally or partially dehydrogenated B rings in their skeletons.

1. (±)-Linaresine and (±)-Dihydrolinaresine

In the ^{1}H-NMR spectrum of racemic linaresine (**18**), signals arising from one methylenedioxy group (6.1 ppm) and two methoxy groups (3.82 and 3.87 ppm) define the aromatic substituents (22). Three benzenoid protons, one singlet (7.28 ppm), and one AB spin system (7.11 and 6.50 ppm, J = 9.1 Hz) agree with a 6,7,4′,5′-substituted cularine. A second AB spin system (8.38 and 7.73 ppm), together with a smaller coupling constant (5.6 Hz), indicates B-ring aromaticity. This has been confirmed by the UV spectrum in neutral (236, 298, and 334 nm) and acid solution (245, 312, and 346 nm). An absorption at 7.27 ppm coupled to a neighboring hydroxy group locates this group at Cα. In addition, NOE studies interrelated most of the protons of **18.**

The spectroscopic properties of (±)-dihydrolinaresine (**19**) are very similar to those of linaresine (22). The main change in the ^{1}H-NMR spectrum is the presence of two high-field multiplets (2.8 and 3.88 ppm) instead of the AB spin system which corresponds to H-3 and H-4 in linaresine (**18**). N-Methylation of dihydrolinaresine gives the expected dihydrolinaresinone (**45**).

18　　　　　**19**　　　　　**45**

2. Gouregine

The UV spectrum of gouregine (**17**) shows absorption bands at 229, 247, 291, and 348 nm in neutral solution. Bathochromic shifts are observed in acid solution (231, 274, 304, and 404 nm) and in basic solution (260, 307, and 379 nm), the latter showing the phenolic nature of **17** (21). Comparison of its ^{1}H-NMR and ^{13}C-NMR spectra with those of the recently isolated aporphine, melosmine (**46**) (46), was fundamental in establishing its structure. The main features of the ^{1}H-NMR spectrum of **17** are the geminal dimethyl group (1.9 ppm in deuterochloroform solution), the pair of doublets of H-3 and H-4 (8.16 and 7.69 ppm, J = 6 Hz), and the appearance of H-4′ as a doublet doublet (6.68 ppm) with one ortho coupling (9Hz) and one meta coupling (3Hz). The ^{1}H-NMR spectrum of the diacetyl derivative of gouregine (**17a**) was used to assign positions 7 and 3′ for the phenolic groups. The structure of gouregine, deduced from spectral analysis, was confirmed by X-ray diffraction of a single crystal of **17a.** Oxidation of melosmine (**46**) by Fenton's reagent yielded gouregine (**17**) (90%) (21).

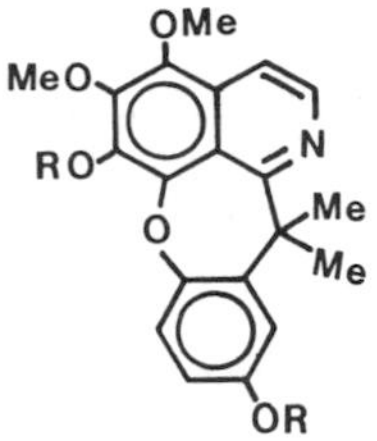

17 R=H
17a R=Ac

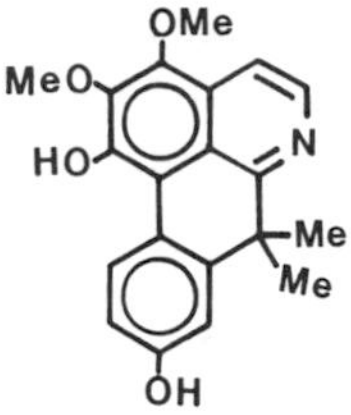

4 6

D. OXOCULARINES

Alkaloids of the yellow oxocularine group are easily recognized by UV absorption maxima at about 400 nm (in neutral solution), by bathochromic shifts in acid media, and by conjugated carbonyl absorption at 1670 cm^{-1} in the IR spectra. Characteristic features of the mass spectra are the appearance of a strong molecular ion (usually the base peak). In the ^{1}H-NMR spectra the absence of aliphatic protons and the presence of a pair of aromatic proton doublets with a reduced coupling constant (5.5–5.7 Hz) for H-3 and H-4, both clearly differentiated from benzenoid protons, are noted. These are the essential spectroscopic data needed for structural determination of oxocularines.

1. 7,3',4'-Oxocularines: Oxocompostelline and Oxocularine

Oxocompostelline (**20**), isolated as yellow crystals, showed a conjugated system in the UV spectrum (208, 254, 292, and 397 nm). The substitution pattern shown in structure **20** was deduced from the ^{1}H-NMR spectrum, which shows one methylenedioxy group (5.99 ppm), one methoxy group (4.08), and six aromatic protons, two as singlets (7.11 and 6.90 ppm) corresponding to a 3',4'-substituted D ring and four as two AB quartets, the lowest field doublet (8.63) being the H-3 proton ($J = 5.7$ Hz). The proposed structure was later confirmed by total synthesis (*23*).

The structure of oxocularine (**21**) is inferred from its ^{1}H-NMR spectrum,

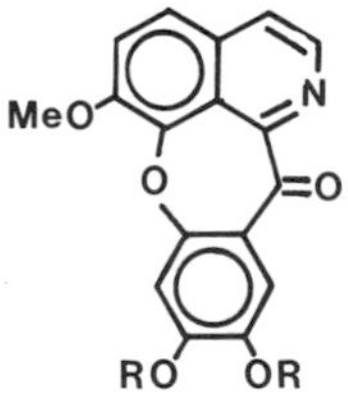

20 R+R=CH$_2$
21 R=Me

showing three methoxy groups (4.14, 3.99, and 3.91 ppm) and six aromatic protons, two as singlets (7.25 and 6.95 ppm) and four as two sets of two doublets, each with a well differentiated coupling constant (J = 9 and 5.5 Hz, respectively) (9, 23). Oxidation of (+)-cularine (11) with lead tetraacetate gives oxocularine in low yield (9). Compound 21 is also obtained by oxidation of synthetic (+)-cularimine (3) with pyridine–chromic acid (35).

2. 7,4′,5′-Oxocularines: Oxosarcocapnidine, Oxosarcophylline, and Oxosarcocapnine

Oxosarcocapnidine (22) shows a bathochromic shift on addition of acid or base. The latter shift indicates the presence of a phenolic group (17). The ^{1}H-NMR spectrum reveals two methoxy groups (3.98 and 3.95 ppm) and three aromatic AB spin systems, corresponding to a 7,4′,5′-substituted skeleton, which is confirmed by O-methylation to oxosarcocapnine (24). The position of the phenolic function at C-5′ has been established by chemical correlation with sarcocapnidine (13). Fremy's salt oxidation of the 5′-O-methoxymethyl derivative of 13, followed by acid deprotection, gives 22 in 58% yield.

Oxosarcophylline (23) displays UV and IR spectra similar to those of oxosarcocapnidine (22) (24). The presence of a phenolic group and the 7,4′,5′-substitution pattern of 23 has been chemically established by methylation to oxosarcocapnine (24). The location of the phenolic group at C-7 has been deduced by comparison of the mass spectrum of 23 with those of other oxocularines carrying a methoxy group at C-7 and confirmed by NOE experiments.

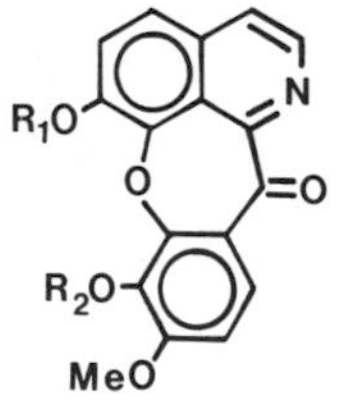

22 R_1=Me, R_2=H
23 R_1=H, R_2=Me
24 R_1=R_2=Me

The UV spectrum (254, 330, and 400 nm) of oxosarcocapnine (24) only shows bathochromic shifts in acid media (266, 298, and 462 nm), and its IR spectrum shows no signals above 3000 cm^{-1} (18). The ^{1}H-NMR spectrum in deuterochloroform and TFA exhibits a pronounced down-field effect on the aromatic protons of the isoquinoline moiety when compared with the spectra of other oxocularines in neutral media. Three methoxy groups (4.23, 4.03, and 3.97 ppm) and six aromatic protons as three AB quartets indicate the substituent positions. Oxidation of sarcocapnine (14) with Fremy's salt gives a 50% yield of

24, which is reduced with Zn–HCl–AcOH to (±)-norsarcocapnine (**47**) (a compound that has not been isolated to date from natural sources). N-Methylation of **47** gave (±)-sarcocapnine (**14**) (*18*).

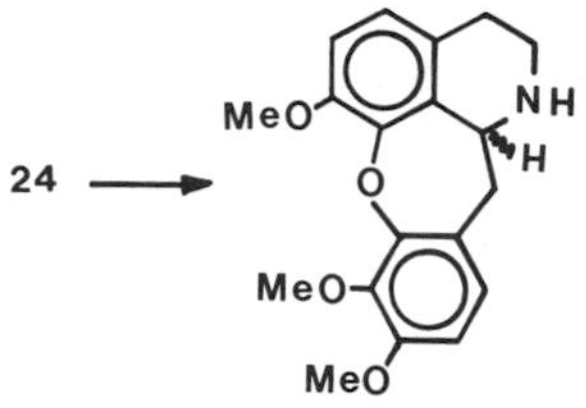

E. 3,4-DIOXOCULARINES

3,4-Dioxocularines are structurally related to 4,5-dioxoaporphines (*47*). As of 1986, only one example had been isolated from natural material.

1. Yagonine

Yagonine (**25**), obtained as red needles, shows a highly conjugated chromophore in the UV spectrum (217, 254, 340, and 435 nm), and its IR spectrum exhibits a broad absorption band at 1680 cm^{-1} (*24*). ^{13}C-NMR and ^{1}H-NMR spectroscopy show the presence of two carbonyl groups (175.24 and 156.96 ppm), three methoxy groups (4.07, 4.09, and 3.90 ppm), and one N-methyl group (3.67 ppm), together with five aromatic protons, one singlet (6.62 ppm) and two pairs of doublets with normal benzenoid coupling constants (8.6 and 7.20 ppm with $J = 8.7$ Hz; 6.90 and 6.76 with $J = 8.7$ Hz). In addition, NOE difference experiments have been used to locate the two methoxy groups at the 4' and 5' positions of the D ring. The aromatic singlet at 6.62 ppm (H-α) increases by 6.1 and 8.2% on monitoring the signal of the N-methyl group and the doublet at 6.9 ppm, respectively. The methoxy group at 4.07 ppm correlates with H-6 (7.20 ppm). Total synthesis of **25** has been achieved by oxidation of the two epimers of 4-hydroxysarcocapnine (**16**) with DDQ (*24, 48*).

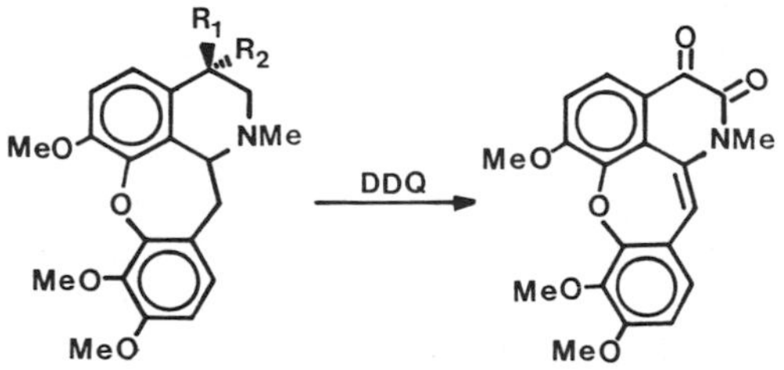

F. ARISTOCULARINES

Only one example of aristocularines, aristoyagonine, is known. This compound is structurally similar to the aristolactams of the aporphine series of alkaloids (49).

1. Aristoyagonine

Aristoyagonine (**26**) is isolated as yellow needles, and no change is observed in the UV spectrum (230, 250, 296, 330, and 410 nm) on addition of acid or base (24). ^{13}C-NMR and ^{1}H-NMR spectra are similar to those obtained for yagonine (**25**), the major difference being the presence of a single carbonyl carbon atom (166.13 ppm). Transformation of **25** to **26** is achieved by a benzylic rearrangement followed by oxidation, reactions used previously to correlate aristolactams with 4,5-dioxoaporphines (50, 51). The reactions with **25**, however, show several differences. Reaction of **25** with barium hydroxide gives **26** in low yield, together with unoxidized **48**. Treatment of **48** with DDQ produced **26** in quantitative yield. The observed replacement of the methoxy group by an ethoxy group at the 4,5-dioxoaporphine C-1 position (50) does not occur at the homologous 3,4-dioxocularine C-7 position (24).

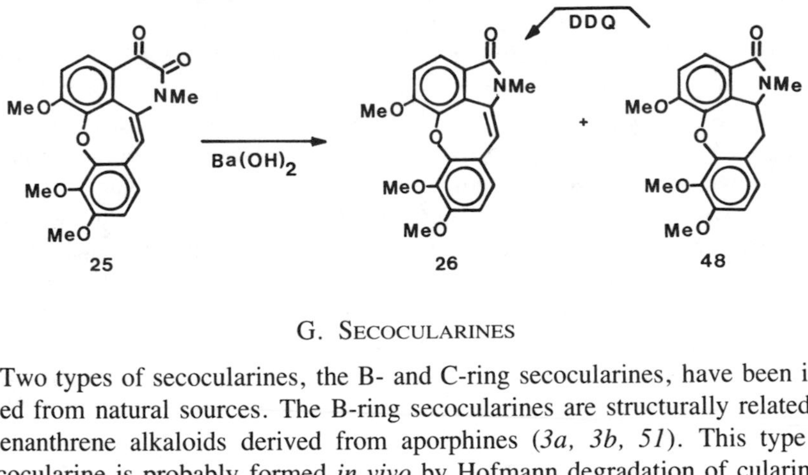

25   26   48

G. SECOCULARINES

Two types of secocularines, the B- and C-ring secocularines, have been isolated from natural sources. The B-ring secocularines are structurally related to phenanthrene alkaloids derived from aporphines (3a, 3b, 51). This type of secocularine is probably formed in vivo by Hofmann degradation of cularines. They can be readily recognized by their ^{1}H-NMR spectra which show a sufficiently characteristic aromatic AB spin system with an abnormally large coupling constant (11.5 Hz), corresponding to H-12 and H-13. In the mass spectrum, the appearance of a base peak at m/e 58 ($CH_2=\overset{+}{N}Me_2$) characterizes the N-dimethyl secocularines, while a base peak at m/e 44 indicates ($CH_2=\overset{+}{N}HMe$) the norsecocularine structure. C-Ring secocularines represent a new type of alkaloid without counterpart among aporphinoid alkaloids.

1. B-Ring Secocularines: Secocularine, Secocularidine,
   and Norsecocularine

Secocularine (**28**), isolated as an amorphous substance (crystallized as perchlorate) shows a characteristic UV spectrum (220, 235, 296, and 320 nm). The
substitution pattern has been established from the ^{1}H-NMR spectrum, which
shows two singlets (6.68 and 6.92 ppm) and an AB quartet with a normal ortho
coupling constant (8.3 Hz) (*25*). Confirmation of the structure has been obtained
by synthesis. Hofmann degradation of cularine methiodide (**49**) with refluxing
sodium hydroxide solution produces material of unknown structure (*52*). The
same reaction, when carried out with sodium ethoxide, gives an 83% yield of a
mixture of compounds **50** and **28** (*25*).

Secocularidine (**27**) shows a UV spectrum similar to that of **28** but exhibits a
bathochromic shift in basic solution. Alkylation with diazomethane gives **28**.
The phenolic group is unambiguously located at C-7, as **27** is identical with the
Hofmann degradation product from cularidine methiodide (**51**). Synthetic **52** has
also been prepared from enneaphylline methiodide (**53**) and proved different
from **27** (*25*).

The ^{1}H-NMR spectrum of norsecocularine (**29**) shows the aromatic part to be
very similar to that of other secocularines (*36*). The presence of a band at 3400
cm^{-1} in the IR spectrum agrees with the presence of a single *N*-methyl group,
confirmed by MS (base peak at *m/e* 44). N-Methylation of **29** with formalde-
hyde–sodium borohydride gave secocularine (**28**) (*36*).

49 $R_1=R_2=R_3=Me$
51 $R_1=H$, $R_2=R_3=Me$
53 $R_1=R_3=Me$, $R_2=H$

28 $R_1=R_2=R_3=Me$
27 $R_1=H$, $R_2=R_3=Me$
52 $R_1=R_3=Me$, $R_2=H$
29 $R_1=R_2=Me$, $R_3=H$

50

2. C-Ring Secocularines: Noyaine

Obtained as an amorphous solid, noyaine (**30**) exhibits a UV spectrum (220,
254, and 312 nm) that does not change on addition of acid or base (*36*). The IR
spectrum displays a complex pattern of carbonyl absorptions. The lactamic system is readily deduced from the two multiplets (3.49 and 2.92 ppm) and the low-
field *N*-methyl group (3.01 ppm) in the ^{1}H-NMR spectrum, together with the

appearance of an mass spectral peak at m/e 190 assigned to the fragment **54**. A second carbonyl absorption in the ^{13}C-NMR spectrum (165.99 ppm) and a characteristic methyl carbon (56 ppm) different from normal Ar—OMe resonances indicate the presence of a carbomethoxy group as a substituent in a tetrasubstituted benzene ring, with two aromatic protons shown as singlets (7.46 and 6.17 ppm). Reduction of **30** with lithium aluminium hydride gives the alcohol **55**. Total synthesis of **30** has been achieved by Ullmann condensation of isoquinoline **56** and bromo derivative **57**, followed by oxidation of the resulting isoquinoline with potassium permanganate (*36*).

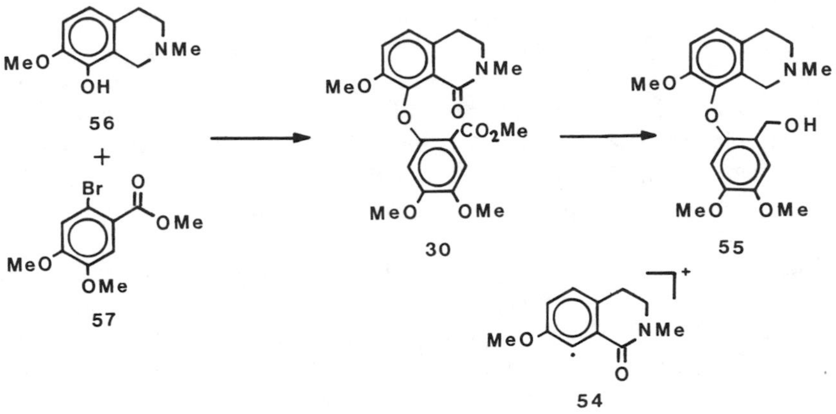

IV. Total Synthesis

Two general approaches have been used for synthesis of cularines. One involves the formation of the diaryl ether linkage at the initial stages and subsequent formation of the B and C rings. The other approach begins with the synthesis of appropriately substituted benzylisoquinolines and reserves construction of the oxepine ring as the last step.

A. FROM SUBSTITUTED DIARYL ETHERS

The interesting approach developed by Kametani *et al.* (*53, 54*) for the synthesis of (±)-cularine (**11**) and (±)-cularimine (**3**) forms the tetracyclic system in a single step by cyclization of dicarboxylic acid **58** to key lactone **59** by polyphosphoric acid. This has now been applied to a synthesis of phenolic cularidine (**8**) (Scheme 2). The major problem, that the protective group must be compatible with the conditions of the cyclization, is resolved by the use of either isobutyl or ethyl groups. However, selective removal of isobutyl or ethyl groups from cularimines **60** and **61** is not possible (*55, 56*).

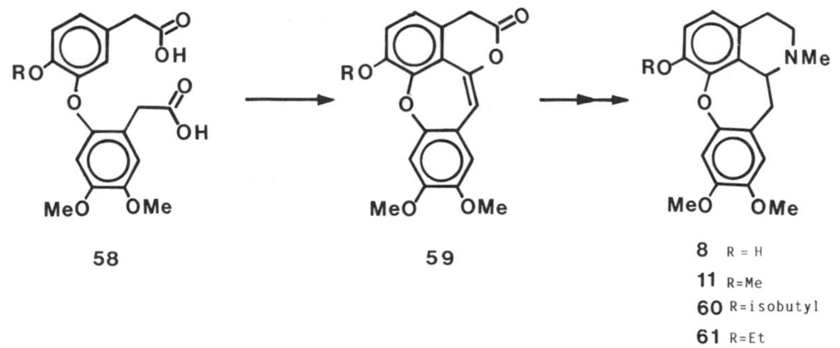

58   59   8 R = H
                                                        11 R=Me
                                                        60 R=isobutyl
                                                        61 R=Et

SCHEME 2

The first synthesis of (±)-cularine (**11**) (*57*), using oxepinones as intermediates, has been modified by Noguchi and MacLean (*26*) for a total synthesis of phenolic cularicine (**6**) (Scheme 3). Attempts to cyclize the acid chloride **62a** under a variety of Friedel–Crafts conditions had led to failure owing to the sensitivity of the phenol protective group to Lewis acid. However, under Bischler–Napieralsky conditions **62b** is cyclized to give the key oxepinone **63** after hydrolysis. Cyclization of the acetal **64** produces 4-ethoxynorcularicine (**65**), which is readily converted to (±)-cularicine (**6**).

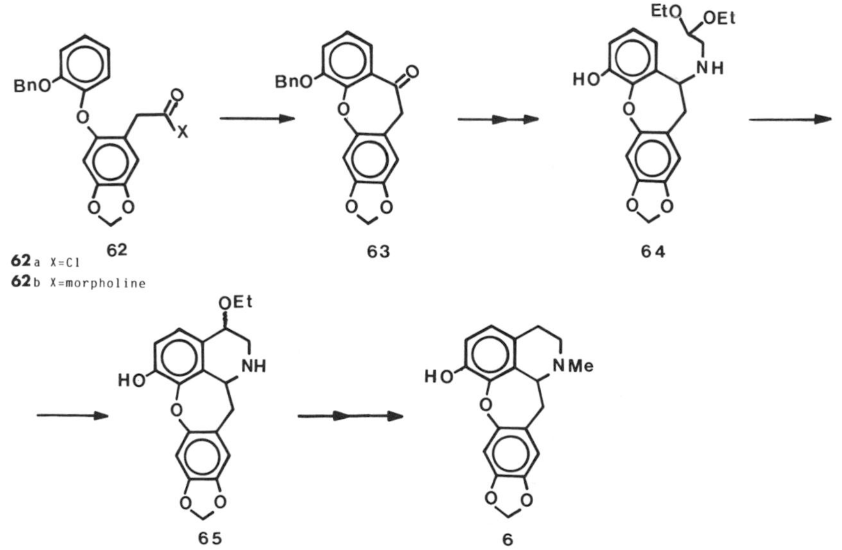

62a X=Cl   62   63   64
62b X=morpholine

65   6

SCHEME 3

B. FROM BENZYLISOQUINOLINES

There are two problems to be solved in perfecting the benzylisoquinoline approach: synthesis of 7,8-substituted benzylisoquinolines with appropriate functional groups and formation of the oxepine ring. Bischler–Napieralsky cyclization of amides (66) is known to afford 6,7-substituted benzylisoquinolines. To avoid this unwanted reaction, two new approaches have been developed, namely, activating position C-2 and blocking position C-6.

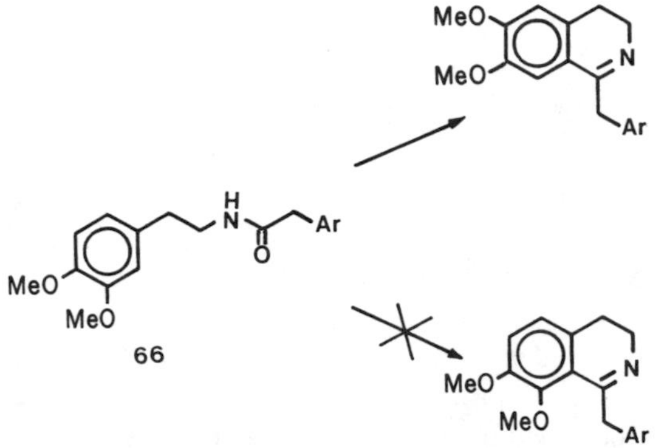

Ishiwata *et al.* (*58*) have succeeded in synthesizing (±)-cularine (11) from amide 67. Ethoxycarbamide selectively orients the cyclization to the para position, giving 68 after N-methylation and reduction. Removal of activating and protecting groups left compound 69 ready for an Ullmann cyclization.

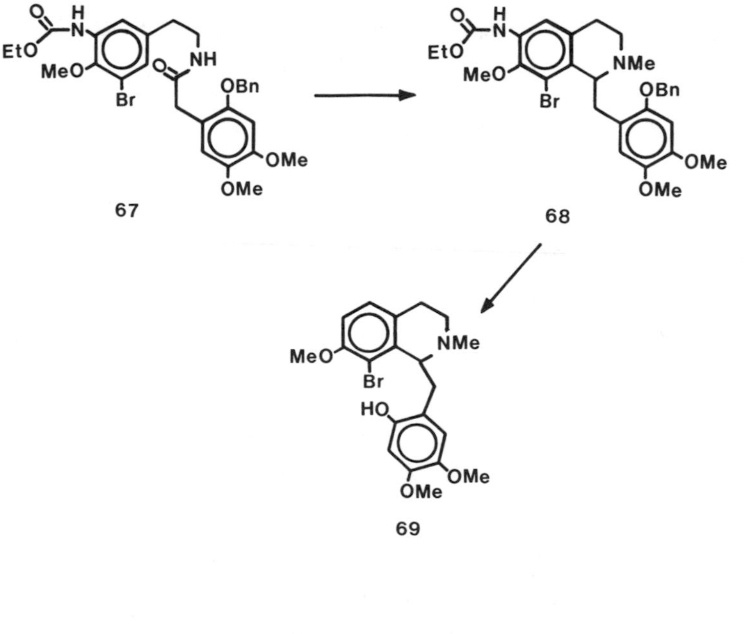

Two variations have been introduced by Iida et al. (30, 59, 60) in order to synthesize several cularine alkaloids. After cyclization, reduction, and N-methylation, the dibrominated amide **70** affords the 7,8-substituted isoquinoline **71**, whose bromine at C-2′ is needed for the subsequent Ullmann cyclization. The protective group and the second bromine are removed after ring closure.

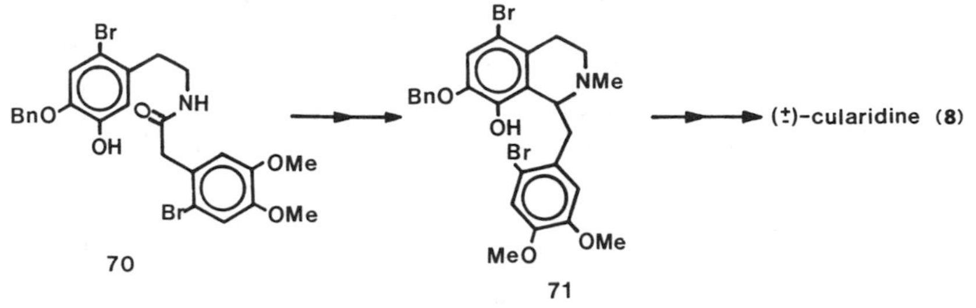

A Pictet–Spengler reaction for the synthesis of benzylisoquinoline **72** has been used by Kametani et al. (43) to synthesize (±)-sarcocapnidine (**13**) and (±)-sarcocapnine (**14**). Again, bromine is used as a blocking group to force cyclization at the desired position.

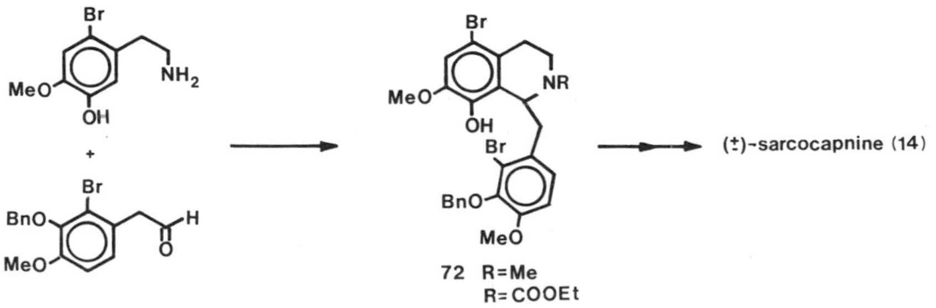

The Ullmann reaction gave better yields with tertiary amines than with amines protected by an ethoxy carbonyl group.

An alternative approach (61) to the synthesis of the strategic benzylisoquinoline **73** uses electrochemical reductive alkylation of isoquinolinium salts.

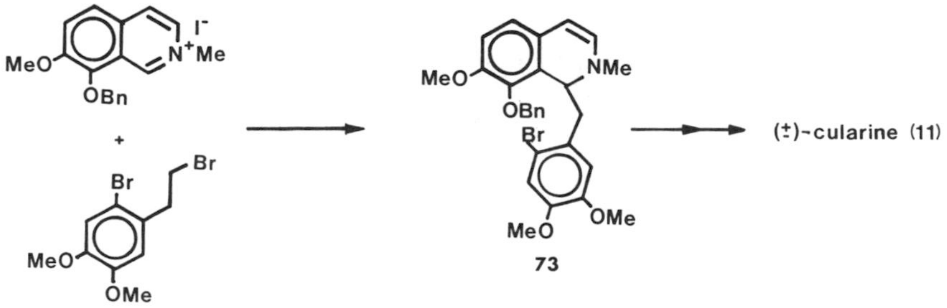

Reduction of **73** and deprotection followed by standard Ullmann reaction give an excellent yield of (±)-cularine (**11**).

An alternative synthesis has been used by Jackson and co-workers to synthesize (±)-cularine (**11**) (*62, 63*). Remarkable improvement in the synthesis of the required (±)-crassifoline (**75**) is obtained by alkylating the Reissert compound **74,** easily prepared from *O*-vanillin by a small modification of the Pomerantz–Fritsch cyclization.

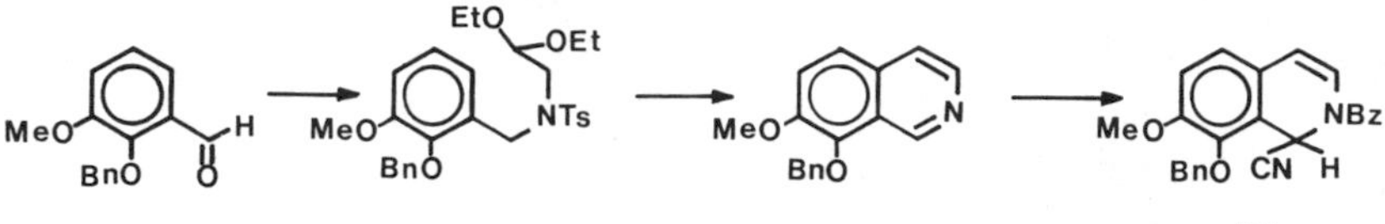

74

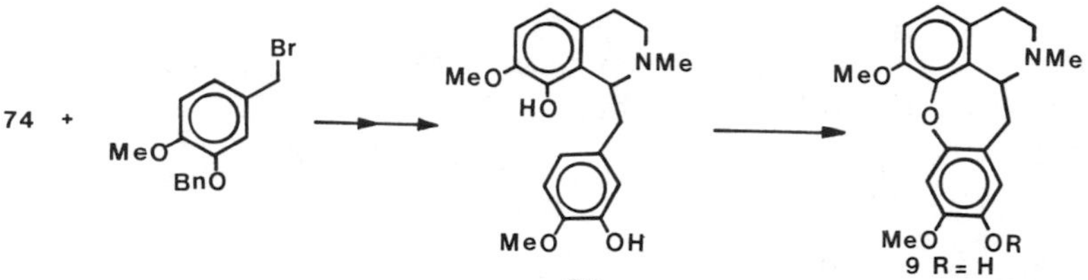

75   9 R = H
          11 R = Me

To form the biaryl ether linkage, a biosynthetic type of oxidative phenolic coupling is used. After trials employing several oxidizing reagents, the best results were obtained with potassium ferricyanide in a two-phase system. Following a diazomethane reaction, low yields of two para coupling products, (±)-cularine (**11**) and (±)-enneaphylline (**9**), are obtained.

Kametani *et al.* (*39, 40*) report some differences in the ferricyanide oxidation of (±)-crassifoline (**75**). Although para coupling occurred, an ortho-coupled product, (±)-sarcocapnidine (**13**), was the main product; however, overall yield of both products was less than 10%.

The biomimetic cyclization of a (±)-crassifoline–borane complex (**76**, R = H), using VOF₃ in TFA, results in a spectacular improvement in the synthesis of (±)-enneaphylline (**9**), now obtained in 40% yield (*64*). Even more interesting is

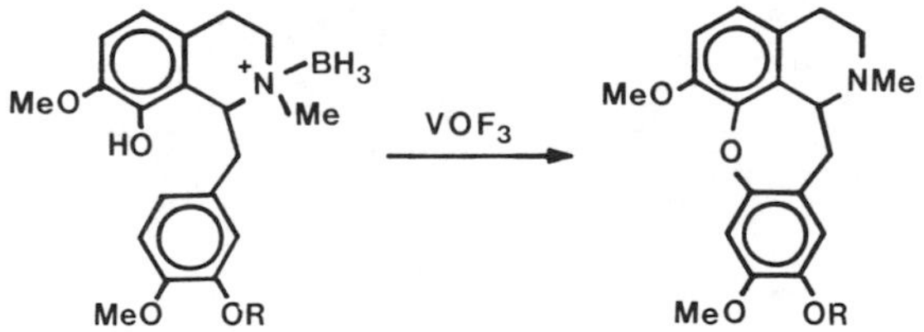

76   9 R=H
          11 R=Me

the reaction of VOF_3 with the borane complex of monophenolic benzyliso-quinoline (**76**, R = Me) which gives a 70% yield of ($\pm$)-cularine (**11**). The absence of ortho coupling is noteworthy and limits the use of this reaction to synthesis of 7,3′,4′-substituted cularines. In theory, if the last reaction could be reversed, introduction of the phenolic group at the 2′ position of the benzyliso-quinoline would permit the synthesis of 7,4′,5′-substituted cularines and greatly simplify the synthesis of the isoquinoline moiety. However, no cyclization is observed with model compound **77** and VOF_3 (*65*).

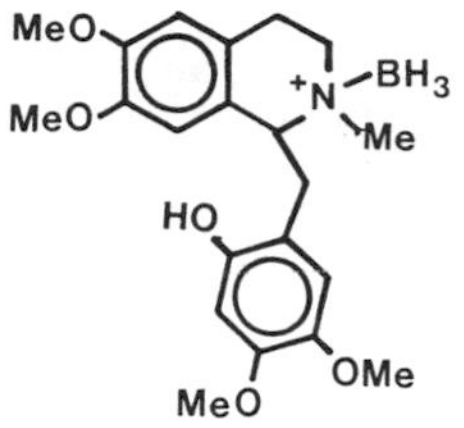

77

Taking advantage of the Reissert approach to the synthesis of 7,8-disubstituted benzylisoquinolines, followed by Ullmann cyclization, Castedo *et al.* have car-ried out convergent synthesis of several cularine alkaloids and obtained excellent yields (*48*). As depicted in Scheme 4, condensation of Reissert compound **74** with benzylchlorides **78** under phase-transfer conditions and hydrolysis gives benzylisoquinolines (**79**). N-Methylation, reduction, and deprotection give **80**, which, after Ullmann reaction, produced the following alkaloids: ($\pm$)-cularine (**11**) from **80a** (91% yield), ($\pm$)-sarcocapnine (**14**) from **80b** (86%), and O-methyl cularicine (**81**) from **80c** (83%).

Benzylisoquinoline **80b** was also used in the first total syntheses of 4-hy-droxysarcocapnine, yagonine, and aristoyagonine (*48*). Lead tetraacetate oxida-tion of **80b** gives *o*-quinol acetate **82**, which after rearrangement gives a racemic mixture (1:9) of epimeric 4-acetoxy derivatives (**83**). Ullmann reaction of **83**

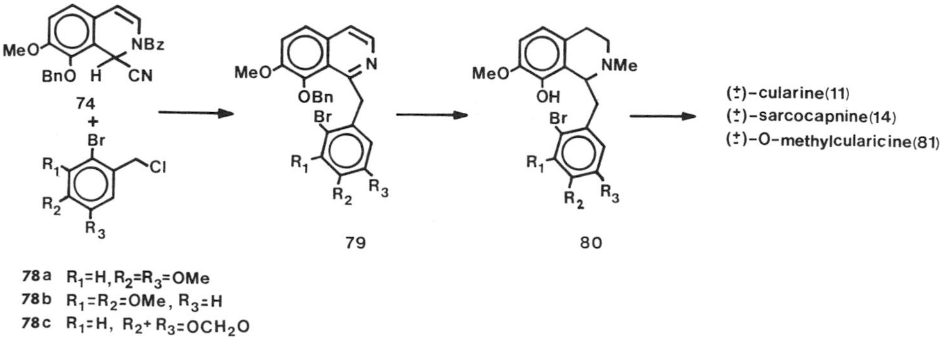

78a R_1=H, R_2=R_3=OMe
78b R_1=R_2=OMe, R_3=H
78c R_1=H, R_2+R_3=OCH$_2$O

SCHEME 4

gives a 92% yield of the epimeric mixture of 4-hydroxysarcocapnine (**16**), with the natural epimer as the minor component. Treatment of **16** with DDQ yields 40% of yagonine (**25**), which, after oxidative decarbonylation, gives aristoyagonine (**26**) (*24*) (Scheme 5). Attempts to directly synthesize oxocularines and tetradehydrocularines by Ullmann cyclization of benzylisoquinolines **84** and **85** have been unsuccessful.

Nucleophilic aromatic substitution, based on an intramolecular attack of the phenoxide ion in the benzyne intermediate **87** (generated by dimsyl sodium treatment of **86**), has been used by Castedo *et al.* (*66*) in a versatile synthesis of cularine alkaloids. However, N-attack competes with O-attack in all instances to give dibenzopyrrocoline derivates (**88**), together with tetradehydrocularines (**89**). The main value of this approach is the easy conversion of **89** to norcularines, by catalitic hydrogenation; to cularines, by quaternization with methyl iodide fol-

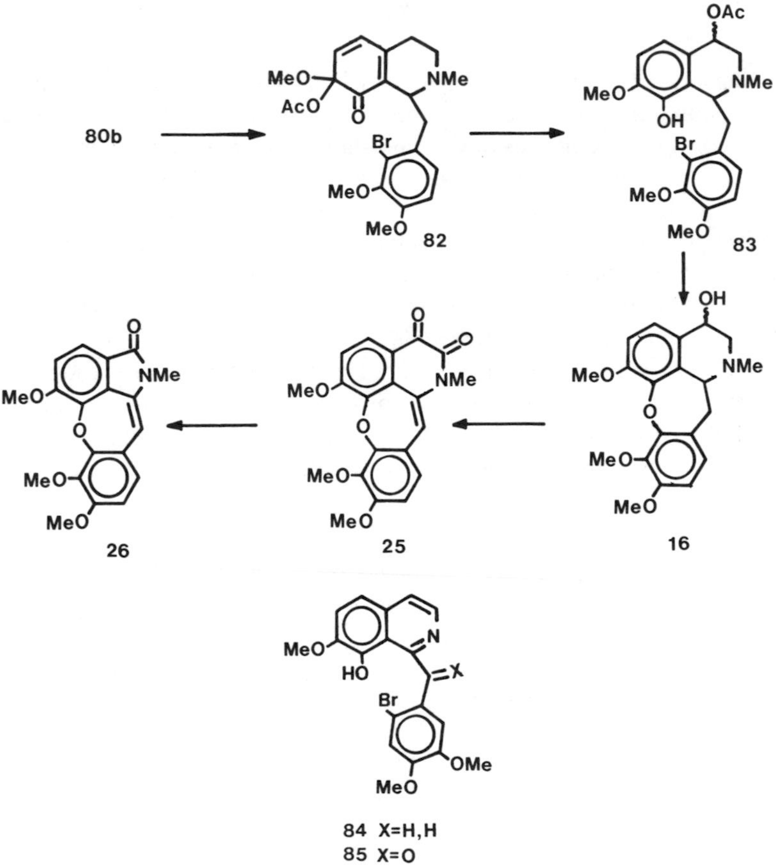

84 X=H,H
85 X=O

SCHEME 5

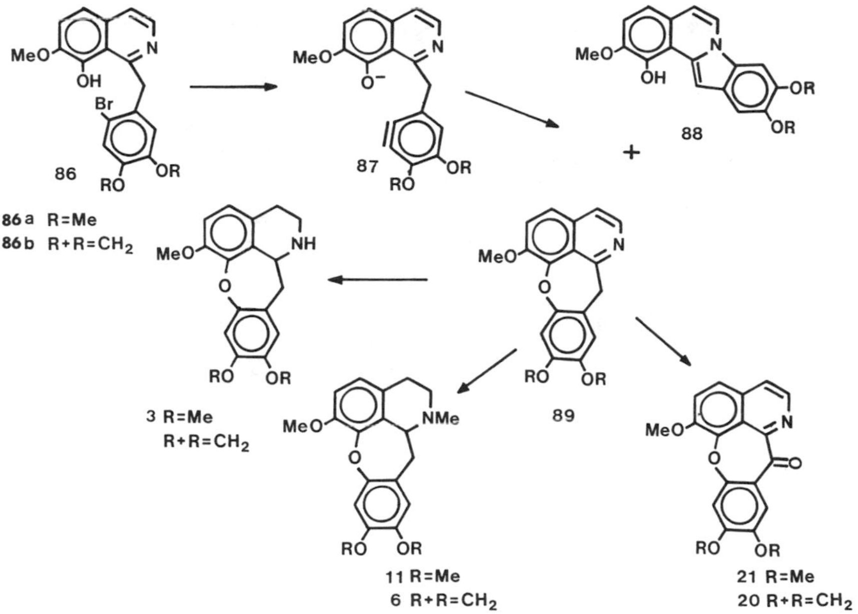

SCHEME 6

lowed by reduction; or to oxocularines, by oxidation with Fremy's salt (*23*) (Scheme 6).

The *N*-methosalt **90** forms the violet derivative **91** in basic media, preventing cyclization. Moreover, the oxobenzylisoquinoline **92,** a direct precursor of ox-

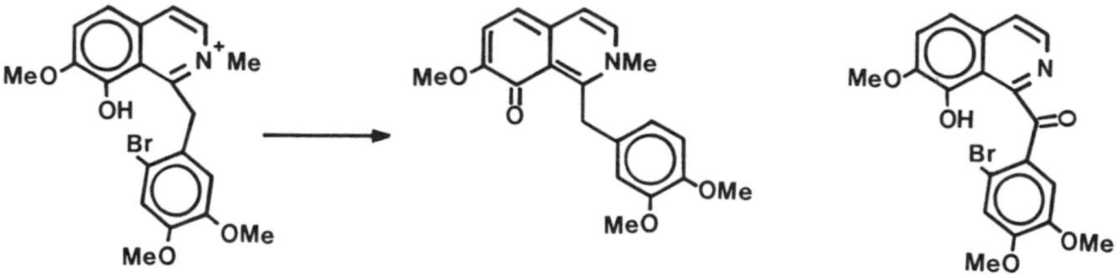

90 91 92

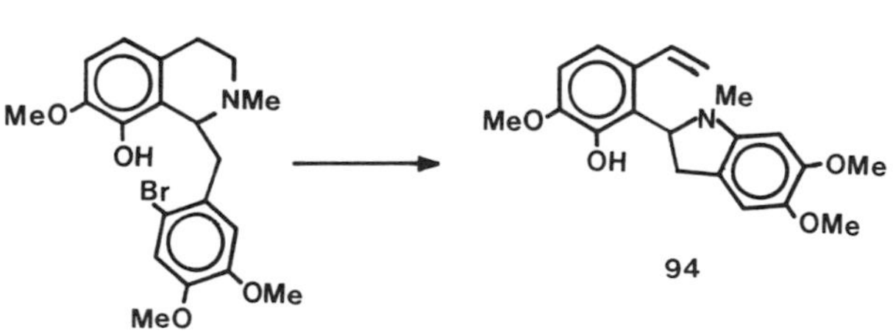

93 94

ocularines, does not cyclize to the corresponding oxocularine when treated with
dimsyl sodium. Furthermore, the indoline **94** is the only product resulting from
N-attack on the benzyne generated from isoquinoline **93.**

Failure was met with when the formation of the dibenzoxepine ring of cularines was attempted by generating an electron-deficient system like that of *ortho*-
quinones or *para*-quinone methides in ring C of an 8-hydroxybenzylisoquinoline
(*66*). Finally, acid treatment of benzylisoquinoline **95** (*66*) gives benzofuran **96,**
which is related to the quettamine class of alkaloids (*67*).

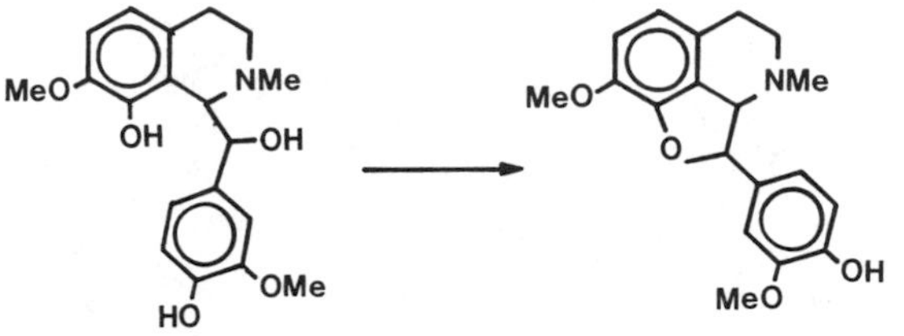

V. Stereochemistry

Originally the absolute configuration of (+)-cularine (**11**) was suggested to be
(*R*) by optical rotatory dispersion measurements of its sodium–liquid ammonia

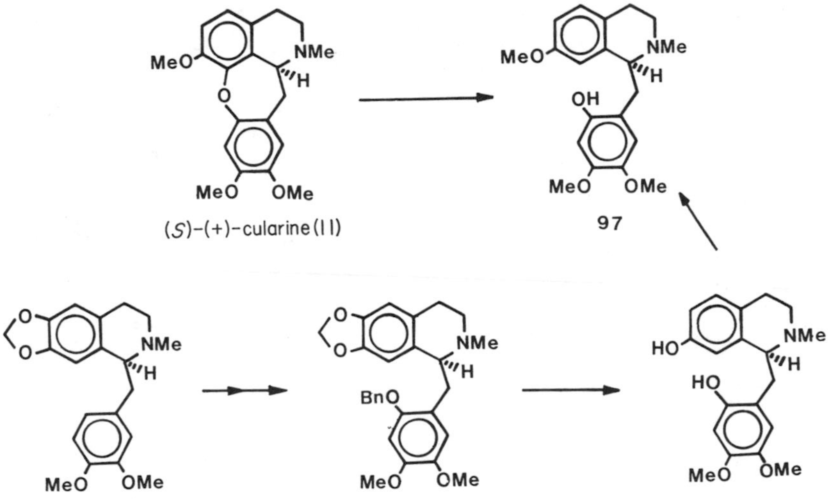

SCHEME 7

reduction product (**97**) (*32*). The opposite configuration, (*S*)-(+)-cularine has subsequently been assigned by chemical correlation of **97** with (*S*)-romneine (**98**), whose stereochemistry had previously been correlated to that of (*S*)-laudanonosine (*74*). The key step of this transformation is the sodium–liquid ammonia reductive cleavage of derivative **99** to give diphenolic benzylisoquinoline **100** (Scheme 7).

Confirmation for the (*S*) of (+)-cularine has been obtained by three-dimensional X-ray diffraction analysis of its methiodide (*75*). The two independent molecules in the crystal lattice have the same conformation, with a twist–boat-shaped dihydrooxepine ring, the oxygen atom being the bow. Since all cularine alkaloids are strongly dextrorotatory like (*S*)-(+)-cularine, and many of them have been chemically correlated with (+)-cularine, it has been concluded that all naturally occurring cularines have the (*S*) configuration at C-1. Similar arguments are advanced to show that the most probable configuration for (+)-limousamine (**15**) and (+)-4-hydroxysarcocapnine (**16**) is (1*S*,4*S*) (*19, 20*).

VI. Biosynthesis

Several routes have been proposed for the biosynthesis of cularine alkaloids, involving one of four types of biogenetic precursors:

A. Bis(phenethyl)amines
B. 1-Benzylisoquinolines
C. Aporphines
D. Protoberberinium salts

Precursors A and B are thought to be the starting compounds for the most common cularines, while C and D are believed to be precursors in the biosynthesis of specific types of cularine alkaloids.

A. FROM BIS(PHENETHYL)AMINES

Phenolic oxidative coupling of the diphenolic bis(phenethyl)amine **101** to the 11-membered ring structure (**102**) controls the formation of the isoquinoline system at the stage of iminium salt **103** and explains the 7,8-substitution pattern in the isoquinoline part of cularine alkaloids (*40, 63*). However, to date, no such precursors have been isolated from natural sources. Moreover, phenolic oxidative coupling of synthetic precursors such as **101,** using a variety of one electron oxidation reagents, gives rise to polymerized compounds or products with the skeleton of erythrine alkaloids (**104**) (*68*).

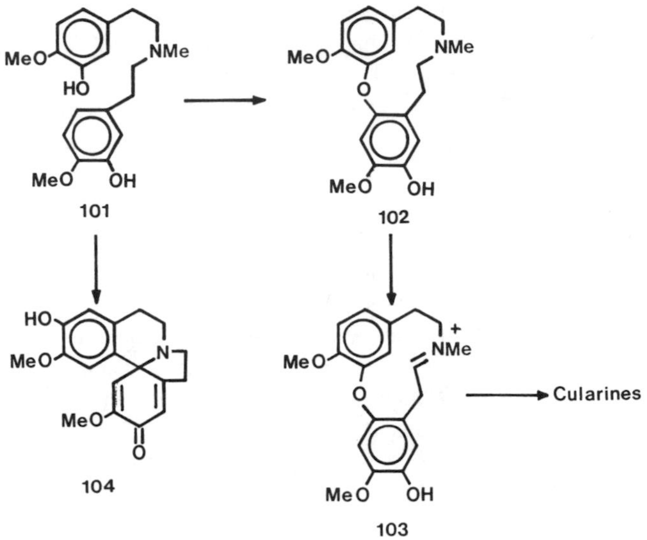

B. FROM BENZYLISOQUINOLINES

Starting from diphenolic tetrahydrobenzylisoquinolines, two different routes of biosynthesis have been postulated. As depicted in Scheme 8, phenolic oxidative coupling of a 8,4'-diphenolic benzylisoquinoline **105** leads to the procularines (**106**), which undergo dienone–phenol rearrangement to give 3',4'-(**107a**) and 4',5'-substituted (**107b**) cularines or which are reduced to **108**, undergoing a dienol–benzene rearrangement, to give cularines (**109**) with loss of an

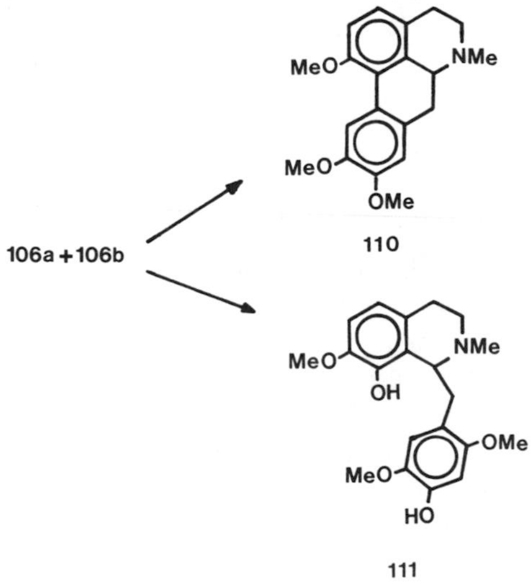

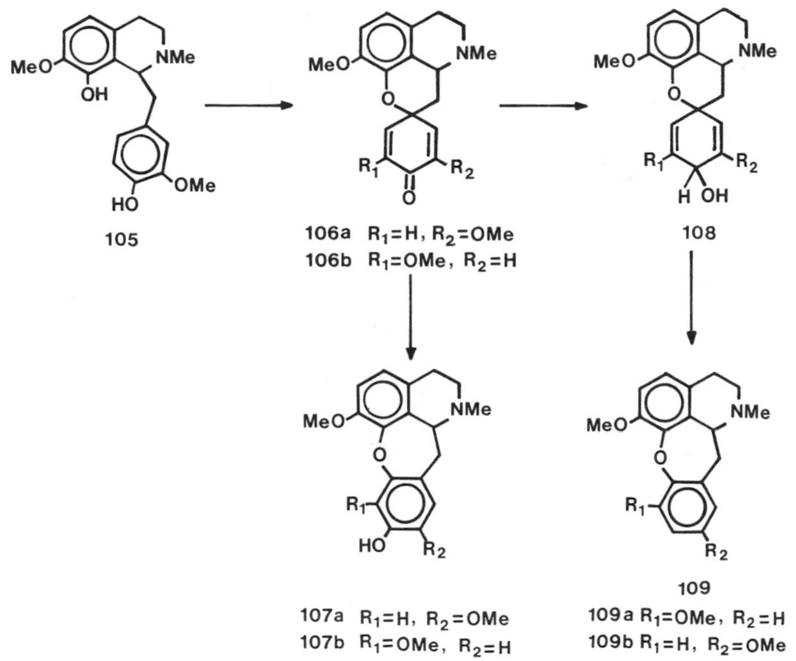

SCHEME 8

oxygenated substituent. However, no natural procularines or simple D-ring monooxygenated cularines have yet been found. Moreover, none of the diastereomeric procularines (106) produced by potassium ferricyanide oxidation of 105 undergo rearrangement to the corresponding cularine (107). Instead, they formed aporphines (110) (69) and C-ring trioxygenated benzylisoquinolines (111) (70).

An alternative biosynthetic pathway originating from tetrahydrobenzylisoquinolines like 112 has been proposed to explain the formation of quettamines (113), a new class of isoquinoline alkaloid (67, 71). In laboratory synthesis, the 1,6-addition to the p-quinonemethide 114 (path a) competes favorably with the 1,4-addition (path b) which leads to the cularine skeleton (66) (Scheme 9).

The second hypothetical route suggests a direct oxidative phenolic coupling of the 8,3'-diphenolic benzylisoquinoline 115. Para and ortho coupling would afford, respectively, 7,3',4'- and 7,4',5'-trioxygenated cularines (Scheme 10). Isolation of (+)-crassifoline (115) and several cularine alkaloids from both Sarcocapnos crassifolia (17) and Corydalis claviculata (10, 15, 17) lends support to this pathway. Conclusive proof that crassifoline is the biogenetic precursor of cularine alkaloids is provided by results of experiments in which tritium-labeled crassifoline was fed to Corydalis claviculata (72). Analysis of the nonphenolic fraction of the plant alkaloids indicated that cularine was the source of most of

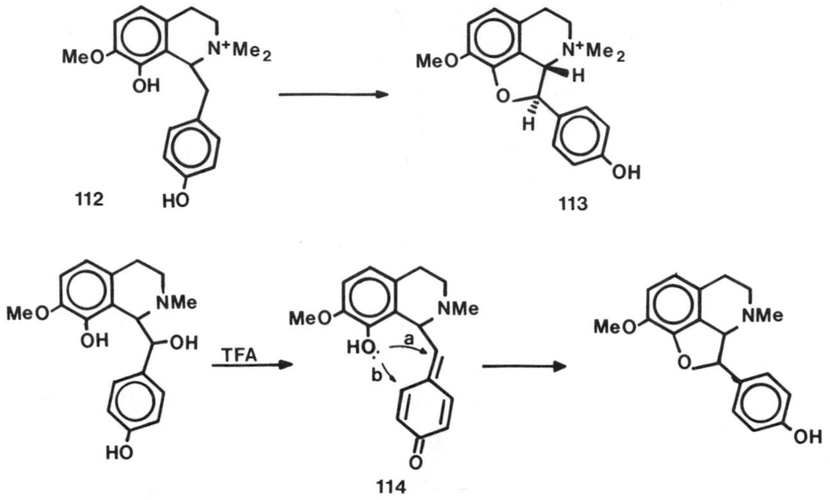

SCHEME 9

the radioactivity. The presence of the protoberberine alkaloids caseamine (**116**) and caseadine (**117**), together with cularines in *Ceratocapnos heterocarpa*, strongly suggests that a benzylisoquinoline such as crassifoline could be a common precursor (*14*). Furthermore, reaction of (±)-crassifoline with postassium ferricyanide produced a mixture of sarcocapnidine (**13** and enneaphylline (**9**) (*39, 40, 62, 63*).

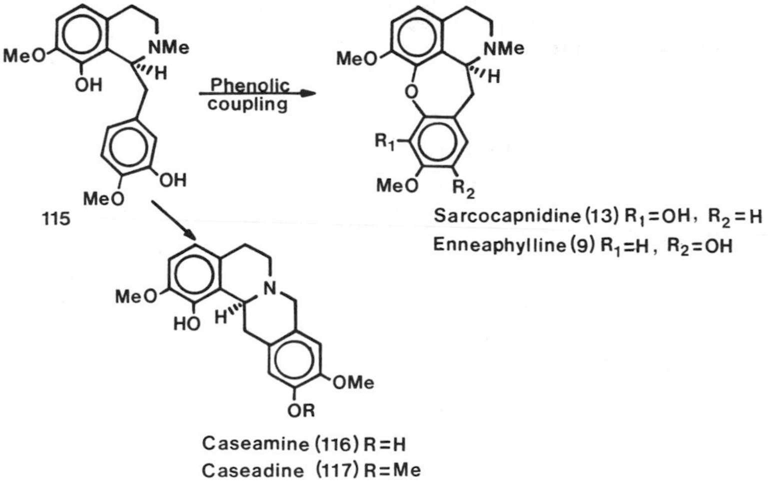

Sarcocapnidine (13) R_1=OH, R_2=H
Enneaphylline (9) R_1=H, R_2=OH

Caseamine (116) R=H
Caseadine (117) R=Me

SCHEME 10

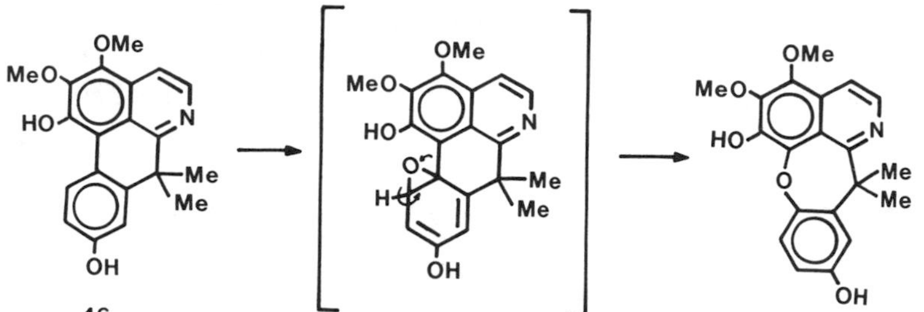

46 17

SCHEME 11

C. FROM APORPHINES

In *Guatteria ouregou* the unusual functional grouping and substitution pattern of gouregine (**17**) and the coproduction of the aporphine melosmine (**46**) leads to the suggestion that **17** may be formed by oxidation of **46** at positions 11–11a or 7a–11a, followed by rearrangement of the biarylic bond, as indicated in Scheme 11. In fact, oxidation of **46** with Fenton's reagent gives **17** in high yield *(21)*.

D. FROM PROTOBERBERINIUM SALTS

Protoberberinium salts have been suggested as starting compounds for the biogenesis of the nonclassical cularine alkaloids linaresine (**18** and dihydrolinaresine (**19**). This hypothesis is based on the copresence of poliberbine

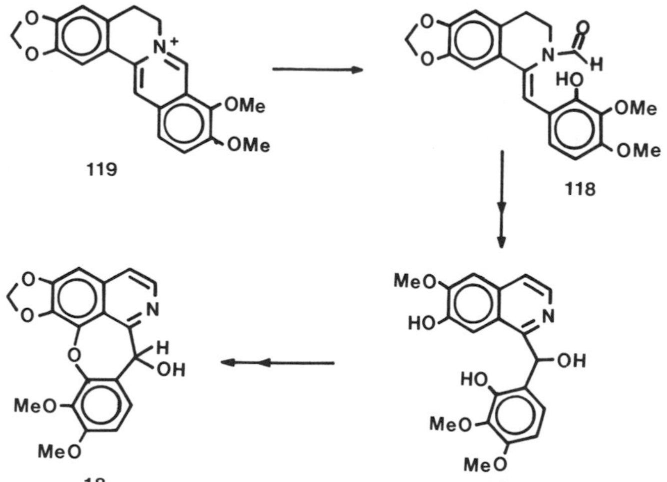

119 118 120 18

SCHEME 12

(**118**) with two cularines in *Berberis valdiviana* (*22*) (Scheme 12). Since oxidation of berberinium salt **119** gives **118,** it is reasonable to assume that its diphenolic analog **120** could undergo intramolecular oxidative phenolic coupling to give **18** after oxidation (*22*).

In conclusion, oxidized cularine alkaloids might be formed by reactions analogous to those occurring in aporphine biosynthesis (*73*), according to the following general sequence:

Crassifoline → 4-Hydroxycrassifoline ← Noradrenaline
↓   ↓
Cularines → 4-Hydroxycularines
↓   ↓
Oxocularines   3,4,-Dioxocularines → Aristocularines

VII. Pharmacological Properties

Because of the small number of cularine alkaloids, whose existence was known until very recently, there are very few published studies about their physiological effects. However, studies of the pharmacological activity of (+)-cularine hydrochloride and some comparison with other isoquinoline alkaloids such as papaverine and hydrastine have been reported (*76*). Dilute solutions of (+)-cularine produced marked anesthesia of the rabbit cornea. The effects on smooth muscle included stimulation of the uterus and depression of intestinal motility. Heart tone and contractility were also increased. When injected intravenously into rabbits, (+)-cularine **11** produced a drop in blood pressure which was followed by recovery to the normal level.

In studies carried out with emetine-resistant mutants of Chinese hamster ovary cells to establish the structural basis for cross-resistance and the common site of action of benzoisoquinoline, phenanthroindolizidine, and phenanthroquinolizidine alkaloids (*77*), no cross-resistance was observed with cularine or other isoquinoline alkaloids.

Finally, it may be pointed out that cularines possess roughly the same N–O–O triangle arrangement that is common to many antineoplastic agents (*78*). This suggests that it may be worthwhile testing their activity in this area.

REFERENCES

1. R. H. F. Manske, *in* "The Alkaloids" (R. H. F. Manske and H. L. Holmes, eds.), Vol. 4, p. 249. Academic Press, New York, 1954.
2. R. H. F. Manske, *in* "The Alkaloids" (R. H. F. Manske and H. L. Holmes, eds.), Vol. 10, p. 463. Academic Press, New York, 1967.
3a. M. Shamma, "The Isoquinoline Alkaloids," Chap. 6, Academic Press, New York, 1972.

3b. M. Shamma and J. L. Moniot, "Isoquinoline Alkaloids Research 1972–1977," Chap. 7, Plenum, New York, 1978.
4a. T. Kametani, "The Chemistry of Isoquinoline Alkaloids," Vol. 1, Chap. 7, Hirokawa Publishing Co., Tokyo, 1968.
4b. T. Kametani, "The Chemistry of Isoquinoline Alkaloids," Vol. 2, Chap. 8, Kinkodo Publishing Co., Sendai, 1974.
5. Specialist Periodical Reports, "The Alkaloids," Vols. 1–13. Chemical Society, London, 1971–1983.
6. B. Gözler and M. Shamma, *J. Nat. Prod.* **47**, 753 (1984).
7. R. Rodrigo, in "The Alkaloids" (R. H. F. Manske and H. L. Holmes eds.) Vol. 14, p. 407. Academic Press, New York, 1973.
8. P. Candau and A. Soler, *Botanica Macaronesica* **8–9**, 147 (1981).
9. D. P. Allais and H. Guinaudeau, *J. Nat. Prod.* **46**, 881 (1983).
10. H. Guinaudeau and D. P. Allais, *Heterocycles* **22**, 107 (1984).
11. R. H. F. Manske, *Can. J. Res.* **16B**, 81 (1938).
12. J. M. Boente, L. Castedo, A. Rodriguez de Lera, J. M. Saá, R. Suau, and M. C. Vidal, *Tetrahedron Lett.*, 1829 (1984).
13. R. H. F. Manske, *Can. J. Chem.* **43**, 989 (1965).
14. R. Suau, M. Valpuesta, and M. V. Silva, unpublished results.
15. G. Blaschke and G. Scriba, *Z. Naturforsch.* **38c** 670 (1983).
16. G. Blaschke and G. Scriba, *Phytochemistry* **24**, 585 (1985).
17. J. M. Boente, L. Castedo, R. Cuadros, A. Rodriguez de Lera, J. M. Saá, R. Suau, and M. C. Vidal, *Tetrahedron Lett.*, 2303 (1983).
18. M. J. Campello, L. Castedo, J. M. Saá, R. Suau, and M. C. Vidal, *Tetrahedron Lett.*, 239 (1982).
19. D. P. Allais and H. Guinaudeau, *Heterocyles* **20**, 2055 (1983).
20. L. Castedo, D. Dominguez, A. Rodriguez de Lera, and E. Tojo, *Tetrahedron Lett.*, 4573 (1984).
21. M. Leboeuf, D. Cortes, R. Hocquemiller, A. Cavé, A. Chiaroni, and C. Riche, *Tetrahedron* **38**, 2889 (1982).
22. S. Firdous, A. J. Freyer, M. Shamma, and A. Urzua, *J. Am. Chem. Soc.* **106**, 6099 (1984).
23. J. M. Boente, L. Castedo, A. Rodriguez de Lera, J. M. Saá, R. Suau, and M. C. Vidal, *Tetrahedron Lett.*, 2295 (1983).
24. M. J. Campello, L. Castedo, D. Dominguez, A. Rodriguez de Lera, J. M. Saá, R. Suau, E. Tojo, and M. C. Vidal, *Tetrahedron Lett.*, 5933 (1984).
25. J. M. Boente, L. Castedo, D. Dominguez, A. Fariña, A. Rodriguez de Lera, and M. C. Villaverde, *Tetrahedron Lett.*, 889 (1984).
26. I. Noguchi and D. B. MacLean, *Can. J. Chem.* **53**, 125 (1975).
27. T. Kametani and S. Shibuya, *J. Chem. Soc.*, 5565 (1965).
28. T. Kametani, S. Shibuya, C. Kibayashi, and S. Sasaki, *Tetrahedron Lett.*, 3215 (1966).
29. R. H. F. Manske, *Can. J. Chem.* **43**, 989 (1965).
30. H. Iida, H. C. Hsu, and T. Kikuchi, *Chem. Pharm. Bull.* **21**, 1001 (1973).
31. L. Castedo, S. Lopez, and E. Tojo, unpublished results.
32. N. S. Bhacca, J. C. Craig, R. H. F. Manske, S. K. Roy, M. Shamma, and W. A. Slusarchyk, *Tetrahedron* **22**, 1467 (1966).
33. D. H. Hughes and D. B. MacLean, in "The Alkaloids" (R. H. F. Manske and R. G. A. Rodrigo, eds.), Vol. 18, p. 217. Academic Press, New York, 1981.
34. M. Ohashi, J. M. Wilson, H. Budzikievicz, M. Shamma, W. A. Slusarchyk, and C. Djerassi, *J. Am. Chem. Soc.* **85**, 2807 (1963).
35. T. Kametani, S. Shibuya, and I. Noguchi, *Yakugaku Zasshi* **85**, 667 (1965).
36. J. Boente, Ph.D. Thesis. Universidad de Santiago. Spain (1985).

37. T. Kametani, S. Shibuya, and S. Sasaki, *Yakugaku Zasshi* **37**, 198 (1967).
38. A. Rodriguez de Lera, C. Villaverde, and L. Castedo, *Heterocycles* **24**, 109 (1986).
39. T. Kametani, K. Fukumoto, and M. Fujihara, *J. Chem. Soc., Chem. Commun.*, 352 (1971).
40. T. Kametani, K. Fukumoto, and M. Fujihara, *Bioorg. Chem.* **1**, 40 (1971).
41. L. Castedo, J. M. Saá, R. Suau, and M. C. Villaverde, *Heterocycles* **9**, 659 (1978).
42. L. Castedo, R. Riguera, and J. Sardina, *An. Quim.* **77C**, 138 (1981).
43. T. Kametani, K. Fukumoto, and M. Fujihara, *Chem. Pharm. Bull.* **20**, 1800 (1972).
44. L. Castedo, R. Riguera, and J. Sardina, S. Garcia-Blanco, and I. Fonseca, *Heterocycles* **19**, 1591 (1982).
45. M. H. Abu Zarga and M. Shamma, *Tetrahedron Lett.*, 3739 (1980).
46. V. Zabel, W. H. Watson, C. H. Phoebe Jr., J. E. Knapp, and P. L. Slatkin, *J. Nat. Prod.* **45**, 94 (1982).
47. T. Kametani and T. Honda, *in* "The Alkaloids" (A. Brossi, ed.), Vol. 24, p. 153, Academic Press, New York, 1985.
48. A. Rodriguez de Lera, R. Suau, and L. Castedo, submitted for publication.
49. D. B. Mix, H. Guinaudeau, and M. Shamma, *J. Nat. Prod.* **45**, 657 (1982).
50. L. Castedo, A. Mouriño, and R. Suau, *Tetrahedron Lett.*, 50 (1976).
51. H. Guinaudeau, M. Leboeuf and A. Cavé, *J. Nat. Prod.* **46**, 761 (1983).
52. R. H. F. Manske, *J. Am. Chem. Soc.* **72**, 55 (1950).
53. T. Kametani, S. Shibuya, S. Seino, and K. Fukumoto, *Tetrahedron Lett.*, 25 (1964).
54. T. Kametani, S. Shibuya, S. Seino, and K. Fukumoto, *J. Chem. Soc.*, 4146 (1964).
55. T. Kametani, H. Iida, and C. Kibayashi, *J. Heterocycl. Chem.* **6**, 61 (1969).
56. T. Kametani, H. Iida, and C. Kibayashi, *J. Heterocycl. Chem.* **7**, 339 (1970).
57. T. Kametani and K. Fukumoto, *J. Chem. Soc.*, 4289 (1963).
58. S. Ishiwata, T. Fujii, N. Miyaji, Y. Satoh, and K. Itakura, *Chem. Pharm. Bull.* **18**, 1850 (1970).
59. H. C. Hsu, T. Kikuchi, S. Aoyagi, and H. Iida, *Yakugaku Zasshi* **92**, 1030 (1972).
60. H. Iida, H. C. Hsu, T. Kikuchi, and K. Kawano, *Yakugaku Zasshi* **92**, 1242 (1972).
61. T. Shono, T. Miyamoto, M. Mizukami, and H. Hamaguchi, *Tetrahedron Lett.*, 2385 (1981).
62. A. H. Jackson and G. W. Stewart, *Chem. Commun.*, 149 (1971).
63. A. H. Jackson, G. W. Stewart, G. A. Charnock, and J. A. Martin, *J. Chem. Soc., Perkin Trans. I*, 1911 (1974).
64. A. Rodriguez de Lera, J. M. Saá, R. Suau and L. Castedo, submitted for publication.
65. R. Suau, M. L. Ruiz, and A. Rodriguez de Lera, unpublished results.
66. A. Rodriguez de Lera, J. M. Saá, R. Suau, and L. Castedo, submitted for publication.
67. M. H. Abu Zarga, G. A. Miana, and M. Shamma, *Tetrahedron Lett.*, 541 (1981).
68. S. F. Dyke and N. S. Quessy, *in* "The Alkaloids" (R. H. F. Manske and R. Rodrigo, eds.), Vol. 18, p. 1. Academic Press, New York, 1981.
69. T. Kametani, K. Fukumoto, and M. Fujihara, *J. Chem. Soc., Perkin Trans. I*, 394 (1972).
70. A. J. Birch, A. H. Jackson, P. V. R. Shannon, and G. W. Stewart, *J. Chem. Soc., Perkin Trans. I*, 2492 (1975).
71. S. Chattopadhyay and M. Shamma, *Heterocycles* **19**, 697 (1982).
72. G. Blaschke and G. Scriba, *Tetrahedron Lett.*, 5855 (1983).
73. M. Shamma and H. Guinaudeau, *Tetrahedron*, **40**, 4795 (1984).
74. J. Kunimoto, K. Morimoto, K. Yamamoto, Y. Yoshikawa, K. Azuma, and K. Fujitani, *Chem. Pharm. Bull.* **19**, 219 (1971).
75. H. Shimanouchi, Y. Sasada, T. Honda, and T. Kametani, *J. Chem. Soc., Perkin Trans. II*, 1226 (1973).
76. A. K. Reynolds, *J. Pharmacol.* **69**, 112 (1940).
77. R. S. Gupta, J. J. Krepinsky, and L. S. Siminovitch. *Pharmacology* **18**, 136 (1980).
78. C. C. Cheng and R. K. Y. Zee-Cheng, *Heterocycles* **15**, 1275 (1981).

INDEX

A

Acetonyldihydrosanguinarine, 16
Adhatodine, 112, 117, 134
Adlumiceine, 61
Adlumidiceine, 12
Adlumine, 54, 61
Aflatrem
  biological properties, 226
  biosynthesis, 227
  structure, 226
Aflavinine, 228
Alborine, 29
Allocryptopine, 21, 27, 43, 46
Alpinigenine, 24, 27
1-Amino-2-nitrocyclopentanecarboxylic acid,
  273
Amurensine, 27
Amurensine methohydroxide, 34
Amurensinine, 19, 27
Amurensinine methohydroxide, 5
Amurine, 23, 27, 35
Amuroline, 28
Ancistine, 152
Ancistrine, 152
Ancistrocladeine, 150
Ancistrocladidine, 153
Ancistrocladine
  occurrence of, 146
  structure, 142
  synthesis, 180
Ancistrocladinine, 153
Ancistrocladisine, 151
Ancistrocladonine, 149
Ancistrocladus alkaloids, 142, 158
Ancistrocline, 148
Ancistrocongine, 151
Ancistrocongolensine, 150
Ancistroealaensine, 148
Ancistroquinone, 163
Ancistrotectorine, 153

Angoline, 43
Aniflorine
  occurrence of, 134
  spectral data, 118
  structure, 112
Anisessine
  occurrence of, 134
  spectral data, 119
  structure, 112
Anisotine
  occurrence of, 134
  spectral data, 119
  structure, 112
Anonaine, 32
Aobamidine, 55
Aobamine, 55
Apoaranotin, 215
Apocavidine, 55
Aporheine, 9
Aporheine methohydroxide, 5
Aporphines, 73
Aranotin, 214
Arboricine, 100
Arborine
  biological properties, 131
  biosynthesis, 128
  occurrence of, 134
  structure, 100
Argemonine methohydroxide, 5
Argemonine N-oxide, 36
Arguosin B, 249
Aristoyagonine, 65
Armepavine, 19, 38
Arosinine, 30
Arylpropanones reductive amination,
  178
Aspergillic acid
  biosynthesis, 191
  chemistry, 190
  occurrence of, 186
Asperopterin A and B, 247

325

Asperlicin, 224
Aspirochlorine, 217
Aspochalasins, 241
Asterriquinone, 229
Aszonalenin, 223
Autotensine, 52

B

Benzylisoquinolines, 67
Berberine, 9, 32
Bicucine, 52
Bicuculline, 52
Bicuculline methohydroxide, 6
Bicucullinine, 61
Bisnorargemonine, 37
Bocconoline, 43
Breoganine
    occurrence of, 290
    O-methylation of, 298
    spectral data, 294
Bulbocapnine, 30
Bulbocapnine methohydroxide, 5
Bulbodione, 52

C

Californidine, 5
Canadine, 11
Canadine methohydroxide, 32
Cancentrine, 64
Candidine, 123, 134
Capaurine, 52
Capaurimine, 55
Capnoidine, 57
Carpoxidine, 56
Caseadine, 52
Caseamine, 52, 320
Caseanadine, 52
Cataline, 30
Celtine
    occurrence of, 290
    spectral data, 294, 298
Celtisine
    occurrence of, 290
    spectral data, 294
    structure, 298
Cheilanthifoline, 36
Chelamidine, 42

Chelamine, 42
Chelerythridimerine, 43
Chelerythrine, 19, 49
Chelidonine, 32, 42
Chelilutine, 49, 64
Chelirubine, 29
Chrysogine
    occurrence of, 132
    structure, 103
Claviculine
    occurrence of, 290
    spectral data, 295, 299
Cochliodinol, 229
Codamine, 16
Codaphniphylline, 267
Codeine
    biosynthesis, 17
    occurrence, 16
Codeine N-oxide, 16
Columbamine, 43
Consperine, 53
Coptisine, 12, 17, 43
Coreximine, 16
Corgoine, 53
Corledine, 61
Corlumine, 55
Corycavamine, 52
Corycavidine, 52
Corydalisol, 54
Corydalispiron, 54
Corydalmine, 54
Corydine, 27, 32
Corynolamine, 54
Corynoline, 54
Corynoloxine, 53
Corypalmine, 52
Corysamine, 31, 38
Corytenchine, 55
Corytuberine, 11, 52
Corytuberine methohydroxide, 5
Crassifolazonine, 53
Crassifoline, 312, 320
Cryptoechinulins B, D, and G, 197
Cryptopine, 29
Culacarine, 53
Culacorine, see Breoganine
Cularicine
    occurrence of, 290
    spectral data, 294, 299
    synthesis, 309

Cularidine
  occurrence of, 290
  spectral data, 294
  synthesis, 311
Cularimine
  occurrence of, 290
  O-demethylation, 297
  spectral data, 294
Cularine
  configuration optical isomers, 316
  occurrence of, 290
  spectral data, 295
  synthesis, 309
Cularine alkaloids, biosynthesis, 322
Cularines, 70
Cyclanoline, 46
Cyclopiamine B, 209
Cyclopiazonic acid, 236
Cytochalasins, 237

D

Daphgraciline
  properties, 266
  spectral data, 282
  structure, 284
Daphmacrine, 267
Daphnigraciline
  properties, 266
  spectral data, 279
Daphnigracine
  properties, 267
  spectral data, 279
Daphnilactone-A, structure, 267, 273
Daphniphylline, 267
Daphniphyllum alkaloids
  chemotaxonomy, 265
  pharmacological properties, 285
  toxicity, 265
Daphniteijsmanine
  properties, 266
  spectral data, 270
Dehydroapocavidine, 53
Dehydrocorybulbine, 53
Dehydrocorydaline, 53
Dehydrocorydine, 29
Dehydronanteine, 57
Dehydronormacrostomine, 13
Dehydroroemerine, 20
Dehydrothalictricavine, 53

Demethylenesanguinarine, 43
Demethylsanguinarine, 43
Densiflorine, 60
Deoxyaniflorine
  occurrence of, 134
  spectral data, 118
  structure, 112
Deoxyaspergillic acids, 188
Deoxybrevianamide E, 202
Deoxyhydroxyaspergillic acid, 188
Deoxypeganidine
  occurrence of, 133
  structure, 112
Deoxyuzurimine
  properties, 266
  spectral data, 277
Deoxyvasicine
  occurrence of, 132
  pharmacological properties, 130
  structure, 112, 115
Dicentrine, 29
Dicentrinone, 31
Dihydrochelerythrine, 39
Dihydrochelilutine, 42
Dihydrolinaresine
  occurrence of, 292
  spectral data, 295, 302
Dihydroparfumidine, 60
Dihydroprotopine, 16
Dihydrosanguinarine, 36, 43
Dihydroxyflavinine, 228
Diketones in napthyl isoquinoline alkaloid synthesis, 167
1,3-Dimethylisoquinolines, 173, 178
Dionochophyllaceae alkaloids, 145
Dipegine
  occurrence of, 134
  structure, 112
Ditryptophenaline, 209
Domesticine, 30
Domestine, 53

E

Echinulin, 192
Emerin, 253
Emestrin, 216
Enneaphylline
  occurrence of, 290
  spectral data, 294
  synthesis, 313

Epialpinine, 28
Epicorazine A, 216
Epicorynoline, 54
Epiglaudine, 20
Epioxodaphnigraciline
   properties, 266
   spectral data, 281
Escholamine, 40
Escholanidine, 5
Escholidine, 3, 39
Escholine, 39
Escholinine, 5, 39
Eschscholtzidine, 39
Eschscholtzine, 39

F

Febrifugine
   antimalarial properties, 131
   biosynthesis, 125
   chemistry, 102
   occurrence of, 134
   structure, 100
Ferriaspergillin, 189
Flavacol, 189
Flavinantine, 19, 35
Flavipucine, 244
Florpavidine, 21
Fumaramidine, 60
Fumaramine, 56, 60
Fumariaceae and subtypes
   alkaloids of, 66
   taxonomy, 1, 51
      Corydalis, 51
      Dicentra, 62
      Fumaria, 60
      Sarcocapnos, 65
      Adlumia, 65
Fumaricine, 61
Fumariflorine, 61
Fumariline, 61
Fumaritine, 61
Fumaritine *N*-oxide, 60
Fumarofine, 60
Fumarophycine, 60
Fumarostelline, 61
Fumitremogen A and B
   chemistry, 207, 208
   structures, 204
Fumivailine, 61
Fumschleicherine, 61

G

Galudine, 12
Glaucamine, 20
Glaudine, 20
Glaufidine, 29
Glaufine, 31
Glaunidine, 30
Glaunine, 31
Glaziovine, 19
Gliotoxin
   biosynthesis, 212
   structure, 211
   synthesis, 212
Glomerine
   occurrence of, 132
   properties, 104
   structure, 100
Glycophymoline
   occurrence of, 134
   properties, 105
   structure, 100
Glycorine
   occurrence of, 132
   properties, 105
   structure, 100
Glycosmicine
   occurrence of, 132
   pharmacology, 131
   structure, 106
Gouregine
   occurrence of, 291
   spectral data, 295, 302

H

Hamatine
   occurrence of, 147
   synthesis, 181
Heliotrine, 29
Hexahydromecambrine B, 11
Hinnuliquinone, 229
Homoglomerine, 104, 133
Homosecodaphniphyllate, 275
Hunnemanine, 46
Hydrastine, 47, 53
Hydrastine methohydroxide, 6
Hydrocotarnine, 16
Hydroxybulbocapnine, 31
Hydroxycodeine, 16

Hydroxydaphnigraciline
  properties, 266
  spectral data, 284
Hydroxyneoaspergillic acid, 186
Hydroxypeganine, 114
Hydroxypyrazines, 186
Hydroxysarcocapnine
  configuration optical isomers, 317
  occurrence of, 291
  spectral data, 295
  synthesis, 300, 313
Hydroxythebaine, 16
Hypecorine, 45
Hypecorinine, 45

I

Indenobenzazepines, 88
Indoloquinazoline alkaloids, 123
Isoargemonine, 37
Isoboldine, 16, 29
Isocochliodinol, 229
Isocorybulbine, 53
Isocorydine, 30
Isocorypalmine, 16, 41
Isocorypalmine methohydroxide, 5, 31
Isocorytuberine, 31
Isocularine, 288
Isodaphnilactone-B
  properties, 266
  spectral data, 274
Isoechinulin, 196
Isofebrifugine, 134
Isoflavipucine, 244
Isonorargemonine, 38
Isopavines, 69
Isopeganidine, occurrence of, 133
Isoquinoline alkaloids, biosynthesis of, 170
Isorhoeadine, 9
Isorhoeagenine glycoside, 9
Isothebaine, 24
Isothebaine methohydroxide, 5
Isotriphophyllline, 155
Izmirine, 61

J

Juziphine, 61

K

Kikemanine, 56

L

Lahoramine, 61
Lahorine, 61
Latericine, 21
Laudanidine, 16
Laudanine, 16
Laudanosine, 16
Laurioscholtzine, 39
Laurotetanine, 40
Ledeborine, 54
Ledecorine, 54, 61
Lederine, 54
Limousamine
  configuration optical isomers, 317
  occurrence of, 291
  spectral data, 295
  synthesis, 301
Linaresine
  biosynthesis, 321
  occurrence of, 291
  spectral data, 295, 302
Lirinidine, 19
Liriodenine, 18, 32
LL-S490 beta, 223
Luduine, 30

M

Macarpine, 39
Macrantaline, 26
Macrantolidine, 26
Macrostomine, 13
Magnaflorine, 3, 46
Malaria, febrifugine in, 131
Marschaline, 55
Mecambridine, 6
Mecambridine methohydroxide, 6
Mecambrine, 19
Mecambroline, 19
Melosmine, 321
Menispermine, 5
Methylcodeine, 16
Morphinans, 77
Morphine
  biosynthesis, 17
  occurrence of, 16, 27
Morphine N-oxide, 16
Muramine, 11, 28
Mutaaspergillic acid, 186

N

N-Benzylisoquinolines, 68
N-Methyladlumine, 6, 61
N-Methylasimilobine, 11, 19
N-Methylcrotonsine, 19
N-Methylhydrastine, 54
N-Methyllaurotetanine, 29
N-Methyllindcarpine, 31
N-Methyloridine, 28
N-Methylsinactine, 5
N-Methyltriphophylline, 156
Nandazurine, 52
Nantenine, 20, 52
Naphthalene, biomimetic type synthesis,
  169
Narceine, 16
Narcotine methohydroxide, 6
Narcotoline, 16
Neoaspergillic acid, 186
Neocochliodinol, 229
Neoechinulin, 194
Neohydroxyaspergillic acid
  biosynthesis, 191
  structure, 186
Neolitsine, 31
Neopine, 16
Neoxaline, 211
Nigerazine B, 243
Nigragillin, 243
Nokoensine, 55
Norargemonine, 36
Norbracteoline, 31
Norbreoganine, 297
Norchelerythrine, 38
Norchelidonine, 42
Norcularicine
  occurrence of, 290
  spectral data, 294
  structure, 297
Norcularidine
  occurrence of, 290
  spectral data, 294
Norisocorydine, 31
Norlumidine, 60
Normorphine, 16
Nornarceine, 16
Norprotosinomenine, 64
Norsarcocapnine, 305

Norsecocularine
  occurrence of, 293
  spectral data, 296, 307
Nortryptoquivaline, 220
Noyaine
  occurrence of, 293
  spectral data, 297
Nuciferine, 19
Nuciferoline, 19
Nudaurine, 27

O

O-Methylancistrocladine
  occurrence of, 147
  synthesis, 180
O-Methylatheroline, 29
O-Methylcularicine
  occurrence of, 290
  spectral data, 294
  synthesis, 313
O-Methyldidehydrotriphophylline,
  157
O-Methylfumarophycine, 60
O-Methyltetradehydrotriphophylline
  occurrence of, 157
  synthesis, 176
O-Methylthalisopavine, 29
O-Methyltriphophylline, 155
Ochotensimine, 55
Ochotensinine, 55
Ochrobirine, 56
Omoflavipucine, 247
Oreodine, 21
Oreogenine, 21
Oridine, 28
Orientalidine, 24, 26
Orientalin, biosynthesis, 3
Oripavine, 20, 27
Oxaline, 224
Oxocompostelline
  occurrence of, 292
  spectral data, 295, 303
Oxocoptisine, 61
Oxocorynoline, 54
Oxocryptopine, 16
Oxocularine
  occurrence of, 292
  spectral data, 295, 303

Oxodaphnigraciline
  properties, 266
  spectral data, 281
Oxodaphnigracine
  properties, 266
  spectral data, 281
Oxoglaucine, 30
Oxonanthenine, 52
Oxoprotopine, 13
Oxosarcocapnidine
  occurrence of, 292
  spectral data, 296, 304
Oxosarcocapnine
  occurrence of, 292
  spectral data, 296, 304
Oxosarcophylline
  occurrence of, 292
  spectral data, 296, 304
Oxyhydrastinine, 38
Oxysanguinarine, 11

P

Palaudine, 16
Palmatine, 19
Pancoridine, 56
Pancorine, 56
Papaveraceae and subtypes
  alkaloids of, 7, 14, 15, 66–91
  historical survey, 10
  quaternary alkaloids, 5
  taxonomy, 1
    Argemonorhoeades, 12
    Argemone, 35
    Bocconia, 43
    Carinatae, 13, 16
    Chelidonium, 42
    Dicranostigma, 40
    Eschscholzia, 39
    Glauca, 14, 18
    Glaucium, 27, 29
    Hunnemannia, 46
    Hypecoum, 44
    Macrantha, 23
    Meconella, 50
    Meconopsis, 33
    Miltantha, 18
    Orthorhoeades, 9
    Pilosa, 21
    Platystemon, 49

Roemeria, 32
Romneya, 49
Sanguinaria, 47
Scapiflora, 27
Stylomecon, 47
Stylophorum, 44, 46
Papaverine, 17, 20
Parfumidine, 61
Parfumine, 61
Paspalicine, 227
Paspaline, 227
Paspalinine, 225
Pavines, 69
Pegamine
  occurrence of, 133
  spectral data, 109
  structure, 100
Peganidine
  occurrence of, 134
  spectral data, 121
  structure, 112
Peganol
  occurrence of, 133
  structure, 112
Peganum Harmala alkaloids, 112
Pentaketones in naphthyl isoquinoline synthesis, 165
Peschawarine, 45
Phomanide, 213
Phtahideisoquinolines, 86
Platycerine methohydroxide, 5
Plumbagine, 162
Pontevedrine, 31
Predicentrine, 52
Proaporphines, 72
Procularines, 319
Promorphinans, 76
Pronuciferine, 11
Protoberberines, 78
Protopine, occurrence of, 9, 19, 21, 27, 32, 44, 46, 47, 49, 51
Protopine methohydroxide, 6
Pseudomonas aeruginosa, quinazoline alkaloids from, 110
Pseudomorphine, 16
Pyrazine metabolites, 186
Pyridoquinazoline alkaloids, 121

Q

Quettamine alkaloids, 316

R

Raddeanamine, 55
Raddeanidine, 55
Raddeanine, 55
Raddeanone, 55
Reframidine, 32
Reframine methohydroxide, 33
Reframoline, 32
Remrefine, 32
Remrefridine, 5, 19
Reohybrine, 32
Reticuline, biosynthesis, 3
Rhoeadine alkaloids, occurrence of, 9, 15, 17, 27, 44
Rhoeadines, 90
Rhoeagenine, 18, 20
Ribasine, 53
Roehybrine, 32
Roemeramine, 32
Roemeridine, 13
Roemerine, 18, 32
Roemerine N-oxide, 18
Roemeroline, 32
Roemeronine, 32
Romneine, 317
Roquefortin, 224

S

S-Govanine, 54
Salutaridine, 24, 26
Sanguidimerine, 49
Sanguinarine, 12, 49
Sanguirubine, 49
Sarcocapnidine
   biosynthesis, 320
   occurrence of, 290
   spectral data, 295, 299
   synthesis, 311
Sarcocapnine
   occurrence of, 290
   spectral data, 295, 299
   synthesis, 313
Scoulerine methohydroxide, 5
Secoberbines, 81
Secocularidine
   occurrence of, 293
   spectral data, 296, 307
Secodaphniphyllline, 267

Secophthalideisoquinolines, 86
Sendaverine, 53
Sessiflorine
   chemistry, 120
   occurrence of, 134
   structure, 112
Sevanine, 13
Sibiricine, 56
Silvaticamide, 249
Sinactine, 20
Sinacutine, 20
Spirobenzylisoquinolines, 89
Stepholidine, 16
Stylopine, 44
Stylopine methohydroxide, 6, 46

T

Tetrahydrocorysamine, 53, 57
Tetrahydrocorysamine methohydroxide, 6
Tetrahydroisoquinoline alkaloids, 3
Tetrahydrojatrorrhizine, 52
Tetrahydropalmatine, 53
Tetrahydropalmatine methohydroxide, 6, 31
Tetrahydropapaverine, 16
Thalictrifoline, 52
Thaliporphine, 30
Thebaine
   biosynthesis, 17
   occurrence of, 9, 19, 21, 27, 32, 44, 46, 47, 49, 51
Thebaine methohydroxide, 5
TR-2, 208
Triphyopeltine, 156
Triphyophylline
   occurrence of, 155
   synthesis, 176
Triphyophyllum alkaloids, 159
Trypethelone, 170
Tryptoquivalines
   biosynthesis, 221
   structures, 217
   synthesis, 221

V

Vaillantine, 61
D-Valyl-L-tryptophan anhydride, 210
Vasicine
   biosynthesis, 126

occurrence of, 132
pharmacological properties, 129
structure, 111
Vasicinoline, 112, 115, 132
Vasicinone
occurrence of, 132
pharmacological properties, 129
structure, 112, 114
Vasicol
occurrence of, 133
spectral data, 114
structure, 112
Vasicoline
occurrence of, 134
spectral data, 117
structure, 112
Vasicolinone, 112
Verruculogen, 207
Versimide, 248
Viguine, 53

W

WF-5239, 253
Wilsonine, 56

X

Xanthoascin, 252
Xanthocillin-X, 252, 254

Y

Yenhusomidine, 55
Yenhusomine, 55
Yuzurimine
chemistry, 266
structure, 275
Yuzurimine-C
chemistry, 266
properties, 275
Yuzurine
properties, 266
spectral data, 278

Z

Zanthoxylum arborescens, alkaloids of, 110
Zwitterionic alkaloid from Daphniphyllum, 273